FAO FOOD AND NUTRITION PAPER 14/4 Rev. 1

manual of food quality control 4. Rev. 1. microbiological analysis

Dr W. Andrews
FAO Consultant
Food and Drug Administration
Washington, D.C., USA

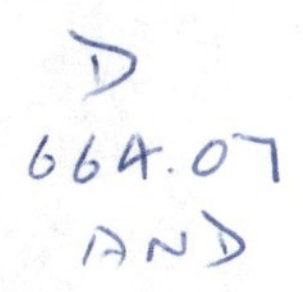

FOOD AND AGRICULTURE ORGANIZATION OF THE UNITED NATIONS
Rome 1992

The designations employed and the presentation of material in this publication do not imply the expression of any opinion whatsoever on the part of the Food and Agriculture Organization of the United Nations concerning the legal status of any country, territory, city or area or of its authorities, or concerning the delimitation of its frontiers or boundaries.

M-82
ISBN 92-5-103189-4

FOREWORD

The control of food safety and quality is an integral part of national programmes for development. National food control systems are designed to protect the health and welfare of the consumer, to promote the development of trade in food and food products, and to protect the interests of the fair and honest food producer, processor or marketer against dishonest and unfair competition. Emphasis is placed on the prevention of chemical and biological hazards which result from contamination, adulteration or simple mishandling of foods. Also important is the maintenance of general food quality.

An integral part of a national food control system is the analytical laboratory service. The typical food control laboratory has both chemical and microbiological analytical capabilities. Each of these technical areas have differing requirements and procedures to establish and maintain high quality analytical work.

This revised manual is a practical handbook on the analysis of foods for microbiological indices of quality and contamination. Its ultimate aim is to assure that a microbiological laboratory produces reliable high quality analytical results using analytical methodology which has been shown to be accurate and reproducible. A quality assurance (QA) manual for the microbiological laboratory has been previously published in the FAO Food and Nutrition Paper Series as number 14/12 "Quality Assurance in the Food Control Microbiological Laboratory". This and the present manual should be used in conjuction with each other for most effective operation of a microbiological laboratory.

FAO wishes to acknowledge the efforts of Dr. Wallace Andrews, Division of Microbiology, Food and Drug Administration (FDA), Washington DC, USA, who prepared the text for this revised manual. The contribution of other FDA staff including Ms Lois Tomlinson for technical editing and Mrs. Alesia Javins for typing the draft manuscript, are also gratefully acknowledged. The text of Chapter 19 was provided by Dr. Philip B. Mislivic of the Division of Microbiology, FDA. FAO further acknowledges the firm of Lehman-Scaffa Art and Photography, Inc., Silver Springs, Maryland, USA, for preparation of the art work.

The manual in draft form was reviewed by Dr. M. Ali Ahmed, Head, Public Health Laboratory, Dubai, United Arab Emirates; Dr. E. Essien, Director, Department of Food and Drug Administration and Control, Lagos, Nigeria; Dr. D. Majumdar, Director, Inspection and Quality Control, Export Inspection Council of India, New Delhi, India; Mr. Philip Palmyre, Chief Technologist, Public Health Laboratory, Victoria, Seychelles; Dr. M. K. Refai, Faculté de Médecine Vétérinaire, Gizeh, Egypt; and Mrs. N. Thongtan, Director, Division of Agricultural Chemistry, Bangkok, Thailand. FAO is grateful for those reviews.

This publication is available to persons and organizations. Comments on this publication and suggestions for possible future editions should be sent to:

The Chief
Food Quality and Standards Service
Food Policy and Nutrition Division
Food and Agriculture Organization of the United Nations
00100 Rome, Italy

SPECIAL NOTE

The Laboratory procedures described in this Manual are designed to be carried out by properly trained personnel in a suitably equipped laboratory. In common with many such procedures, they may involve hazardous materials.

For the correct and safe execution of these procedures, it is essential that laboratory personnel follow standard safety precautions.

While the greatest care has been exercised in the preparation of this information, FAO expressly disclaims any liability to users of these procedures for consequential damages of any kind arising out of or connected with their use.

The analytical procedures detailed herein are also not to be regarded as official because of their inclusion in this Manual. They are simply procedures which have been found to be accurate and reproducible in a variety of laboratories.

MICROBIOLOGICAL ANALYSIS IN THE FOOD CONTROL LABORATORY

TABLE OF CONTENTS

ANNEXES

FIGURES

ABBREVIATIONS USED IN THIS MANUAL

cm	-	Centimetre(s)
g	-	Gram(s)
hr	-	Hour(s)
id	-	Inside diameter
in	-	inch(es)
kg	-	Kilogram(s)
l	-	Litre
lb	-	Pound
LT	-	Heat labile
mg	-	Milligram(s)
μg	-	Microgram(s)
μm	-	Micrometre(s)
min	-	Minute(s)
ml	-	Millilitre(s)
mm	-	Millimetre(s)
MPN	-	Most probable number
N	-	Normal
nm	-	Nanometre(s)
ppm	-	Parts per million
psi	-	Pounds per square inch
rpm	-	Revolutions per minute
s	-	Second(s)
ST	-	Heat stable
UV	-	Ultraviolet
°C	-	Degrees Celsius
%	-	Parts per hundred (percent)

NOTE: Chemical formulae in standard notation (e.g. NaCl for sodium chloride) are used throughout this Manual. These are not abbreviations and are not included in the above listing.

CHAPTER 1

FOOD SAMPLING

Because interpretation about a large consignment of food is based on a relatively small sample of the lot, established sampling procedures must be applied uniformly. A representative sample is essential when the microorganisms are sparsely distributed within the food. Moreover, the condition of the sample received for analysis is of primary importance. If samples are improperly collected and mishandled or are not representative of the sampled lot, the laboratory results will be meaningless.

Whenever possible, submit samples to the laboratory in the original unopened containers. If products are in bulk or in containers too large for submission to the laboratory, transfer representative portions to sterile containers under aseptic conditions. Sterilize one-piece stainless steel spoons, forceps, spatulas, and scissors in an autoclave or dry-heat oven. Use of a propane torch or dipping the instrument in alcohol and igniting is dangerous and may be inadequate for sterilizing equipment.

Use containers that are clean, dry, leak-proof, wide-mouthed, sterile and of a size suitable for samples of the product. Containers such as plastic jars or metal cans that are leak-proof may be used. Whenever possible, avoid glass containers which may break and contaminate the food product. For dry materials, use sterile metal boxes, cans, bags, or packets with suitable closures. Sterile plastic bags (for dry, unfrozen materials only) or plastic bottles are also useful containers. Take care not to overfill bags or puncture bags by wire closure. Identify each sample unit (defined later) with a properly marked strip of masking tape. Do not use a felt pen on plastic because the ink might penetrate the container. Whenever possible, obtain at least 100 g for each sample unit.

Deliver samples to the laboratory promptly with the original storage conditions maintained as nearly as possible. When collecting liquid samples, take an additional sample as a temperature control. Check the temperature of the control sample at the time of collection and on receipt at the laboratory. Make a record for all samples of the times and dates of collection and of arrival at the laboratory. Dry or canned foods that are not perishable and are collected at ambient temperatures need not be refrigerated. Transport frozen or refrigerated products in approved insulated containers of rigid construction so that they will arrive at the laboratory unchanged. Collect frozen samples in pre-chilled containers. Place containers in a freezer long enough to chill them thoroughly. Keep frozen samples solidly frozen at all times. Cool refrigerated samples in ice at 0-4°C and transport them in a sample chest with suitable refrigerant capable of maintaining the sample at 0-4°C until arrival at the laboratory. Do not freeze refrigerated products. Unless otherwise specified, refrigerated samples should not be analyzed more than 36 hr after collection. Sampling plans for various microorganisms are presented below:

SAMPLING PLANS

1. <u>Salmonella</u>

   a. Sample collection

Because of the continuing occurrence of <u>Salmonella</u> in foods, sampling plans for these organisms have received considerable attention. In general, the number of samples from a particular lot of food varies according to the sampling category to which a food is assigned. Generally, the assignment to a sampling or food category depends on (1) the sensitivity of the consumer group (e.g., the aged, the infirm, and infants) and (2) the possibility that the food may have undergone a processing step lethal to Salmonella during manufacture or in the home. For the Salmonella sampling plan discussed here, 3 categories of foods are identified:

Food Category I. - Foods that would not normally be subjected to a process lethal to Salmonella between the time of sampling and consumption and are intended for consumption by the aged, the infirm, and infants;

Food Category II. - Foods that would not normally be subjected to a process lethal of Salmonella between the time of sampling and consumption and are intended for consumption by the normal adult population;

Food Category III. - Foods that would normally be subjected to a process lethal to Salmonella between the time of sampling and consumption.

A sample unit consists of a minimum of 100 g and is usually a consumer-size container of product. Take sample units at random to ensure that the sample is representative of the lot. When using sample containers, submit a control consisting of one empty sample container that has been exposed to the same conditions as those under which the sample was collected. Collect more than one sample unit from large institutional or bulk containers when the number of sample units required exceeds the number of containers in the lot. A sample unit will consist of more than one container when containers are smaller than 100 g (e.g., four 25 g containers could constitute a sample unit).

The numbers of sample units which should be collected in each food category are as follows: Food Category I, 60 sample units; Food Category II, 30 sample units; Food Category III, 15 sample units. Submit all collected samples to the laboratory for analysis. Advise laboratory in advance of perishable sample shipments.

b. Sample analysis

The laboratory should analyze each sample for the presence of Salmonella according to methods described in this manual. Take 25 g analytical unit at random from each 100 g sample unit. When a sample unit consists of more than 1 container, aseptically mix the contents of each container before taking the 25 g analytical unit. To reduce the analytical workload, the analytical units may be composited. The maximum size of a composite unit is 375 g or 15 analytical units. The minimum number of composite units to be tested for each food category is as follows: Food Category I, 4 composite units; Food Category II, 2 composite units; Food Category III, 1 composite unit. For each 375 g composite, the entire amount of 375 g is analyzed for Salmonella.

Keep the remainder of the sample unit in a sterile container as a reserve. Refrigerate perishable samples and those samples supporting microbial growth. An analytical control is required for each sample tested. The sampled lot is acceptable only if analyses of all composite units are negative for Salmonella. If one or more composite units are positive for Salmonella, the lot should be rejected, provided that the environmental control is negative for Salmonella. A lot should not be resampled unless the environmental control for Salmonella is positive.

c. Classification of food products for sampling purposes

Examples of foods that would be classified into the 3 sampling categories described above are provided below:

Food Category I. - Foods that would not normally be subjected to a process lethal to Salmonella between the time of sampling and consumption and are intended for consumption by the aged, the infirm, and infants. Examples are:

- Dried milk for direct consumption

- Baby and geriatric foods for direct consumption

Food Category II. - Foods that would not normally be subjected to a process lethal to Salmonella between the time of sampling and consumption and are intended for consumption by the normal adult population. Examples are:

- Milled grain products not cooked before consumption (bran and wheat germ)

- Bread, rolls, buns, sugared breads, crackers, custard- and cream-filled sweet goods, and icings

- Breakfast cereals and other ready-to-eat breakfast foods

- Pretzels, chips, and other snack foods

- Butter and butter products; pasteurized milk and raw fluid milk and fluid milk products for direct consumption; pasteurized and unpasteurized concentrated liquid milk products for direct consumption; dried milk and dried milk products for direct consumption; casein, sodium caseinate, and whey

- Cheese and cheese products

- Ice cream from pasteurized milk and related products that have been pasteurized; raw ice cream mix and related unpasteurized products for direct consumption

- Pasteurized and unpasteurized imitation dairy products for direct consumption

- Pasteurized eggs and egg products from pasteurized eggs; unpasteurized eggs for consumption without further cooking

- Canned and cured fish, vertebrates, and other fish products; fresh and frozen raw shellfish and crustacean products for direct consumption; smoked fish, shellfish, and crustaceans for direct consumption

- Meat and meat products, poultry and poultry products, and gelatin (flavored and unflavored bulk)

- Fresh, frozen, and canned fruits and juices, concentrates, and nectars; dried fruits for direct consumption; jams, jellies, preserves, and butters

- Nuts, nut products, edible seeds, and edible seed products for direct consumption

- Vegetable juices, vegetable sprouts, and vegetables normally eaten raw

- Oils consumed directly without further processing; oleomargarine

- Dressings and condiments (including mayonnaise), salad dressing, and vinegar

- Spices, flavors, and extracts

- Soft drinks and water

- Beverage bases

- Coffee and tea

- Candy (with and without chocolate; with and without nuts) and chewing gum

- Chocolate and cocoa products

- Pudding mixes not cooked before consumption and gelatin products

- Syrups, sugars, and honey

- Ready-to-eat sandwiches, stews, gravies, and sauces

- Soups

- Prepared salads

- Nutrient supplements, such as vitamins, minerals, proteins, and dried inactive yeasts

Food Category III: Foods that would normally be subjected to a process lethal to Salmonella between the time of sampling and consumption. Examples are:

- Whole grain, milled grain products that are cooked before consumption (corn meal and all types of flour) and starch products for human use

- Prepared dry mixes for cakes, cookies, breads, and rolls
- Macaroni and noodle products
- Fresh and frozen fish; vertebrates (except those eaten raw); fresh and frozen shellfish and crustaceans (except raw shellfish and crustaceans for direct consumption); other aquatic animals (including frog legs, marine snails, and squid)
- Vegetable protein products (simulated meats) normally cooked before consumption
- Fresh vegetables, frozen vegetables, dried vegetables, cured and processed vegetable products normally cooked before consumption
- Vegetable oils, oil stock, and vegetable shortening
- Dry dessert mixes, pudding mixes, and rennet products that are cooked before consumption

2. Listeria

a. Sample collection

For crabmeat, shrimp, processed imitation seafood (surimi), crayfish and lobster (cooked or parboiled), langostinos (cooked), smoked or salted fish, cheese, milk, ice cream, and other dairy products, collect ten 250 g subsamples (or retail packages) at random. Do not break or cut larger retail packages to obtain a 250 g subsample. Collect the intact retail unit as the subsample even if it is larger than 250 g. Make 2 composites from the 10 subsamples. Prepare each composite by removing 100 g from each of 5 subsamples. Each composite will contain a total of 500 g.

For cheese units that weigh 2 kg or more, collect 2 units per sample. From each of the units of cheese, prepare one composite as follows: Divide the cheese unit in half. Remove a 250 g plug which includes both surfaces from each half of the cheese unit to obtain a 500 g composite. Obtain a second composite for analysis by repeating this process on the other unit of cheese. Two composites are analyzed per cheese sample.

b. Sample analysis

To each 500 g composite in a sterile stomacher bag or blender, add 500 ml enrichment broth for Listeria, and stomach or blend for 2 min. Remove 50 g portion from bag and add to flask containing 200 ml enrichment broth for Listeria. Swirl thoroughly and continue with standard Listeria methodology as described in Chapter 11.

3. **Aerobic plate counts, total coliforms, fecal coliforms, Escherichia coli, Staphylococcus spp., Vibrio spp., Shigella spp., Campylobacter spp., Yersinia spp., Bacillus cereus, and Clostridium perfringens**

a. Sample collection

From any lot of food, collect appropriate number (see Chapter 22) 250 g subsamples (or retail packages) at random. Do not break or cut larger retail packages to obtain a 250 g subsample. Collect the intact retail unit as the subsample even if it is larger than 250 g.

b. Sample analysis.

Analyze each subsample individually as recommended elsewhere in this manual.

EQUIPMENT AND MATERIALS

1. **Mechanical blender**. Several types are available. Use blender that has several operating speeds or rheostat. The term "high-speed blender" designates mixer with 4 canted, sharp-edge, stainless steel blades rotating at bottom of 4 lobe jar at 10,000-12,000 rpm or with equivalent shearing action. Suspended solids are reduced to fine pulp by action of blades and by lobular container, which swirls suspended solids into blades. Waring blender, or equivalent, meets these requirements.

2. **Stomacher** and sterile stomacher bags (for Listeria)

3. Sterile glass or metal **high-speed blender jar**, 1000 ml, with cover, resistant to autoclaving for 60 min at 121°C

4. **Balance**, with weights; 2,000 g capacity, sensitivity of 0.1 g

5. **Sterile beakers**, 250 ml, low-form, covered with aluminum foil

6. **Sterile graduate pipets**, 1.0 and 10.0 ml

7. **Sterile knives, forks, spatulas, forceps, scissors, tablespoons, and tongue depressors** (for sample handling)

MEDIA AND REAGENTS

Butterfield's phosphate buffer (R3), sterilized in bottles to yield final volume of 90 $\pm$ 1 ml

RECEIPT OF SAMPLES

1. **Arrival at the laboratory**. As soon as the sample arrives at the laboratory, the analyst should note its general physical condition. If the sample cannot be analyzed immediately, it should be stored as described later.

2. **Condition of sampling container**. Check sampling containers for gross physical defects. Carefully inspect plastic bags and bottles for tears, pin holes, and puncture marks. If

sample units were collected in plastic bottles, check bottles for fractures and loose lids. If plastic bags were used for sampling, be certain that twist wires did not puncture surrounding bags. Any cross-contamination resulting from one or more of the above defects would invalidate the sample.

3. **Labeling and records.** Be certain that each sample is labeled appropriately and is accompanied by a completed copy of all records. Assign each sample unit an individual unit number and analyze as a discrete unit unless the sample is composited as described previously in this chapter.

4. **Storage.** If possible, examine samples immediately upon receipt. If analysis must be postponed, however, store frozen samples at -20°C until examination. Refrigerate unfrozen perishable samples at 0-4°C for not longer than 36 hr. Store nonperishable, canned, or low-moisture foods at room temperature until analysis.

THAWING

Use aseptic technique when handling product. Before handling or analysis of sample, clean immediate and surrounding work areas. In addition, swab immediate work area with commercial germicidal agent. Preferably, do not thaw frozen sample before analysis. If necessary to temper frozen sample to obtain analytical portion, thaw it in original container or in container in which it was received in the laboratory. Whenever possible, avoid transferring sample to second container for thawing. Normally, a sample can be thawed at 2-5°C within 18 hr. If rapid thawing is desired, thaw sample at less than 45°C for not more than 15 min. When thawing sample at elevated temperatures, agitate sample continuously in thermostatically controlled water bath.

MIXING

Various degrees of non-uniform distribution of microorganisms are to be expected in any food sample. To ensure more even distribution, shake liquid samples thoroughly and, if practical, mix dried samples with sterile spoons or other utensils before withdrawing the analytical unit from a sample of 100 g or greater. Use a 50 g analytical unit of liquid or dry food to determine aerobic plate count value and most probable number (MPN) of coliforms. Other analytical unit sizes (e.g., 25 g for Salmonella) may be recommended, depending on specific analysis to be performed. Use analytical unit size and diluent volume recommended in this manual for appropriate method being used. If contents of package are obviously not homogeneous (e.g., a frozen dinner), withdraw the analytical unit from macerate of the entire contents of the package or, preferably, analyze each different food portion separately, depending on purpose of test.

WEIGHING

Tare high-speed blender jar; then aseptically and accurately ($\pm$ 0.1 g) weigh unthawed food (if frozen) into jar. If entire sample weighs less than the required amount, weigh portion equivalent to one-half of sample and adjust amount of diluent or broth accordingly. Total volume in blender must completely cover blades.

BLENDING AND DILUTING FOR SAMPLES REQUIRING ENUMERATION OF MICROORGANISMS

1. **All foods other than nut meat halves and larger pieces, and nut meal.** Add 450 ml Butterfield's phosphate buffer to blender jar containing 50 g analytical unit and blend 2 min. This results in a dilution of 10^{-1}. Make dilutions of original homogenate promptly, using pipets that deliver required volume accurately. Do not deliver less than 10% of total volume of pipet. For example, do not use pipet with capacity greater than 10 ml to deliver 1 ml volumes; for delivering 0.1 ml volumes, do not use pipet with capacity greater than 1.0 ml. Prepare all decimal dilutions with 90 ml of sterile diluent plus 10 ml of previous dilution, unless otherwise specified. Shake all dilutions vigorously 25 times in 30 cm (1 ft) arc in 7 s. Not more than 15 min should elapse from the time sample is blended until all dilutions are in appropriate media.

2. **Nut meat halves and larger pieces.** Aseptically weigh 50 g analytical unit into sterile screw-cap jar. Add 50 ml diluent (H-1, above) and shake vigorously 50 times through 30 cm arc to obtain 10^{0} dilution. Let stand 3-5 min and shake 5 times through 30 cm arc to resuspend just before making serial dilutions and inoculations.

3. **Nut meal.** Aseptically weigh 10 g analytical unit into sterile screw-cap jar. Add 90 ml of diluent (H-1, above) and shake vigorously 50 times through 30 cm arc to obtain 10^{-1} dilution. Let stand 3-5 min and shake 5 times through 30 cm arc to resuspend just before making serial dilutions and inoculations.

CHAPTER 2

AEROBIC PLATE COUNT

The aerobic plate count is designed to provide an estimate of the total number of aerobic organisms in a particular food. A series of dilutions of the food homogenate is mixed with an agar medium and incubated at 35°C for 48 hr. It is assumed that each visible colony is the result of multiplications of a single cell on the surface of the agar.

The total aerobic plate count is useful for indicating the overall microbiological quality of a product and, thus, is useful for indicating potential spoilage in perishable products. The aerobic plate count is also useful for indicating the sanitary conditions under which the food was produced and/or processed.

A. EQUIPMENT AND MATERIALS

1. Mechanical blender (see Chapter 1).

2. Blender jar (see Chapter 1).

3. Balance with capacity of $\geq$ 2 kg and sensitivity of 0.1 g.

4. Petri dishes, glass (15 x 100 mm) or plastic (15 x 90 mm)

5. Sterile pipets, 1, 5, and 10 ml, graduated in 0.1 ml units.

6. Dilution bottles, 160 ml, made of borosilicate glass, with rubber stopper or plastic screw caps equipped with Teflon liners.

7. Pipet and Petri dish containers, adequate for protection.

8. Water bath, for tempering agar, thermostatically controlled to 48 $\pm$ 1°C.

9. Incubator, 35 $\pm$ 1°C.

10. Colony counter, dark-field, Quebec, or equivalent, with suitable light source and grid plate.

B. MEDIA AND REAGENTS

1. Dilution blanks, 90 $\pm$ 1 ml Butterfield's phosphate buffer (R3).

2. Plate count agar (M77).

C. PROCEDURE FOR FOODS OTHER THAN NUT MEAT HALVES AND LARGER PIECES, AND NUT MEAL

1. Add 450 ml Butterfield's phosphate buffer to blender jar containing 50 g analytical unit and blend 2 min at 10,000 - 12,000 rpm. This results in a dilution of 10^{-1}.

2. Using separate sterile pipets, prepare decimal dilutions of 10^{-2}, 10^{-3}, 10^{-4}, and others, as appropriate, of food homogenate transferring 10 ml of previous dilutions to 90 ml of diluent. Shake all dilutions 25 times in 30 cm arc within 7 s.

3. Pipet 1 ml of each dilution into separate, duplicate, appropriately marked Petri plates.

4. Add 20 ml of plate count agar (cooled to 44 - 46 °C) to each plate within 15 min of original dilution.

5. Immediately mix sample dilutions and agar medium thoroughly and uniformly.

6. Allow agar to solidify, invert Petri plates, and incubate promptly for 48 $\pm$ 2 hr at 35°C.

7. After incubation, count duplicate plates having 25 - 250 colonies. Duplicate plates of at least 1 of 3 dilutions should be in 25 - 250 colony range. When only one dilution is in appropriate range, compute average count per g for dilution and report as aerobic plate count per g. When 2 dilutions are in appropriate range, determine average count dilution before averaging 2 dilution counts to obtain aerobic plate count per g. Round off count to 2 significant figures only at time of conversion to aerobic plate counts. When rounding off numbers, raise second digit to next higher number only when third digit from left is 5 or greater, and replace dropped digit with zero. If third digit is 4 or less, replace third digit with zero and leave second digit the same.

D. PROCEDURE FOR NUT MEAT HALVES, NUT MEAT LARGER PIECES, AND NUT MEAL

See Chapter 1

E. FLOW SHEET

See Figure 1.

F. RECORD SHEET

See Annex 1.

G. PRECAUTIONS AND LIMITATIONS OF METHOD

In determining the aerobic plate count values of foods, the following points should be noted:

1. Microorganisms sometimes appear as clumps or clusters in food and may not be dispersed, even when the food sample is blended. There is no assurance that all organisms will be distributed as single cells. Thus, each colony on a plate of agar may arise from a cluster of cells rather than individual cells.

2. Autoclaving of plate count agar under crowded conditions may lead to inadequate sterilization of media.

3. Butterfield's phosphate buffer dilution blanks must be adjusted to 90 ml after autoclaving.

4. If the temperature of the melted agar is not carefully monitored at 44 - 46°C, then a temperature higher than 46°C can lead to cell shock or cell injury and thereby give a lower than true count.

5. Analyst differences in making dilutions, pipetting, and counting plates may lead to a variation in results. Extreme care should be taken to insure that errors due to analyst variability are kept to an absolute minimum.

6. In making dilutions, pipet tip should not extend more than 2.5 cm below level of buffer.

OTHER ACCEPTABLE METHODS

In addition to the conventional plate count method, there are rapid methods and/or methods of convenience which are acceptable for determining the aerobic plate count. These methods are: (The section numbers refer to AOAC's Official Methods of Analysis (1) which contains details of the procedures.

1. Spiral plate method, section 977.27.
2. Hydrophobic grid membrane filter method, section 986.32.
3. Pectin gel method, section 988.18.
4. Dry rehydratable film method, sections 986.33, 989.10, and 990.12.

REFERENCE

1. Association of Official Analytical Chemists. 1990. Official Methods of Analysis of the Association of Official Analytical Chemists, Volume 1, 15th ed., K. Helrich (Ed.). Association of Official Analytical Chemists, Arlington, VA.

Aerobic Plate Count

1. BLEND
food sample

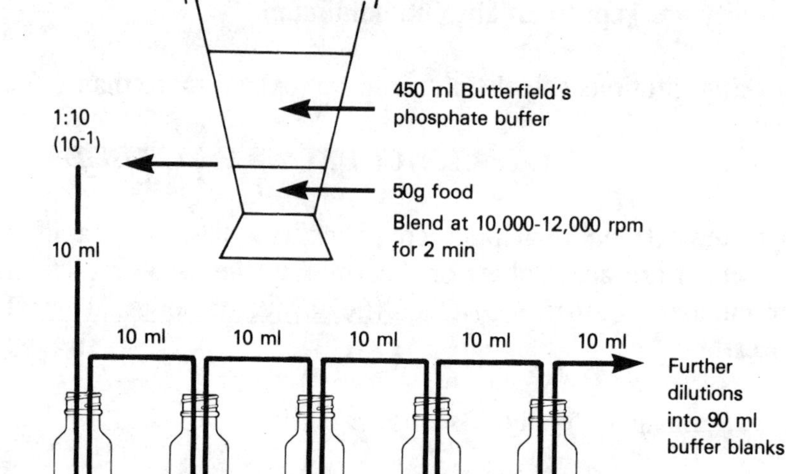

2. DILUTE
food sample homogenate

3. PIPET
1 ml volumes into sterile plates

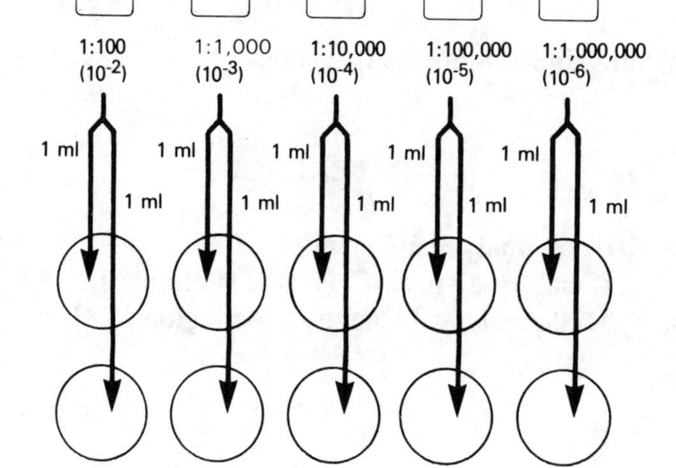

4. POUR
20 ml of plate count agar

5. INCUBATE
at 35°C for 48 hr

6. COUNT
plates containing 25-250 colonies and multiply the number of colonies by the reciprocal of the dilution; express aerobic plate count as number of organisms per g of food

Figure 1

CHAPTER 3

ESCHERICHIA COLI AND OTHER COLIFORMS

Members of the total coliform and fecal coliform groups are referred to as indicator organisms since a quantitation of their presence is used to indicate the potential presence of pathogens in foods. It is believed by some investigators that the higher the number of coliforms, the greater the possibility of pathogenic organisms being present. This indicator/pathogen relationship, however, is scientifically debatable, and by no means accepted unanimously by the scientific community. By definition, the total coliform group contains aerobic and facultatively anaerobic, Gram-negative nonsporeforming rods that ferment lactose in brilliant green lactose bile broth within 48 hr at 35°C.

An elevated temperature is used to separate organisms of the coliform group into those of fecal origin and those of nonfecal origin. By definition, the fecal coliform group contains aerobic and facultatively anaerobic Gram-negative nonsporeforming rods that ferment lactose in EC medium within 48 hr at 45.5°C. The fecal coliform group is restricted to organisms that grow in the gastrointestinal tract of humans and other warm-blooded animals and includes members of at least 3 genera: Escherichia, Klebsiella, and Enterobacter.

It was once assumed, perhaps somewhat questionably, that all members of the fecal coliform group possess equal sanitary significance. Because of the recent involvement of E. coli in several cases of food poisoning, it has been suggested that perhaps this organism, rather than the fecal coliform group, should be used as an indicator of sanitary quality.

The strains capable of causing illness in humans are placed in one of 4 categories of enteropathogenic E. coli:

1. Classical enteropathogenic strains associated with outbreaks of diarrhea in infants and young children.

2. Facultatively enteropathogenic strains which are associated with sporadic diarrhea and include many serogroups associated with normal intestinal microflora.

3. Enterotoxigenic strains which produce a heat-stable (ST) enterotoxin, a heat-labile (LT) enterotoxin, or both.

4. Enteroinvasive strains which cause illness by their invasive infection of the gastrointestinal tract.

In analyzing foods for the presence of enteropathogenic E. coli, the sample is enriched and streaked to selective plating agars to obtain isolated colonies. The picked colonies are subjected to a series of biochemical tests to screen enteropathogenic E. coli from the organisms of related species. The isolates are then serologically characterized. Finally, the enteropathogenicity of an organism is demonstrated by determining its invasiveness and its production of LT or ST.

Methods for determining the presence of enteropathogenic E. coli are not within the scope of this manual. Appropriate references, however, are provided later in this chapter (1-15).

CONVENTIONAL METHOD FOR ENUMERATING TOTAL COLIFORMS, FECAL COLIFORMS AND E. COLI

A. EQUIPMENT AND MATERIALS

1. Incubator, 35 ± 1°C
2. Covered water bath, with circulating system to maintain temperature of 45.5 ± 0.2°C. Water level should be above that of medium in immersed tubes
3. Immersion-type thermometer, 1-55°C, with 0.1°C subdivisions, certified by National Institute of Standards, or equivalent
4. Balance with capacity of ≥ 2 kg and sensitivity of 0.1 g
5. Mechanical blender (see Chapter 1)
6. Blender jar (see Chapter 1)
7. Sterile pipets, 1, 5, and 10 ml, graduated in 0.1 ml units
8. Sterile utensils for sample handling (see Chapter 1)
9. Dilution bottles, 160 ml, made of borosilicate glass, with stopper or polyethylene screw caps equipped with Teflon liners

B. MEDIA AND REAGENTS

1. Brilliant green bile 2% (BGB) broth (M13)
2. Lauryl tryptose (LT) broth (M41)
3. EC medium (M26)
4. Levine's eosin-methylene blue (L-EMB) agar (M43)
5. Tryptone (tryptophane) broth, 1% (M111)
6. MR-VP medium (M61)
7. Koser citrate broth (M38)
8. Plate count agar (PCA) (M77)
9. Butterfield's phosphate buffer (R3)
10. Kovacs' reagent (R16)
11. Voges-Proskauer (VP) reagents (R40)

12. Methyl red indicator (R19)

13. Gram stain reagents (S6)

C. PRESUMPTIVE TEST FOR TOTAL COLIFORMS AND FECAL COLIFORMS

1. Weigh 50 g unthawed (if frozen) sample into sterile high-speed blender jar. Add 450 ml Butterfield's phosphate buffer (R3) and blend 2 min at 10,000-12,000 rpm. If necessary to temper frozen sample to remove 50 g portion, hold 18 hr at 2-5°C. If entire sample consists of <50 g, weigh portion equivalent to 1/2 sample and add volume of Butterfield's phosphate buffer (R3) to make 1:10 dilution.

2. Prepare all decimal dilutions with 90 ml Butterfield's phosphate buffer (R3) plus 10 ml from previous dilution unless otherwise specified. The dilutions to be prepared depend on the anticipated coliform density. Shake all dilutions 25 times in 30 cm arc for 7 s. Do not use pipets to deliver < 10% of their total volume. Transfer 1 ml portions to 3 LT tubes for each dilution. Hold pipet at an angle so that its lower edge rests against tube. Let pipet drain 2-3 s. Not more than 15 min should elapse from time sample is blended until all dilutions are in appropriate media.

3. Incubate tubes 48 ± 2 hr at 35°C. Examine tubes at 24 ± 2 hr for gas, i.e., displacement of medium in fermentation vial or effervescence when tubes are gently agitated.

4. Reincubate negative tubes for additional 24 hr.

5. Examine a second time for gas.

6. Perform a confirmed test on all presumptive positive (gassing) tubes.

D. CONFIRMED TEST FOR TOTAL COLIFORMS

1. Gently agitate each gassing LT tube and transfer loopful of suspension to tube of BGB broth. Hold LT tube at angle and insert loop to avoid transfer of pellicle (if present).

2. Incubate BGB tubes 48 ± 2 hr at 35°C.

3. Examine for gas production and record.

4. Calculate MPN (Chapter 21) of total coliforms based on combination of confirmed gassing LT tubes for 3 consecutive dilutions.

E. CONFIRMED TEST FOR FECAL COLIFORMS

1. Gently agitate each gassing LT tube and transfer loopful of each suspension to tube of EC medium.

2. Incubate EC medium tubes 48 ± 2 hr at 45.5 ± 0.2°C. Examine for gas production at 24 ± 2 hr; if negative, examine again at 48 ± 2 hr.

3. Calculate MPN (Chapter 21) of fecal coliforms based on proportion of confirmed gassing EC medium tubes for 3 consecutive dilutions.

F. **CONFIRMED TEST FOR E. COLI**

1. Streak loopful of suspension from each gassing EC medium tube to L-EMB agar.

2. Incubate plates 18-24 hr at 35°C.

3. Examine plates for typical E. coli colonies, i.e., dark centered with or without metallic sheen. Pick 2 typical colonies from each L-EMB plate and transfer to PCA slants for morphological and biochemical tests.

4. Incubate PCA slants 18-24 hr at 35°C.

5. If typical colonies are not present, pick 2 or more colonies most likely to be E. coli. Pick 2 colonies from every plate.

6. Perform Gram stain on each culture.

7. For all cultures appearing as Gram-negative, short rods or cocci, continue as described below:

   a. **Indole production**. Inoculate tube of tryptone or tryptophane broth and incubate 24 ± 2 hr at 35°C. Test for indole by adding 0.2-0.3 ml Kovacs' reagent. Appearance of distinct red color in the upper layer is positive test.

   b. **Voges-Proskauer-reactive compounds**. Inoculate tube of MR-VP medium and incubate 48 ± 2 hr at 35°C. Transfer 1 ml to 13 X 100 mm tube. Add 0.6 ml alpha-naphthol solution and 0.2 ml 40% KOH, and shake. Add a few crystals of creatine. Shake and let stand for 2 hr. Test is positive if eosin pink color develops.

   c. **Methyl red-reactive compounds**. Incubate MR-VP tube additional 48 ± 2 hr at 35°C after performing Voges-Prokauer test. Add 5 drops methyl red solution to each tube. A distinct red color is a positive test. A yellow color is a negative reaction.

   d. **Utilization of citrate**. Lightly inoculate tube of Koser citrate broth. Incubate 96 ± 2 hr at 35°C. Development of distinct turbidity is a positive reaction.

   e. **Production of gas from lactose**. Inoculate tube of LT broth and incubate 48 ± 2 hr at 35°C. Displacement of medium from inner vial or effervescence after gentle agitation is a positive reaction.

   f. **Interpretation**. All cultures that (a) ferment lactose with production of gas within 48 hr at 35°C, (b) appear as Gram-negative nonsporeforming rods or cocci, and (c) give IMViC patterns of ++-- (Biotype 1) or -+-- (Biotype 2) are considered to be E. coli. Calculate MPN (Chapter 21) of E. coli based on proportion of EC medium tubes in 3 successive dilutions which have been shown to contain E. coli.

G. FLOW SHEET

See Figures 2 and 3.

H. RECORD SHEET

See Annex 2.

I. PRECAUTIONS AND LIMITATIONS OF METHOD

In enumerating total coliforms and fecal coliforms in foods by the conventional method, the following points should be noted:

1. In the total coliform and fecal coliform group tests, only the lactose fermentation reactions of total bacterial populations are quantitated by the MPN techniques. Individual organisms are not isolated in pure culture, transferred to lactose broth, nor Gram stained. Thus, conformity with the definitions of the total coliform and fecal coliform groups is not absolute.

2. Some samples tend to foam during homogenization, and this foam should be avoided when pipetting.

3. After receiving dilutions of the homogenized sample, the inner walls of the tubes of lauryl tryptose broth should be "washed down" by slowly rotating tube.

4. Anaerogenic coliform cultures, or cultures not capable of fermenting lactose, will cause false negative reactions.

5. An adequate water level must be maintained in the water baths at all times. The level of water should be at least as high as the medium in the inner fermentation vial or the medium in the outer tube, whichever is higher. Failure to maintain an adequate level of water may result in false positive reactions.

6. Fresh E. coli and Enterobacter aerogenes control cultures must be used for each sequential 24-hr incubation period in the water bath. Failure of the E. coli culture to ferment lactose indicates that the water bath temperature may have gone excessively above 45.5°C. If E. aerogenes ferments lactose, this indicates that the water bath temperature may have gone too low.

7. The EC medium must be dispensed into 16 X 150 mm tubes, rather than 20 X 150 mm tubes, to allow for the more rapid attainment of a temperature of 45.5°C when tubes are placed in the water bath.

In enumerating E. coli in foods by the conventional method, the following points should be noted:

1. Because LT broth contains the surfactant, sodium lauryl sulfate, no resuscitation is provided for injured or stressed E. coli cells. Thus, severely damaged E. coli cells may not be recovered by this procedure.

2. Some biotypes of E. coli ferment lactose slowly or not at all, and these colonies will have an atypical appearance on L-EMB agar.

3. Koser citrate broth is used to determine an organism's ability to use citrate as a sole carbon source. Because of the possibility of introducing other nutrients from the plate count agar subculture, extreme care must be taken to use a very light inoculum. Koser citrate broth should be inoculated before the other IMViC media to minimize the possibility of nutrient "carry over".

4. Recent data indicate that the IMViC profile may be insufficient to distinguish E. coli from other members of the Enterobacteriaceae. Other enteric organisms such as E. fergusoni, E. ewingii, E. vulneris, E. hermanii, Plesiomonas shigelloides, Klebsiella Taxon 53, Klebsiella Taxon 62, Klebsiella Taxon 68, Hafnia alvei, and Aeromonas hydrophila may have an IMViC pattern of ++-- or -+--. Thus, it is essential that the Gram stain be performed in addition to the IVMiC determination. If the Gram stain indicates Gram negative bacteria with a morphology not typical for E. coli, then the following biochemical tests should be performed: motility, cytochrome oxidase, growth in KCN medium, malonate reaction, urease production, and fermentation of glucose, lactose, mannitol, sorbitol, cellobiose, and adonitol.

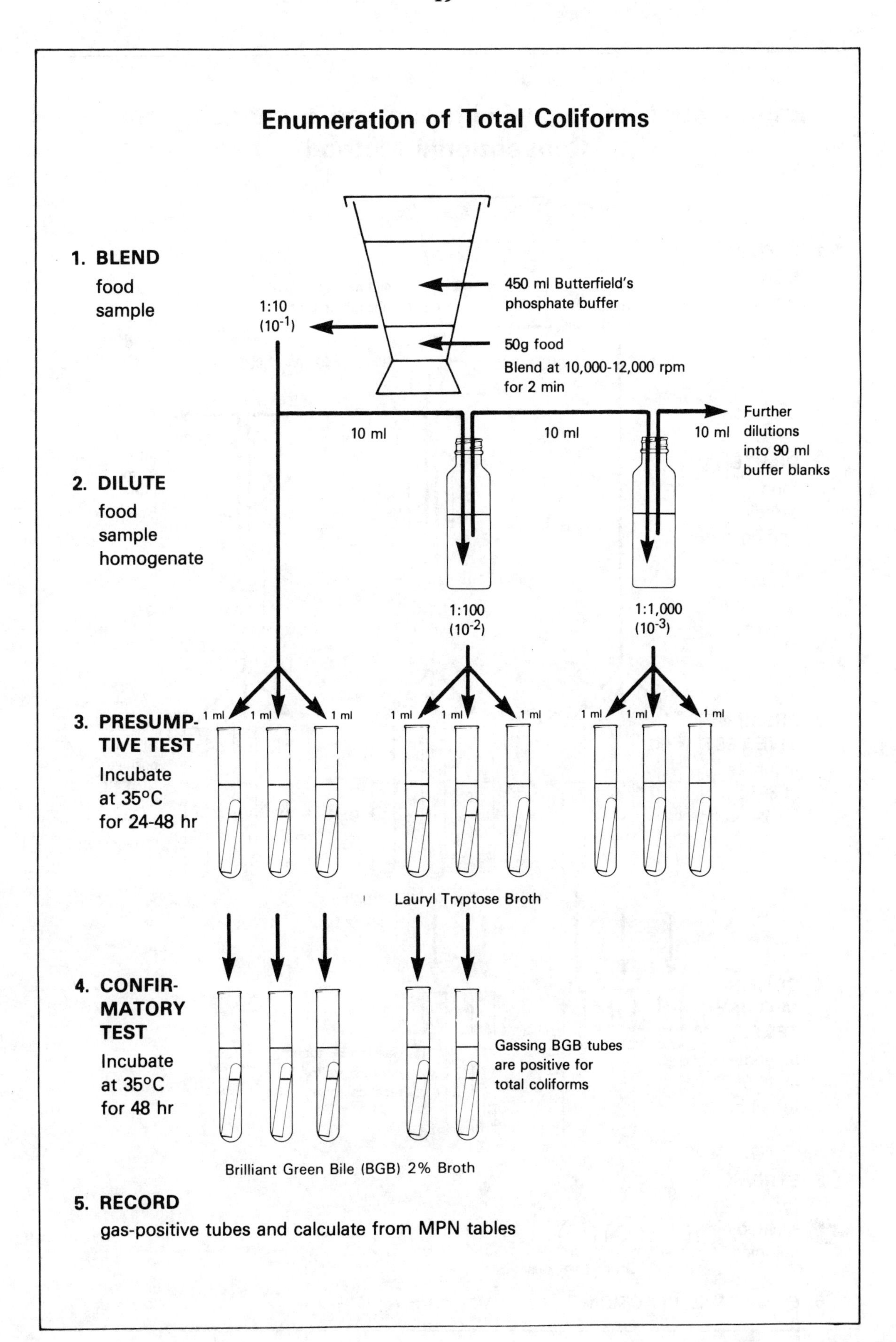

Enumeration of Total Coliforms
1. BLEND
food
sample
1:10
(10^{-1})
450 ml Butterfield's
phosphate buffer
50g food
Blend at 10,000-12,000 rpm
for 2 min
10 ml
10 ml
10 ml
Further
dilutions
into 90 ml
buffer blanks
2. DILUTE
food
sample
homogenate
1:100
(10^{-2})
1:1,000
(10^{-3})
3. PRESUMP-
TIVE TEST
Incubate
at 35°C
for 24-48 hr
1 ml
Lauryl Tryptose Broth
4. CONFIR-
MATORY
TEST
Incubate
at 35°C
for 48 hr
Gassing BGB tubes
are positive for
total coliforms
Brilliant Green Bile (BGB) 2% Broth
5. RECORD
gas-positive tubes and calculate from MPN tables

Figure 2

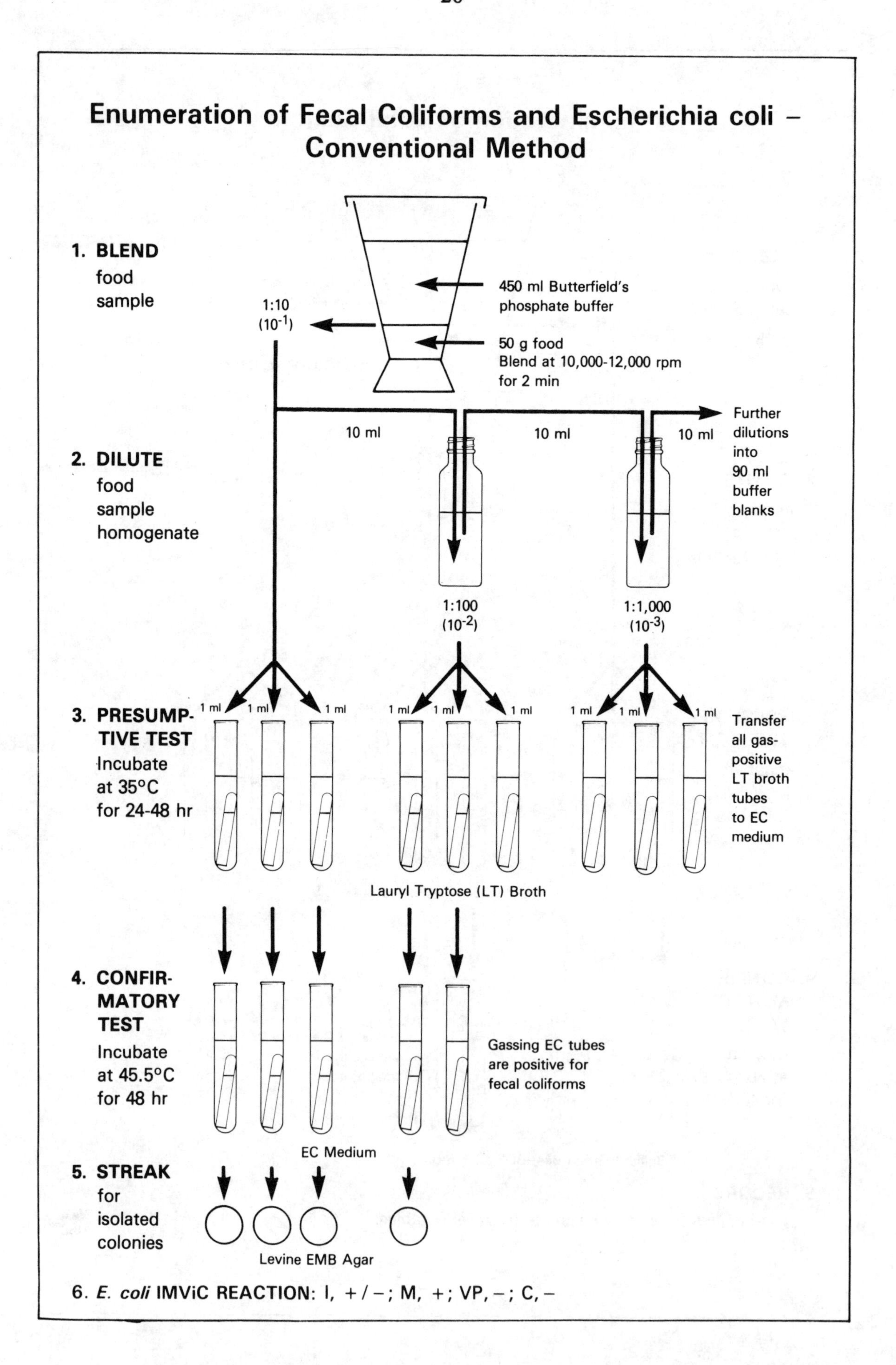
Enumeration of Fecal Coliforms and Escherichia coli – Conventional Method
1. BLEND food sample
450 ml Butterfield's phosphate buffer
1:10 (10-1)
50 g food
Blend at 10,000-12,000 rpm for 2 min
10 ml
10 ml
10 ml
Further dilutions into 90 ml buffer blanks
2. DILUTE food sample homogenate
1:100 (10-2)
1:1,000 (10-3)
3. PRESUMPTIVE TEST Incubate at 35°C for 24-48 hr
1 ml
Transfer all gas-positive LT broth tubes to EC medium
Lauryl Tryptose (LT) Broth
4. CONFIRMATORY TEST Incubate at 45.5°C for 48 hr
Gassing EC tubes are positive for fecal coliforms
EC Medium
5. STREAK for isolated colonies
Levine EMB Agar
6. E. coli IMViC REACTION: I, +/–; M, +; VP, –; C, –

Figure 3

MUG METHOD FOR RAPID ENUMERATION OF E. COLI IN CHILLED OR FROZEN FOODS EXCLUSIVE OF SHELLFISH

Glucuronidase is present in most strains of E. coli but absent in most other enteric microorganisms. Thus, an assay for this enzyme is useful for determining the presence of E. coli. The substrate, 4-methylumbelliferyl beta-D-glucuronide (MUG) is incorporated into LT broth. The inoculated tubes are incubated under specified conditions and examined under longwave UV light for the presence of a fluorogenic glucuronidase end product. Fluoresence in the LT-MUG medium is indicative of the presence of E. coli.

A. EQUIPMENT AND MATERIALS

1. Incubator, 35 ± 1°C
2. Balance with capacity of ≥ 2 kg and sensitivity of 0.1 g
3. Mechanical blender (see Chapter 1)
4. Blender jar (see Chapter 1)
5. Sterile pipets, 1, 5, and 10 ml, graduated in 0.1 ml units
6. Sterile utensils for sample handling (See Chapter 1)
7. Dilution bottles, 160 ml, made of borosilicate glass, with stopper or polyethylene screw caps equipped with Teflon liners

B. MEDIA AND REAGENTS

1. Lauryl tryptose-MUG (LT-MUG) broth (M42)
2. Lauryl tryptose (LT) broth (M41)
3. Levine's eosin-methylene blue (L-EMB) agar (M43)
4. MR-VP medium (M61)
5. Koser citrate broth (M38)
6. Plate count agar (PCA) (M77)
7. Butterfield's phosphate buffer (R3)
8. Kovacs' reagent (R16)
9. Voges-Proskauer test reagents (R40)
10. Methyl red indicator (R19)
11. Gram stain reagents (S6)

C. PRESUMPTIVE LT-MUG TEST FOR E. COLI

1. Prepare food samples as described in the Conventional Method for Enumerating Total Coliforms, Fecal Coliforms, and E. coli, above.

2. Prepare serial 10-fold dilutions as described in the Conventional Method for Enumerating Total Coliforms, Fecal Coliforms, and E. coli, above.

3. Inoculate 1 ml portions into 3 LT-MUG tubes for each dilution for appropriate number of dilutions to obtain determinant MPN.

4. Incubate tubes for 24 ± 2 hr at 35°C and examine each tube for growth, gas, and fluorescence. To observe fluorescence, examine tubes in the dark under a longwave (366 nm) UV lamp. A bluish fluorescence is a positive presumptive test for E. coli.

D. CONFIRMED LT-MUG

1. Streak loopful of suspension from each fluorescent tube to L-EMB agar and incubate 24 ± 2 hr at 35°C. Follow protocols outlined in the sections on Conventional Method for Enumerating Total Coliforms, Fecal Coliforms and E. coli, above, for performing Gram stains, IMViC tests, and determination of production of gas from lactose to confirm E. coli.

2. Interpretation. All cultures that (a) fluoresce in LT-MUG broth, (b) ferment lactose with production of gas within 48 hr at 35°C, (c) appear as Gram-negative nonsporeforming rods or cocci, and (d) give IMViC patterns of ++-- (Biotype 1) or -+-- (Biotype 2) are considered to be E. coli.

3. Calculate MPN (Chapter 21) of E. coli based on combination of fluorescent tubes in 3 successive dilutions that contain E. coli.

E. FLOW SHEET

See Figure 4.

F. RECORD SHEET

See Annex 3.

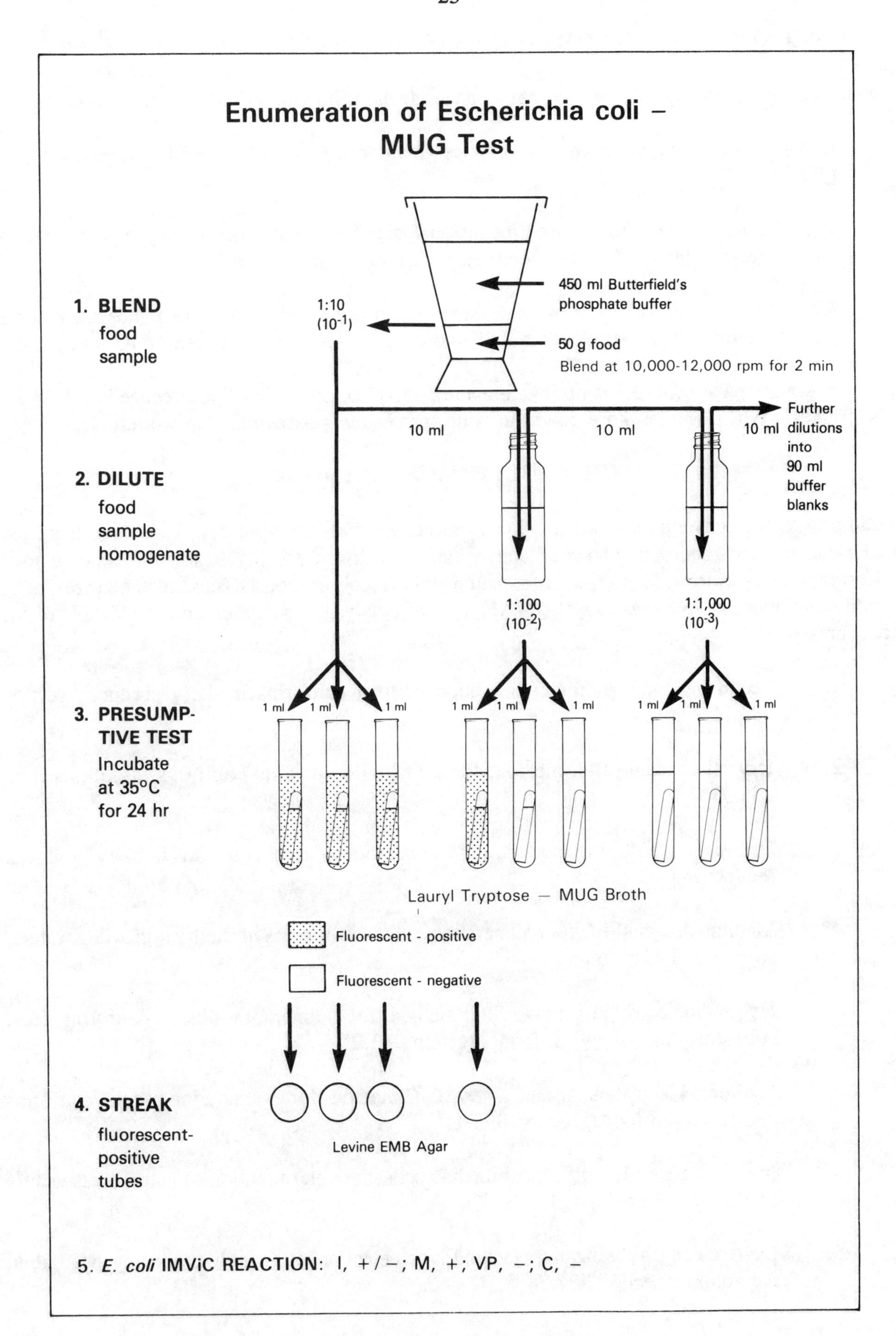
Enumeration of Escherichia coli – MUG Test
1. BLEND food sample
1:10 (10⁻¹)
450 ml Butterfield's phosphate buffer
50 g food
Blend at 10,000-12,000 rpm for 2 min
10 ml
10 ml
10 ml
Further dilutions into 90 ml buffer blanks
2. DILUTE food sample homogenate
1:100 (10⁻²)
1:1,000 (10⁻³)
1 ml
3. PRESUMPTIVE TEST Incubate at 35°C for 24 hr
Lauryl Tryptose – MUG Broth
Fluorescent - positive
Fluorescent - negative
4. STREAK fluorescent-positive tubes
Levine EMB Agar
5. E. coli IMViC REACTION: I, +/–; M, +; VP, –; C, –

Figure 4

G. PRECAUTIONS AND LIMITATIONS OF METHOD

In enumerating E. coli in foods by the MUG method, the following points should be noted:

1. About 5% of E. coli strains do not produce glucuronidase and, thus, will be negative by the MUG test.

2. The MUG test should not be used for the analysis of shellfish since this food type contains glucuronidase, thus causing the occurrence of false positive reactions.

3. All test tubes used for the MUG test should be screened for auto-fluorescence since some glass contains cerium oxide which will fluoresce and interfere with the MUG test.

4. A 6-watt, hand held UV lamp is safe and effective for observing fluorescence in LT-MUG tubes. When using a more powerful lamp, protective glasses must be worn.

OTHER ACCEPTABLE METHODS

In addition to the conventional fermentation tube method and the rapid MUG method, there are other rapid methods and/or methods of convenience which are acceptable for enumerating total coliforms, fecal coliforms, E. coli, or enteropathogenic E. coli in foods. These methods are: (The section numbers refer to AOAC's Official Methods of Analysis (16) which contains details of the procedures.)

1. Pectin gel method, for enumeration of total coliforms in dairy products, section 989.11

2. Dry rehydratable film method, for enumeration of total coliforms in milk, section 986.33

3. Dry rehydratable film method, for enumeration of total coliforms in dairy products, section 989.10

4. Medium A-1 method, for enumeration of fecal coliforms in shellfish-growing waters, section 978.23

5. Hydrophobic grid membrane filter method, for enumeration of total coliforms, fecal coliforms, and E. coli in foods, section 983.25

6. Hydrophobic grid membrane filter/MUG method, for enumeration of total coliforms and E. coli in foods, section 990.11

7. Invasiveness by E. coli of mammalian cells, for determination of enteroinvasiveness of E. coli cultures, section 982.36

8. DNA colony hybridization method, for detection of E. coli producing heat-labile enterotoxin, section 984.34

9. DNA colony hybridization method using synthetic oligodeoxyribonucleotides and paper filters, for detection of enterotoxigenic E. coli, section 986.34

10. Mouse adrenal cell and suckling mouse assays, for detection of E. coli enterotoxins, section 984.35

REFERENCES

1. **Burrows, W., and G. M. Musteikis**. 1966. Cholera infection and toxin in the rabbit ileal loop. J. Infect. Dis. 116:183-190.

2. **Dean, A. G., Y. Ching, R. G. Williams, and L. B. Harden**. 1972. Test for Escherichia coli enterotoxin using infant mice. Application in a study of diarrhea in children in Honolulu. J. Infect. Dis. 125:407-411.

3. **Donta, S. T., H. W. Moon, and S. C. Whipp**. 1974. Detection of heat-labile Escherichia coli enterotoxin with the use of adrenal cells in tissue culture. Science 183:334-336.

4. **Evans, D. G., D. J. Evans, Jr., and N. F. Pierce**. 1973. Differences in the response of rabbit small intestine to heat-labile and heat-stable enterotoxins of Escherichia coli. Infect. Immun. 7:873-880.

5. **Evans, D. J., Jr., D. G. Evans, and S. E. Gorbach**. 1973. Production of vascular permeability factor by enterotoxigenic Escherichia coli isolated from man. Infect. Immun. 8:725-730.

6. **Formal, S. B., H. L. Dupont, R. Hornick, M. J. Snyder, J. Libonati, and E. H. LaBrec**. 1971. Experimental models in the investigation of the virulence of dysentery bacilli and Escherichia coli. N. Y. Acad. Sci. 176:190-196.

7. **Gianella, R. A**. 1976. Suckling mouse model for detection of heat-stable Escherichia coli enterotoxin. Characteristics of the model. Infect. Immun. 14:95-99.

8. **Guerrant, R. L., L. L. Brunton, T. C. Schnaitman, L. I. Rebhun, and A. G. Gilman**. 1974. Cyclic adenosine monophosphate and alteration of Chinese hamster ovary cell morphology: A rapid, sensitive in vitro assay for the enterotoxins of Vibrio cholerae and Escherichia coli. Infect. Immun. 10:320-327.

9. **LaBrec, E. H., H. Schneider, T. J. Magnani, and S. B. Formal**. 1964. Epithelial cell penetration as an essential step in the pathogenesis of bacterial dysentery. J. Bacteriol. 88:1503-1518.

10. **Lovett, J., and J. T. Peeler**. 1983. Detection of E. coli enterotoxin using mouse adrenal cell and suckling mouse assay: Collaborative study. 97th Annual International Meeting and Exposition, AOAC, Washington, DC, (Abstract 45).

11. **Mackel, D. C., L. F. Langley, and L. A. Venice**. 1961. The use of the guinea pig conjunctivae as an experimental model for the study of virulence of Shigella organisms. Am. J. Hyg. 73:219-223.

12. **Mehlman, I. J., E. L. Eide, A. C. Sanders, M. Fishbein, and C. C. G. Aulisio**. 1977. Methodology for recognition of invasive potential of Escherichia coli. J. Assoc. Off. Anal. Chem. 60:546-562.

13. **Sack, D. A., and R. B. Sack**. 1975. Test for enterotoxigenic Escherichia coli using Yl adrenal cells in miniculture. Infect. Immun. 11:334-336.

14. **Scherer, W. F., J. T. Syverton, and G. O. Gey**. 1953. Studies on the propagation in vitro of poliomyelitis viruses. IV. Viral multiplication in a stable strain of human malignant epithelial cells (strain HeLa) derived from an epidermoid carcinoma of the cervix. J. Exp. Med. 97:695-710.

15. **Sereny, B**. 1955. Experimental Shigella keratoconjunctivitis. Acta Microbiol. Acad. Sci. Hung. 2:293-296.

16. Association of Official Analytical Chemists. 1990. Official Methods of Analysis of the Association of Official Analytical Chemists, Volume 1, 15th ed., K. Helrich (Ed.). Association of Official Analytical Chemists, Arlington, VA.

CHAPTER 4

SALMONELLA

Members of the genus Salmonella are infectious pathogens capable of initiating clinical symptoms in humans. The annual incidence of reported isolations of Salmonella from human sources has steadily increased for each year for the past several years.

Salmonella organisms were first detected in clinical specimens. Consequently, the earliest methods were directed towards isolating these pathogens from human feces. However, it was only a matter of time until certain foods were suspected of being contaminated with this pathogen. At the time, it seemed logical to apply these clinical methods, sometimes with modifications, to the analysis of food samples. It soon became evident, however, that this practice was not advisable due to several factors. First, Salmonella is usually present in much lower numbers in food samples than in clinical specimens. Furthermore, these organisms in foods have usually been subjected to processing and the surviving organisms are often injured or debilitated. Adding to the problem is the inherent viability of the foods encountered during analysis. Many foods not only have the ability to alter the physical nature of enriching cultivation media, but also can affect their modes of action. Thus, a long, laborious search began for improved methods of isolating Salmonella from foods. Methods were developed specifically for those foods most often involved or incriminated in salmonellosis outbreaks.

CONVENTIONAL METHOD FOR THE DETECTION OF SALMONELLA

A. EQUIPMENT AND MATERIALS

1. Mechanical blender (see Chapter 1)

2. Blender jar (see Chapter 1)

3. Sterile, 500 ml wide-mouth, screw-cap jars, sterile 500 ml Erlenmeyer flasks, sterile 250 ml beakers, sterile glass or paper funnels of appropriate size, and, optionally, containers of appropriate capacity to accommodate composited samples

4. Sterile, bent-glass spreader rods

5. Balance, with weights; 2,000 g capacity, sensitivity of 0.1 g

6. Balance, with weights; 120 g capacity, sensitivity of 5 mg

7. Incubator, 35 ± 1°C

8. Water baths, 37 ± 0.5°C and 48-50°C

9. Sterile spoons or other appropriate instruments for transferring food samples

10. Sterile culture dishes, 15 x 100 mm, glass or plastic

11. Sterile pipets, 1 ml, with 0.01 ml graduations; 5 and 10 ml, with 0.1 ml graduations

12. Inoculating needle and inoculating loop (about 3 mm id), nichrome, platinum-iridium, or chromel wire

13. Sterile test or culture tubes, 16 x 150 mm and 20 x 150 mm; serological tubes, 10 x 75 mm or 13 x 100 mm

14. Test or culture tube racks

15. Vortex mixer

16. Sterile shears, large scissors, scalpel, and forceps

17. Lamp (for observing serological reactions)

18. Fisher or Bunsen burner

19. pH test paper (pH range 6-8) with maximum graduations of 0.4 pH units per color change

20. pH meter

(NOTE: Items 21 to 23 are needed in the analysis of frog legs and rabbit carcasses)

21. Plastic bags, 28 x 37 cm, sterile, with resealable tape.

22. Plastic beakers, 4 liter, autoclavable, for holding plastic bag during shaking and incubation

23. Mechanical shaker, any model that can be adjusted to give 100 excursions/min with a 4 cm stroke, such as the Eberback shaker with additional 33 and 48 cm clamp bars

B. MEDIA AND REAGENTS

1. Lactose broth (M39)

2. Nonfat dry milk (reconstituted) (M65)

3. Selenite cystine (SC) broth (M87)

4. Tetrathionate (TT) broth (M96)

5. Rappaport-Vassiliadis (RV) medium (M84)

6. Xylose lysine desoxycholate (XLD) agar (M121)

7. Hektoen enteric (HE) agar (M34)

8. Bismuth sulfite (BS) agar (M9)

9. Triple sugar iron (TSI) agar (M100)

10. Tryptone (tryptophane) broth (M111)

11. Trypficase (tryptic) soy broth (M104)

12. Lauryl tryptose (LT) broth (M41)

13. Trypticase soy-tryptose broth (M110)

4. MR-VP medium (M61)

15. Simmons citrate agar (M91)

16. Urea broth (M116)

17. Urea broth (rapid) (M117)

18. Malonate broth (M53)

19. Lysine iron agar (Edwards and Fife) (M50)

20. Lysine decarboxylase broth (M49)

21. Motility test medium (semi-solid) (M59)

22. Potassium cyanide (KCN) medium (M79)

23. Phenol red carbohydrate broth (M74)

24. Purple carbohydrate broth (M81)

25. MacConkey agar (M52)

26. Nutrient broth (M68)

27. Brain heart infusion (BHI) broth (M12)

28. Gelatinase solution, 5% (M30)

29. Potassium sulfite powder, anhydrous

30. Mercuric chloride solution, 0.1% (R18)

31. Ethanol, 70% (R6)

32. Kovacs' reagent (R16)

33. Voges-Proskauer (VP) test reagents (R40)

34. Creatine phosphate crystals

35. Potassium hydroxide solution, 40% (R30)

36. 1 N sodium hydroxide solution (R37)

37. 1 N hydrochloric acid, (R14)

38. Methyl red indicator (R19)

39. Sterile distilled water

40. Tergitol Anionic 7 (R38)

41. Triton X-100 (R39)

42. Physiological saline solution (sterile) (R28)

43. Formalinized physiological saline solution (R9)

44. Brilliant green dye solution, 1% (S2)

45. Bromcresol purple dye solution, 0.2% (S3)

46. Salmonella polyvalent somatic (O) antiserum

47. Salmonella polyvalent flagellar (H) antiserum

48. Salmonella somatic group (O) antisera: A, B, C_1, C_2, C_3, D_1, D_2, E_1, E_2, E_3, E_4, F, G, H, I, Vi, and other groups, as appropriate

49. Salmonella Spicer-Edwards flagellar (H) antisera

C. PREPARATION OF FOODS FOR ISOLATION OF SALMONELLA

The following methods are based on the analysis of a 25 g analytical unit at a 1:9 sample/broth ratio. Depending on the extent of compositing, add enough broth to maintain this 1:9 ratio, unless otherwise indicated. For samples not analyzed on an exact weight basis, e.g., frog legs, refer to the specific method for instructions.

1. **Dried egg yolk, dried egg whites, dried whole eggs, frozen eggs and egg products, prepared powdered mixes (cake, cookie, doughnut, biscuit, and bread), infant formula, and oral or tube feedings containing egg**. Preferably, do not thaw frozen samples before analysis. If frozen samples must be tempered to obtain analytical portion, thaw suitable portion as rapidly as possible to minimize increase in number of competing organisms or to reduce potential to injured Salmonella organisms. Thaw below 45°C for $\leq$ 15 min with continuous agitation in thermostatically controlled water bath or thaw within 18 hr at 2-5°C. Aseptically weigh 25 g sample into sterile, wide-mouth, screw-cap jar (500 ml) or other appropriate container. For nonpowdered samples, add 225 ml sterile lactose broth. If product is powdered, add about 15 ml sterile lactose broth and stir with glass rod, spoon, or tongue depressor to smooth suspension. Add 3 additional portions of lactose broth, 10, 10, and 190 ml, for total of 225 ml. Stir thoroughly until sample is suspended without lumps. Cap jar securely and let stand 60 min at room temperature. Mix well by swirling and determine pH with test paper. Adjust pH, if necessary, to 6.8 $\pm$ 0.2 with sterile 1 N

NaOH or 1 N HC1. Cap jar securely and mix well before determining final pH. Loosen jar cap about 1/4 turn and incubate 24 ± 2 hr at 35°C. Continue as in D, 1-12, below.

2. **Fresh liquid eggs**. Wash fresh eggs with stiff brush and drain. Soak in 0.1% mercuric chloride solution for 1 hr. Protective gloves should be worn when handling this solution. Pour off mercuric chloride solution and replace with 70% ethanol. Soak 30 min. Crack eggs aseptically. Retain yolks only. Aseptically weigh 25 g egg yolk into sterile 500 ml Erlenmeyer flask. Add 225 ml lactose broth and swirl thoroughly. Let stand 60 min at room temperature. Mix well by swirling and determine pH with test paper. Adjust pH, if necessary, to 6.8 ± 0.2. Incubate 24 ± 2 hr at 35°C. Continue as in D, 1-12, below.

3. **Nonfat dry milk**

   a. Instant. Aseptically weigh 25 g sample into sterile beaker (250 ml) or other appropriate container. Using sterile glass or paper funnel (made with tape to withstand autoclaving), pour 25 g analytical unit gently and slowly over surface of 225 ml brilliant green water contained in sterile 500 ml Erlenmeyer flask or other appropriate container. Alternatively, 25 g analytical units may be composited and poured over the surface of proportionately larger volumes of brilliant green water. Prepare brilliant green water by adding 2 ml 1% brilliant green dye solution per 1,000 ml sterile distilled water. Let container stand undisturbed for 60 ± 5 min. Incubate loosely capped container, without mixing or pH adjustment, for 24 ± 2 hr at 35°C. Continue as in D, 1-12, below.

   b. Noninstant. Examine as described for instant nonfat dry milk, except 25 g analytical units may not be composited.

4. **Dry whole milk**. Aseptically weigh 25 g sample into sterile, wide-mouth, screw-cap jar (500 ml) or other appropriate container. Add 225 ml sterile distilled water and mix well by swirling. Cap jar securely and let stand 60 min at room temperature. Mix well by swirling and determine pH with test paper. Adjust pH, if necessary, to 6.8 ± 0.2. Cap jar securely and mix well before determining final pH. Add 0.45 ml 1% aqueous brilliant green dye solution and mix well. Loosen jar cap 1/4 turn and incubate jar 24 ± 2 hr at 35°C. Continue as in D, 1-12, below.

5. **Casein**. Aseptically weigh 25 g sample into sterile blender jar. Add 225 ml sterile lactose broth to sample and blend for 2 min. Aseptically transfer blended homogenate into sterile, wide-mouth, screw-cap jar (500 ml) or other appropriate container. Cap jar securely and let stand 60 min at room temperature. Mix well by shaking and determine pH with test paper. Adjust pH, if necessary, to 6.8 ± 0.2. Cap jar securely and mix well before determining final pH. Loosen jar cap 1/4 turn and incubate jar 24 ± 2 hr at 35°C. Continue as in D, 1-12, below.

6. **Soy flour**. Aseptically weigh 25 g sample into sterile beaker (250 ml) or other appropriate container. Using sterile glass or paper funnel (made with tape to withstand autoclaving), pour 25 g analytical unit gently and slowly over surface of 225 ml lactose broth contained in sterile 500 ml Erlenmeyer flask or other appropriate container. Analytical units weighing 25 g may not be composited. Let container stand undisturbed for 60 ± 5 min. Incubate loosely capped container, without mixing or pH adjustment, for 24 ± 2 hr at 35°C. Continue as in D, 1-12, below.

7. **Egg-containing products (noodles, egg rolls, macaroni, spaghetti), cheese, dough, prepared salads (ham, egg, chicken, tuna, turkey), fresh, frozen, or dried fruits and vegetables, nut meats, crustaceans (shrimp, crab, crayfish, langostinos, lobster), and fish**. Preferably, do not thaw frozen samples before analysis. If frozen sample must be tempered to obtain analytical portion, thaw below 45°C for $\leq$ 15 min with continuous agitation in thermostatically controlled water bath or thaw within 18 hr at 2-5°C. Aseptically weigh 25 g sample into sterile blending container. Add 225 ml sterile lactose broth and blend 2 min. Aseptically transfer homogenized mixture to sterile, wide-mouth, screw-cap jar (500 ml) or other appropriate container and let stand 60 min at room temperature with jar securely capped. Mix well by swirling and determine pH with test paper. Adjust pH, if necessary, to 6.8 $\pm$ 0.2. Mix well and loosen jar cap about 1/4 turn. Incubate 24 $\pm$ 2 hr at 35°C. Continue as in D, 1-12, below.

8. **Dried yeast**. Aseptically weigh 25 g sample into sterile, wide-mouth, screw-cap jar (500 ml) or other appropriate container. Add 225 ml sterile trypticase (tryptic) soy broth. Mix well to form smooth suspension. Let stand 60 min at room temperature with jar securely capped. Mix well by swirling and determine pH with test paper. Adjust pH, if necessary, to 6.8 $\pm$ 0.2, mixing well before determining final pH. Loosen jar cap 1/4 turn and incubate 24 $\pm$ 2 hr at 35°C. For dried inactive yeast, continue as in D, 1-12, below. For dried active yeast, mix incubated sample and transfer 1 ml each to 10 ml lauryl tryptose broth and to 10 ml tetrathionate (TT) broth. Incubate selective enrichment broths 24 $\pm$ 2 hr at 35°C. Vortex sample mixtures thoroughly and steak 3 mm loopful of each broth to each of 3 selective agars as described in D, 3 and 4, below. Continue as in the same procedure, 5-12.

9. **Frosting and topping mixes**. Aseptically weigh 25 g sample into sterile, wide-mouth, screw-cap jar (500 ml) or other appropriate container. Add 225 ml nutrient broth and mix well. Cap jar securely and let stand 60 min at room temperature. Mix well by swirling and determine pH with test paper. Adjust pH, if necessary, to 6.8 $\pm$ 0.2. Loosen jar cap about 1/4 turn and incubate 24 $\pm$ 2 hr at 35°C. Continue as in D, 1-12, below.

10. **Spices**

    a. Black pepper, white pepper, celery seed or flakes, chili powder, cumin, paprika, parsley flakes, rosemary, sesame seed, thyme, and vegetable flakes. Aseptically weigh 25 g sample into sterile, wide-mouth, screw-cap jar (500 ml) or other appropriate container. Add 225 ml sterile trypticase soy broth and mix well. Cap jar securely and let stand 60 min at room temperature. Mix well by swirling and determine pH with test paper. Adjust pH, if necessary, to 6.8 $\pm$ 0.2. Loosen jar cap about 1/4 turn and incubate 24 $\pm$ 2 hr at 35°C. Continue as in D, 1-12, below.

    b. Onion flakes, onion powder, garlic flakes, and garlic powder. Aseptically weigh 25 g sample into sterile, wide-mouth, screw-cap jar (500 ml) or other appropriate container. Preenrich sample in trypticase soy broth with added K_2SO_3 (5 g K_2SO_3 per 1000 ml trypticase soy broth, resulting in final 0.5% K_2SO_3 concentration). Add K_2SO_3 to broth before autoclaving 225 ml volumes in 500 ml Erlenmeyer flasks at 121°C for 15 min. After autoclaving, aseptically determine, and, if necessary, adjust final volume to 225 ml. Add 225 ml sterile trypticase soy broth with added K_2SO_3 to sample and mix well. Continue as in 10a, above.

c. Allspice, cinnamon, cloves, and oregano. At this time there are no known methods for neutralizing the toxicity of these 4 spices. Dilute them beyond their toxic levels to examine them. Examine allspice, cinnamon, and oregano at 1:100 sample/broth ratio, and cloves at 1:1,000 sample/broth ratio. Examine leafy condiments at sample/broth ratio greater than 1:10 because of physical difficulties encountered by absorption of broth by dehydrated product. Examine these spices as described in 10a, above, maintaining recommended sample/broth ratios.

11. **Candy and candy coating (including chocolate)**. Aseptically weigh 25 g sample into sterile blending container. Add 225 ml sterile, reconstituted nonfat dry milk and blend 2 min at 10,000-12,000 rpm. Aseptically transfer homogenized mixture to sterile, wide-mouth, screw-cap jar (500 ml) or other appropriate container and let stand 60 min at room temperature with jar securely capped. Mix well by swirling and determine pH with test paper. Adjust pH, if necessary, to 6.8 $\pm$ 0.2. Add 0.45 ml 1% aqueous brilliant green dye solution and mix well. Loosen jar caps 1/4 turn incubate. Continue as in D, 1-12, below.

12. **Coconut**. Aseptically weigh 25 g sample into sterile, wide-mouth, screw-cap jar (500 ml) or other appropriate container. Add 225 ml sterile lactose broth, shake well, and let stand 60 min at room temperature with jar securely capped. Mix well by swirling and determine pH with test paper. Adjust pH, if necessary, to 6.8 $\pm$ 0.2. Add up to 2.25 ml steamed (15 min) Tergitol Anionic 7 and mix well. Alternatively, steamed (15 min) Triton X-100 may be used. Limit use of these surfactants to minimum quantity needed to initiate foaming. For Triton X-100 this quantity may be as little as 2 or 3 drops. Loosen jar cap about 1/4 turn and incubate 24 $\pm$ 2 hr at 35°C. Continue as in D, 1-12, below.

13. **Food dyes and food coloring substances**. For dyes with pH 6.0 or above (10% aqueous suspension), use method described for dried whole eggs (1, above). For laked dyes or dyes with pH below 6.0, aseptically weigh 25 g sample into sterile, wide-mouth, screw-cap jar (500 ml) or other appropriate container. Add 225 ml TT broth without brilliant green dye. Mix well and let stand 60 min at room temperature with jar securely capped. Using pH meter, adjust pH to 6.8 $\pm$ 0.2. Add 2.25 ml 0.1% brilliant green dye solution and swirl thoroughly. Loosen jar cap about 1/4 turn and incubate 24 $\pm$ 2 hr at 35°C. Continue as in D, 3-12 below.

14. **Gelatin**. Aseptically weigh 25 g sample into sterile, wide-mouth, screw-cap jar (500 ml) or other appropriate container. Add 225 ml sterile lactose broth and 5 ml 5% aqueous gelatinase solution and mix well. Cap jar securely and let stand 60 min at room temperature. Mix well by swirling and determine pH with test paper. Adjust pH, if necessary, to 6.8 $\pm$ 0.2. Loosen jar cap about 1/4 turn and incubate 24 $\pm$ 2 hr at 35°C. Continue as in D, 1-12, below.

15. **Meats, meat substitutes, meat by-products, animal substances, glandular products, and meals (fish, meat, bone)**. For heated, processed, and dried products, aseptically weigh 25 g sample into sterile blending container. Add 225 ml sterile lactose broth and blend 2 min at 10,000-12,000 rpm. Aseptically transfer homogenized mixture to sterile wide-mouth, screw-cap jar (500 ml) or other appropriate container and let stand 60 min at room temperature with jar securely capped. If mixture is powder or is ground or comminuted, blending may be omitted. Mix well by swirling and determine pH with test paper. Adjust pH, if necessary, to 6.8 $\pm$ 0.2. Add up to 2.25 ml steamed (15 min) Tergitol Anionic 7

and mix well. Alternatively, use steamed (15 min) Triton X-100. Limit use of these surfactants to minimum quantity needed to initiate foaming. Actual quantity will depend on composition of test material. Surfactants will not be needed in analysis of powdered glandular products. Loosen jar caps 1/4 turn and incubate sample mixtures 24 ± 2 hr at 35°C. Continue as in D, 1-12, below.

For raw or highly contaminated products, aseptically weigh 25 g portion of product into sterile blending containers. If product is powder or is ground or comminuted, blending may be omitted and product weighed directly into 500 ml Erlenmeyer flasks or other appropriate container. Add 225 ml lactose broth to blender jar and blend 2 min at 10,000-12,000 rpm. For samples that do not require blending, add lactose broth and mix thoroughly. Aseptically transfer blended sample to sterile 500 ml Erlenmeyer flask or other appropriate container. Let blended and nonblended samples stand 60 min at room temperature with jar securely capped. Mix well by swirling and determine pH with test paper. Adjust pH, if necessary, to 6.8 ± 0.2. Loosen jar caps 1/4 turn and incubate sample mixture 24 ± 2 hr at 35°C. Continue as in D, 1-12, below.

16. **Frog legs**. Place 15 pairs of frog legs into a sterile plastic bag and cover with sterile lactose broth (see A, 21-23, above). If single legs are estimated to average 25 g or more, examine only one leg of each of 15 pairs. Place bag in large plastic beaker or other suitable container and shake 15 min on mechanical shaker set for 100 excursions/min with stroke of 4 cm. Pour off lactose broth from bag into another sterile plastic bag and add more lactose broth to total volume of 3,500 ml. Mix well and let stand 60 min at room temperature. Adjust pH, if necessary, to 6.8 ± 0.2, using pH paper. Place plastic bag containing the lactose broth into plastic beaker or other suitable container. Incubate 24 ± 2 hr at 35°C. Continue as in D, 1-12, below.

17. **Rabbit carcasses**. Place each of 3 rabbit carcasses into sterile plastic bag and cover with sterile lactose broth (see A, 21-23, above). Place bag in large plastic beaker or other suitable container and shake 15 min on mechanical shaker set for 100 excursions/min with stroke of 4 cm. Composite lactose broth rinsings by pouring into another sterile container and add more lactose broth to total volume of 3,500 ml. Mix well and let stand 60 min at room temperature. Adjust pH, if necessary, to 6.8 ± 0.2, using pH paper. Incubate 24 ± 2 hr at 35°C. Continue examination as in D, 1-12, below.

D. ISOLATION OF SALMONELLA

1. Tighten lid and gently shake incubated sample (except shrimp) mixture; transfer 1 ml mixture to 10 ml selenite cystine (SC) broth and another 1 ml mixture to 10 ml TT broth. For incubated shrimp sample mixtures only, gently shake sample and transfer 0.1 ml to 10 ml Rappaport-Vassiliadis (RV) medium and another 1 ml to 10 ml TT broth.

2. Incubate SC and TT broths 24 $\pm$ 2 hr at 35°C. Incubate RV medium 24 $\pm$ 2 hr at 43°C.

3. Mix (vortex, if tube) and streak 3 mm loopful incubated TT broth on bismuth sulfite (BS) agar, xylose lysine desoxycholate (XLD) agar, and Hektoen enteric (HE) agar. Prepare BS plates the day before streaking and store in dark at room temperature until streaked.

4. Repeat with 3 mm loopful of SC broth (for samples other than shrimp) and of RV medium (for samples of shrimp).

5. Incubate plates 24 $\pm$ 2 hr at 35°C.

6. Examine plates for presence of colonies suspected to be Salmonella:

   a. Hektoen enteric (HE) agar. Blue-green to blue colonies with or without black centers. Many cultures of Salmonella may produce colonies with large, glossy black centers or may appear as almost completely black colonies. Atypically, a few Salmonella species produce yellow colonies with or without black centers.

   b. Bismuth sulfite (BS) agar. Typical Salmonella colonies may appear brown, gray, or black; sometimes they have a metallic sheen. Surrounding medium is usually brown at first, but may turn black in time with increased incubation, producing the so-called halo effect. Some strains may produce green colonies with little or no darkening of surrounding medium.

   c. Xylose lysine desoxycholate (XLD) agar. Pink colonies with or without black centers. Many cultures of Salmonella may have large, glossy black centers or may appear as almost completely black colonies. Atypically, a few Salmonella species produce yellow colonies with or without black centers.

7. Select 2 or more colonies typical or suspected to be Salmonella from each selective agar. Inoculate into triple sugar iron (TSI) agar and lysine iron (LI) agar. If BS agar plates have no colonies typical or suspected to be Salmonella, or no growth whatsoever, incubate them an additional 24 hr.

8. Lightly touch the very center of the colony to be picked with sterile inoculating needle and inoculate TSI agar slant by streaking slant and stabbing butt. Without flaming, inoculate LI agar slant by stabbing butt twice and then streaking slant. Since lysine decarboxylation reaction is strictly anaerobic, the LI agar slants must have deep butt (4 cm). Store picked selective agar plates at 5-8°C.

9. Incubate TSI agar and LI agar slants at 35°C for 24 $\pm$ 2 hr and 48 $\pm$ 2 hr, respectively. Cap tubes loosely to maintain aerobic conditions while incubating slants to prevent excessive H_2S production. Salmonella in culture typically produces alkaline (red) slant and acid

(yellow) butt, with or without production H_2S (blackening) in TSI agar. In LI agar, Salmonella typically produces alkaline (purple) reaction in butt of tube. Consider only distinct yellow in butt of tube as acidic (negative) reaction. Do not eliminate cultures that produce discoloration in butt of tube solely on this basis. Most Salmonella cultures produce H_2S in LI agar.

10. Re-examine 48 hr BS agar plates for colonies suspected to be Salmonella. Pick 2 or more of these colonies, as described in 8, above, and continue procedure.

11. All cultures that give an alkaline butt in LI agar, regardless of TSI agar reaction, should be retained as potential Salmonella isolates and submitted for biochemical and serological tests. Cultures that give an acid butt in LI agar and an alkaline slant and acid butt in TSI agar should also be considered potential Salmonella isolates and should be submitted for biochemical and serological tests. Cultures that give an acid butt in LI agar and an acid slant and acid butt in TSI agar may be discarded as not being Salmonella. Test retained, presumed-positive TSI agar cultures as directed in 12, below, to determine if they are Salmonella species, including S. arizonae. If TSI agar cultures fail to give typical reactions for Salmonella (alkaline slant and acid butt), pick additional suspicious colonies from selective medium plate not giving presumed-positive culture, and inoculate TSI agar and LI agar slants as described in 8, above.

12. Apply biochemical and serological identification tests to:

    a. Three presumptive TSI agar cultures recovered from set of plates streaked from SC broth, if present, and 3 presumptive TSI agar cultures recovered from plates streaked from TT broth, if present.

    b. If 3 presumptive-positive TSI culture are not isolated from 1 set of agar plates, test other presumptive-positive TSI agar cultures, if isolated, by biochemical and serological tests. Examine a minimum of 6 TSI cultures for each 25 g analytical unit.

E. IDENTIFICATION OF SALMONELLA

1. **Mixed cultures**. Streak TSI agar cultures, that appear to be mixed, on MacConkey agar, HE agar, or XLD agar. Incubate plates 24 $\pm$ 2 hr at 35°C. Examine plates for presence of colonies suspected to be Salmonella.

    a. MacConkey agar. Typical colonies appear transparent and colorless, sometimes with dark center. Colonies of Salmonella will clear areas of precipitated bile caused by other organisms sometimes present.

    b. Hektoen enteric (HE) agar. See D, 6a, above.

    c. Xylose lysine desoxycholate (XLD) agar. See D, 6c, above.

2. **Pure cultures**

a. Urease test (conventional). With sterile needle, inoculate growth from each presumed-positive TSI agar slant culture into tubes of urea broth. Since occasional, uninoculated tubes of urea broth turn purple-red (positive test) on standing, include uninoculated tube of this broth as control. Incubate 24 $\pm$ 2 hr at 35°C.

b. Optional urease test (rapid). Transfer two 3 mm loopfuls of growth from each presumed-positive TSI agar slant culture into tube of rapid urea broth. Incubate 2 hr in 37 $\pm$ 0.5°C water bath. Discard all cultures giving positive test. Retain for further study all cultures that give negative test (no change in color of medium).

3. **Serological polyvalent flagellar (H) test**

a. Perform the polyvalent flagellar (H) test at this point, or later, as directed in E-5, below. Inoculate growth from each urease-negative TSI agar slant into (1) brain heart infusion broth and incubate 4-6 hr at 35°C until visible growth occurs (to test on same day); or (2) trypticase soy-tryptose broth and incubate 24 $\pm$ 2 hr at 35°C (to test on following day). Add 2.5 ml formalinized physiological saline solution to 5 ml of either broth culture.

b. Select 2 formalinized broth cultures and test with Salmonella polyvalent flagellar (H) antisera. Place 0.5 ml of appropriately diluted Salmonella polyvalent flagellar (H) antiserum in 10 x 75 mm or 13 x 100 mm serological test tube. Add 0.5 ml antigen to be tested. Prepare saline control by mixing 0.5 ml formalinized physiological saline solution with 0.5 ml formalinized antigen. Incubate mixtures in 48-50°C water bath. Observe at 15 min intervals and read final results in 1 hr. Positive--agglutination in test mixture and no agglutination in control. Negative--no agglutination in test mixture and no agglutination in control. Nonspecific--agglutination in both test mixture and control. Test the cultures giving such results with Spicer-Edwards antisera.

4. **Spicer-Edwards serological test**. Use this test as an alternative to the polyvalent flagellar (H) test. It may be used with cultures giving nonspecific agglutination in polyvalent flagellar (H) test. Perform Spicer-Edwards flagellar (H) antisera test as described in 3b, above. Perform additional biochemical tests (5a-c, below) on cultures giving positive flagellar test results. If both formalinized broth cultures are negative, perform serological tests on 4 additional broth cultures (3a-(1), or (2), above). If possible, obtain 2 positive cultures for additional biochemical testing (5a-c, below). If all urease-negative TSI agar cultures from sample give negative serological flagellar (H) test results, perform additional biochemical tests (5a-c, below).

5. **Testing of urease-negative cultures.**

a. Lysine decarboxylase broth. If LI agar test was satisfactory, it need not be repeated. Use lysine decarboxylase broth (M49) for final determination of lysine decarboxylase if culture gives doubtful LI agar reaction. Inoculate broth with small amount of growth from TSI agar slant suspicious for Salmonella. Replace cap tightly and incubate 48 $\pm$ 2 hr at 35°C but examine at 24 hr intervals. Salmonella species cause alkaline reaction indicated by purple color throughout medium. Negative test is indicated by yellow color throughout medium. If medium appears discolored (neither purple nor yellow) add a few drops of 0.2% bromcresol purple dye and re-read tube reactions.

b. Phenol red dulcitol broth or purple broth base with 0.5% dulcitol. Inoculate broth with small amount of growth from TSI agar culture. Replace cap loosely and incubate 48 ± 2 hr at 35°C, but examine after 24 hr. Most Salmonella species give a positive test, indicated by gas formation in the inner fermentation vial and acid pH (yellow) of medium. Production of acid should be interpreted as a positive reaction. Negative test is indicated by no gas formation in inner fermentation vial and red (with phenol red as indicator) or purple (with bromcresol purple as indicator) color throughout medium.

c. Tryptone (or tryptophane) broth). Inoculate broth with small amount of growth from TSI agar culture. Incubate 24 ± 2 hr at 35°C and proceed as follows:

1) Potassium cyanide (KCN) medium. Transfer 3 mm loopful of 24 hr tryptophane broth culture to KCN medium. Heat rim of tube so that good seal is formed when tube is stoppered with wax-coated cork. Incubate 48 ± 2 hr at 35°C but examine after 24 hr. Interpret growth (indicated by turbidity) as positive. Most Salmonella species do not grow in this medium, as indicated by lack of turbidity.

2) Malonate broth. Transfer 3 mm loopful of 24 hr tryptone broth culture to malonate broth. Since occasional uninoculated tubes of malonate broth turn blue (positive test) on standing, include uninoculated tube of this broth as control. Incubate 48 ± 2 hr at 35°C, but examine after 24 hr. Most Salmonella species cultures give negative test (green or unchanged color) in this broth.

3) Indole test. Transfer 5 ml of 24 hr tryptophane broth culture to empty test tube. Add 0.2-0.3 ml Kovacs' reagent. Most Salmonella cultures give negative test (lack of deep red color at surface of broth). Record intermediate, varying shades of orange and pink as ±.

4) Serological flagellar (H) tests for Salmonella. If either polyvalent flagellar (H) test (3, above) or the Spicer-Edwards flagellar (H) test tube test (4, above) has not already been performed, then either test may be performed here.

5) Discard, as not Salmonella, any culture that shows either positive indole test and negative serological flagellar (H) test, or positive KCN test and negative lysine decarboxylase test.

6. **Serological somatic (O) tests for Salmonella**. (Pre-test all antisera to Salmonella with known cultures.)

a. Polyvalent somatic (O) test. Using wax pencil, mark off 2 sections about 1 x 2 cm each on inside of glass or plastic Petri dish (15 x 100 mm). Commercially available sectioned slides may be used. Place one-half of 3 mm loopful of culture from 24-48 hr TSI agar slant on dish in upper portion of each rectangular, crayon-marked section. Add 1 drop saline solution to lower part of one section only. Add 1 drop Salmonella polyvalent somatic (O) antiserum to other section only. With clean sterile transfer loop or needle, emulsify culture in saline solution for one section and repeat for other section containing antiserum. Tilt mixtures in back-and-forth motion for 1 min and observe against dark background in good illumination. Consider any degree of agglutination as a positive reaction. Classify polyvalent somatic (O) test results as follows: Positive--agglutination in test mixture; no agglutination in saline control. Negative--no agglutination in test mixture; no agglutination in saline control. Nonspecific--agglutination in test and in control mixtures. Perform further biochemical and serological tests as described in Edwards and Ewing's Identification Enterobacteriaceae (1).

b. Somatic (O) group tests. Test as in E-6a, above, using individual group somatic (O) antisera including Vi, if available, in place of Salmonella polyvalent somatic (O) antiserum. Record cultures that give positive agglutination with individual somatic (O) antiserum as positive for that group. Record cultures that do not react with individual somatic (O) antisera as negative for that group.

7. **Additional biochemical tests**. Classify, as Salmonella, those cultures which exhibit typical Salmonella reactions for test Nos. 1-11, shown in Table 1. If one TSI agar culture from a 25 g analytical unit is classified as Salmonella, further testing of other TSI cultures from the same 25 g analytical unit is unnecessary. Cultures that contain demonstratable Salmonella antigens as shown by positive Salmonella flagellar (H) test, but do not have biochemical characteristics of Salmonella, should be purified (1, above) and retested, beginning with 2, above. Perform the following additional tests on cultures that do not give typical Salmonella reactions for test Nos. 1-11 in Table 1 and that consequently do not classify as Salmonella.

a. Phenol red lactose broth or purple lactose broth.

1) Inoculate broth with small amount of growth from unclassified 24-48 hr TSI agar slant. Incubate 48 ± 2 hr at 35°C, but examine after 24 hr. Positive--acid production (yellow color) and gas production in inner fermentation vial. Consider production of acid only as positive reaction. Most cultures of Salmonella give negative test result, indicated by no gas formation in inner fermentation vial and red (with phenol red as indicator) or purple (with bromcresol purple as indicator) color throughout medium.

2) Discard, as not Salmonella, cultures that give positive lactose tests, except cultures that give acid slants in TSI agar and positive reactions in LI agar, or cultures that give positive malonate broth reactions. Perform further tests on these cultures to determine if they are S. arizonae.

b. Phenol red sucrose broth or purple sucrose broth. Follow procedure described in E, 7a-1), above. Discard, as not Salmonella, cultures that give positive sucrose tests, except those that give acid slants in TSI agar and positive reactions in LI agar.

c. MR-VP broth. Inoculate medium with small amount of growth from each unclassified TSI agar slant suspected to contain Salmonella. Incubate 48 ± 2 hr at 35°C.

1) Perform Voges-Proskauer (VP) test at room temperature as follows: Transfer 1 ml 48 hr culture to test tube and incubate remainder of MR-VP broth additional 48 hr at 35°C. Add 0.6 ml alpha-naphthol and shake well. Add 0.2 ml 40% KOH solution and shake. To intensify and speed reaction, add a few crystals of creatine. Read results after 4 hr: development of pink-to-ruby red color throughout medium is positive test. Most cultures of Salmonella are VP-negative, indicated by absence of development of pink-to-red color throughout broth.

2) Perform methyl red test as follows: To 5 ml of 96 hr MR-VP broth culture, add 5-6 drops of methyl red indicator. Read results immediately. Most Salmonella cultures give positive test, indicated by diffuse red color in medium. A distinct yellow color is negative test. Discard as not Salmonella, cultures that give positive KCN and VP tests and negative methyl red test.

d. Simmons citrate agar. Inoculate this agar, using needle containing growth from unclassified TSI agar slant. Inoculate by streaking slant and stabbing butt. Incubate 96 ± 2 hr at 35°C. Read results as follows. Positive--presence of growth, usually accompanied by color change from green to blue. Most cultures of Salmonella cultures are citrate-positive. Negative--no growth or very little growth and no color change.

8. **Classification of cultures**. Classify, as Salmonella, cultures that have reaction patterns in Table 1. Discard, as not Salmonella, cultures that give results listed in any subdivision of Table 2. Classify, by performing additional tests described in Edwards and Ewing's Identification of Enterobacteriaceae (1), any culture that is not clearly identified as Salmonella by classification scheme in Table 1 or not eliminated as not being Salmonella by test reactions in Table 2. If neither of the 2 TSI cultures carried through biochemical tests confirms the isolate as Salmonella, perform biochemical tests, beginning with E-5, on remaining urease-negative TSI cultures from same 25 g analytical unit.

9. **Presumptive generic identification of Salmonella**. As an alternative to the conventional biochemical tube system, use any of 4 commercial biochemical systems (API 20E, Enterotube II, Enterobacteriaceae II, or MICRO-ID) for presumptive generic identification of foodborne Salmonella. Choose a commercial system based on demonstration in analyst's own laboratory of adequate correlation between commercial system and biochemical tube system delineated in this identification section. Commercial biochemical kits should not be used as a substitute for serological tests. Assemble supplies and prepare reagents required for the kit. Inoculate each unit according to directions supplied by manufacturer, incubating for time and temperature specified. Add reagents, observe, and record results. For presumptive identification, classify cultures, using flow charts and tables supplied by manufacturer, as Salmonella or not.

For confirmation of cultures presumptively identified as Salmonella, perform the Salmonella serological somatic (O) test (6, above) and the Salmonella serological flagellar (H) test (3, above) or the Spicer-Edwards flagellar (H) test (4, above) and classify cultures according to the following guidelines:

a. Confirm, as Salmonella, those cultures classified as presumptive Salmonella with commercial biochemical kits when the culture demonstrates positive Salmonella somatic (O) test and positive Salmonella (H) test.

b. Discard cultures, presumptively classified as not Salmonella with commercial biochemical kits, when cultures conform to manufacturer's criteria for classifying cultures as not Salmonella.

c. For cultures which do not conform to a or b, above, classify according to additional tests specified in 2-7, above, or additional tests as specified by Ewing (1), or send to reference typing laboratory for definitive serotyping and identification.

10. **Treatment of cultures giving negative flagellar (H) test**. If biochemical reactions of certain flagellar (H)-negative culture strongly suggest that it is Salmonella, the negative flagellar agglutination may be the result of nonmotile organisms or insufficient development of flagellar antigen. Proceed as follows: Inoculate motility test medium in Petri dish, using small amount of growth from TSI slant. Inoculate by stabbing medium once about 10 mm from edge of plate to depth of 2-3 mm. Do not stab to bottom of plate or inoculate any other portion. Incubate 24 hr at 35°C. If organisms have migrated 40 mm or more, retest as follows: Transfer 3 mm loopful of growth that migrated farthest to trypticase soy-tryptose broth. Repeat either polyvalent flagellar (H) (3, above) or Spicer-Edwards (4, above) serological tests. If cultures are not motile after the first 24 hr, incubate an additional 24 hr at 35°C; if still not motile, incubate up to 5 days at 25°C. Classify culture as nonmotile if above tests are still negative. If flagellar (H)-negative culture is supected of being a species of Salmonella on the basis of its biochemical reactions, submit the culture to a reference laboratory for serotyping.

F. FLOW SHEET

See Figure 5.

G. RECORD SHEET

See Annex 4.

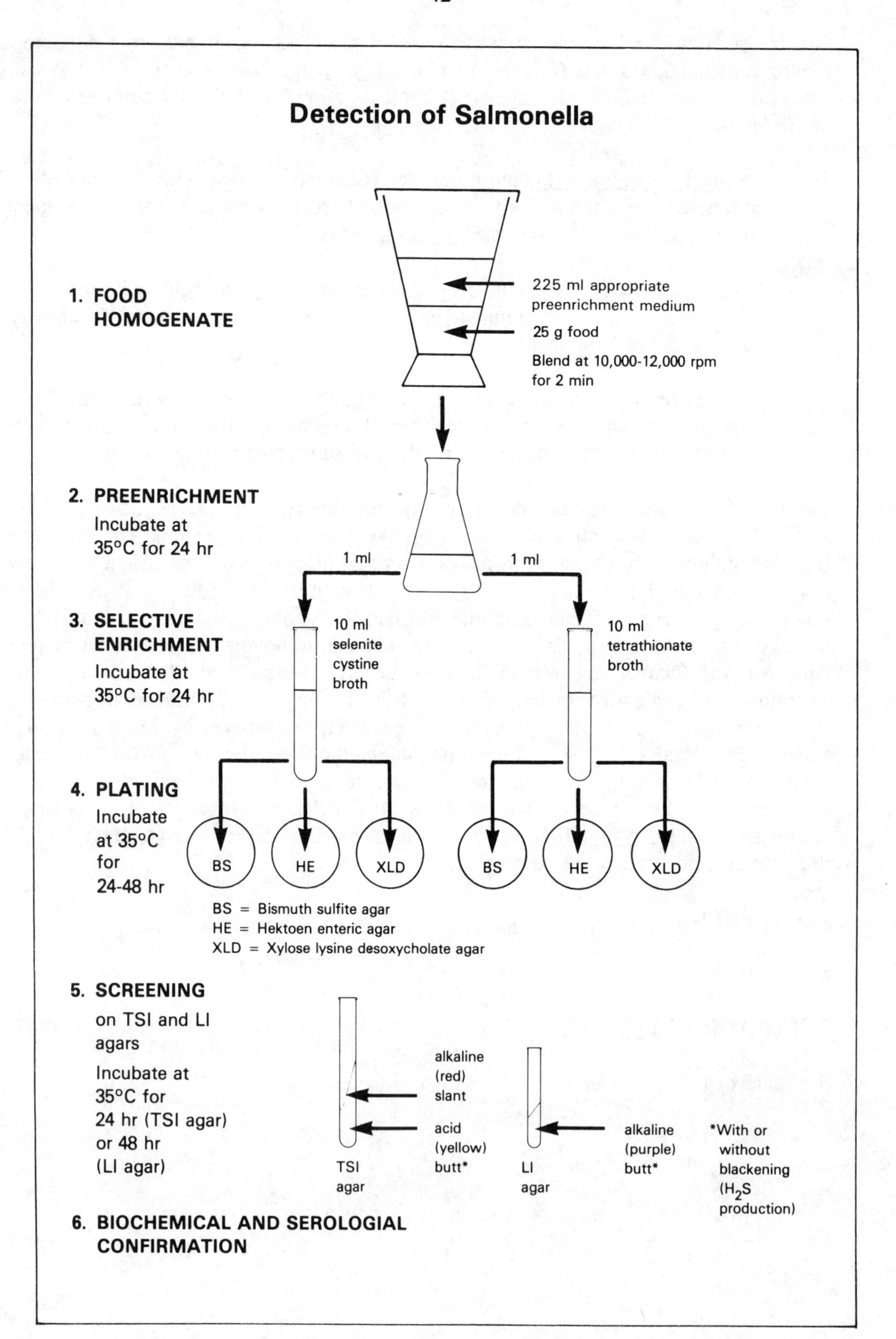
Detection of Salmonella
1. FOOD HOMOGENATE
225 ml appropriate preenrichment medium
25 g food
Blend at 10,000-12,000 rpm for 2 min
2. PREENRICHMENT
Incubate at 35°C for 24 hr
1 ml
1 ml
3. SELECTIVE ENRICHMENT
Incubate at 35°C for 24 hr
10 ml selenite cystine broth
10 ml tetrathionate broth
4. PLATING
Incubate at 35°C for 24-48 hr
BS
HE
XLD
BS
HE
XLD
BS = Bismuth sulfite agar
HE = Hektoen enteric agar
XLD = Xylose lysine desoxycholate agar
5. SCREENING
on TSI and LI agars
Incubate at 35°C for 24 hr (TSI agar) or 48 hr (LI agar)
alkaline (red) slant
acid (yellow) butt*
TSI agar
LI agar
alkaline (purple) butt*
*With or without blackening (H_2S production)
6. BIOCHEMICAL AND SEROLOGIAL CONFIRMATION

Figure 5

H. PRECAUTIONS AND LIMITATIONS OF METHOD

In determining the presence of Salmonella in foods, the following points should be observed:

1. The requisite number of sample units, based on a statistically valid sampling plan, should be analyzed.

2. Samples of instant nonfat dry milk, examined by the soak method, may be composited up to 375 g. Samples of noninstant nonfat dry milk may also be examined by the soak method but must not be composited.

3. For the analysis of coconut and fatty meats, Tergitol 7 is added to the sample homogenate after pH adjustment rather than before the autoclaving of the lactose broth.

4. For the analysis of dried yeast, LT broth is used instead of SC broth as a selective enrichment.

5. Foods composed of several mixed ingredients present a problem as to selection of the most sensitive analytical procedure. The ingredient present in the largest amount usually determines which method is used.

6. The efficiency of the SC broth is greatest when there is a reduced oxidation-reduction potential. Thus, it is essential that the 10 ml volumes be dispensed into 16 X 150 mm, rather than 20 X 150 mm, tubes.

7. TT broth contains an excess amount of calcium carbonate and sodium thiosulfate which will form a precipitate. Thus, this medium must be swirled frequently when dispensing 10-ml volumes into test tubes.

8. Because freshly prepared BS agar plates are inhibitory to some strains of Salmonella, these plates should be made the day before use. The prepared plates should be stored in the dark at room temperature until ready to be used.

9. After being examined at 24 hr, BS agar plates must be reincubated an additional 24 hr. Plate of HE and XLD agar are incubated for 24 hr only.

10. An analysis should not be concluded after observation of colonies typical for Salmonella on selective agar plates. All suspect colonies must be confirmed biochemically and serologically.

11. A sufficient number of colonies should be picked from the selective agar plates to maintain the sensitivity of the method.

12. Lactose-positive Salmonella strains will appear atypical on HE and XLD agars. It is only on BS agar that these strains will appear typical for Salmonella.

13. Any culture that appears to be mixed, or contaminated, in TSI agar must be purified on MacConkey, XLD, or HE agar before proceeding with biochemical screening and serological identification.

14. Do not exclude TSI culture that appears to be bacterium other than Salmonella if reaction in LI agar is typical (alkaline butt) for Salmonella. These cultures may be S. arizonae or other atypical strains of Salmonella.

15. Since lysine decarboxylation occurs anaerobically, LI agar slant must have a deep butt. Allow medium to solidify in slanted position to form 4 cm butt and 2.5 cm slant.

16. Consider only distinct yellow color in butt of LI agar tube as acidic (negative) reaction. Do not eliminate cultures that produce discoloration in butt of tube solely on this basis.

17. Occasionally, Salmonella cultures showing atypical biochemical results (e.g., hydrogen sulfide negative, dulcitol negative, or malonate positive reactions) may be isolated. It should be realized that the classification of an isolate as Salmonella depends ultimately on its antigenic structure rather than exclusively on its biochemical characteristics.

18. KCN medium must be inoculated from a broth. Inoculation of this medium from a slant culture may produce a false positive reaction.

19. Since occasional, uninoculated tubes of malonate broth turn blue (positive test) on standing, include uninoculated tube of this broth as a control.

RAPID SCREENING METHODS FOR SALMONELLA

In addition to the conventional culture method for the isolation and identification of Salmonella, there are other rapid methods which are acceptable.

Acceptable methods for the rapid isolation of Salmonella are:

1. Fluorescent antibody technique, section 975.54
2. Hydrophobic grid membrane filter method, section 985.42
3. DNA hybridization (isotopic) method, section 987.10
4. DNA hybridization (colorimetric) method, section 990.13
5. Monoclonal enzyme immunoassay (colorimetric), sections 986.35 and 987.11
6. Monoclonal enzyme immunoassay (fluorogenic), section 989.15
7. Polyclonal enzyme immunoassay (colorimetric), section 989.14
8. Immunodiffusion method, section 989.13

Acceptable methods for the rapid identification of Salmonella are:

1. API 20E, section 978.24
2. Enterobacteriaceae II Set, section 978.24

3. Enterotube II. section 978.24

4. MICRO-ID System, section 989.12

The above section numbers refer to AOAC's Official Methods of Analysis (2) which contains details of the above procedures.

REFERENCES

1. **Ewing, W. H.** 1986. Edwards and Ewing's Identification of Enterobacteriaceae, 4th ed. Elsevier Science Publishing, New York, NY.

2. Association of Official Analytical Chemists. 1990. Official Methods of Analysis of the Association of Official Analytical Chemists, Volume 1, 15th ed., K. Helrich (Ed.). Association of Official Analytical Chemists, Arlington, VA.

Table 1

Biochemical and serological reactions of Salmonella

	Test or substrate	Positive	Negative	Salmonella species reaction[a]
1.	Glucose (TSI)	yellow butt	red butt	+
2.	Lysine decarboxylase (LI)	purple butt	yellow butt	+
3.	H_2S (TSI and LI)	blackening	no blackening	+
4.	Urease	purple-red color	no color change	-
5.	Lysine decarboxylase broth	purple color	yellow color	+
6.	Phenol red dulcitol broth	yellow color and/or gas	no gas; no color change	+[b]
7.	KCN medium	growth	no growth	-
8.	Malonate broth	blue color	no color change	-[c]
9.	Indole test	violet color at surface	yellow color at surface	-
10.	Polyvalent flagellar test	agglutination	no agglutination	+
11.	Polyvalent somatic test	agglutination	no agglutination	+
12.	Phenol red lactose broth	yellow color and/or gas	no gas; no color change	-[c]
13.	Phenol red sucrose broth	yellow color and/or gas	no gas; no color change	-
14.	Voges-Proskauer test	pink-to-red color	no color change	-
15.	Methyl red test	diffuse red color	diffuse yellow color	+
16.	Simmons citrate	growth; blue color	no growth; no color change	v

[a] +, 90% or more positive in 1 or 2 days; -, 90% or more negative in 1 or 2 days; v, variable.

[b] Majority of S. arizonae cultures are negative.

[c] Majority of S. arizonae cultures are positive.

Table 2

Criteria for discarding non-Salmonella cultures

	Test or substrate	Results
1.	Urease	positive (purple-red color)
2.	Indole test	positive (violet color at surface)
	Polyvalent flagellar (H) test or Spicer-Edwards flagellar test	negative (no agglutination)
3.	Lysine decarboxylase	negative (yellow color)
	KCN medium	positive (growth)
4.	Phenol red lactose broth	positive (yellow color and/or gas)[a,b]
5.	Phenol red sucrose broth	positive (yellow color and/or gas)[b]
6.	KCN medium	positive (growth)
	Voges-Proskauer test	positive (pink-to-red color)
	Methyl red test	negative (diffuse yellow color)

[a] Test malonate broth positive cultures further to determine if they are S. arizonae.

[b] Do not discard positive broth cultures if corresponding LI agar cultures give typical Salmonella reactions; test further to determine if they are Salmonella species.

CHAPTER 5

SHIGELLA

Shigellosis is an infectious disease spread most commonly by person to person transmission. Young children with poor personal hygiene and those of low socio-economic status living in conditions of crowding and/or inadequate sanitation are the most frequently involved in the transmission of this disease.

Shigella may also be spread by contaminated food and water. Food becomes contaminated by being exposed to Shigella-contaminated fecal material, usually on human hands. Foods most commonly incriminated in shigellosis have been salad, chopped meat, sandwich spread, and seafood.

The infective dose for shigellosis is quite low, perhaps as low as 10 organisms. This situation is made even more serious because of the relative insensitivity of present methods for detecting Shigella in foods.

The genus consists of 4 species: S. dysenteriae (Subgroup A), S. flexneri (Subgroup B), S. boydii (Subgroup C) and S. sonnei (Subgroup D). These species are distinguishable by their biochemical and serological reactions.

A. EQUIPMENT AND MATERIALS

1. Same as for Salmonella, see Chapter 4, A

2. Water baths maintained at 42 ± 0.2°C and 44.0 ± 0.2°C

3. Anaerobic jar with catalyst

B. MEDIA AND REAGENTS

1. Shigella broth with novobiocin (M89)

2. MacConkey agar (M52)

3. Triple sugar iron (TSI) agar (M100)

4. Urea broth (M116)

5. Motility test medium (M59)

6. Potassium cyanide (KCN) medium (M79)

7. Malonate broth (M53)

8. Tryptone (tryptophane) broth (M111)

9. MR-VP medium (M61)

10. Koser citrate broth (M38)

11. Veal infusion agar (M118)

12. Bromcresol purple broth (M14) supplemented with the following carbohydrates, each at a level of 0.5%: (a) adonitol, (b) salicin, (c) glucose, (d) inositol, (e) lactose, (f) mannitol, and (g) sucrose

13. Decarboxylase basal medium (arginine, lysine, ornithine) (M25)

14. Kovacs' reagent (R16)

15. Voges-Proskauer test reagents (R40)

16. Methyl red indicator (R19)

17. Physiological saline solution, 0.85% (sterile) (R28)

18. Novobiocin

19. Polyvalent Shigella antisera for groups A, A_1, B, C, C_1, C_2, D and Alkalescens-Dispar biotypes (1-4)

20. Gram stain reagents (S6)

C. **ENRICHMENT**

1. Enrichment of Shigella sonnei. Aseptically weigh 25 g sample into 225 ml Shigella broth to which novobiocin has been added to a level of 0.5 ug/ml. Hold suspension 10 min at room temperature and shake periodically. Pour eluate into sterile 500 ml Erlenmeyer flask. Place flask in anaerobic jar with fresh catalyst, insert GasPak, and activate by adding water. Because of high humidity within jar, heat catalyst as recommended after each use. Incubate jars in 44°C water bath for 20 hr. Agitate suspension and streak on MacConkey agar plates. Incubate 20 hr at 35°C.

2. Enrichment of other Shigella species. Proceed as above, but incorporate novobiocin to a level of 3 μg/ml and incubate anaerobically in 42.0°C water bath.

D. **ISOLATION OF SHIGELLA**

Examine MacConkey agar plates. Shigella colonies are slightly pink and translucent. Inoculate suspicious colonies into the following media: glucose broth, TSI agar slant, lysine decarboxylase broth, motility agar, and tryptone. Incubate at 35°C for 48 hr, but examine at 20 hr. Discard all cultures showing motility, H_2S, gas formation, lysine decarboxylation, and fermentation of sucrose or lactose. With respect to formation of indole, discard positive cultures from 44°C enrichment. All suspicious isolates from 42°C enrichment may be either positive or negative and consequently should be retained.

E. BIOCHEMICAL IDENTIFICATION

Inoculate cultures giving satisfactory screening reactions to the other recommended biochemicals. Incubate inoculated biochemical media aerobically at 35°C and examine at 24 hr and 48 hr. The behavior of Shigella is summarized as follows: negative for H_2S, urease, glucose (gas), motility, lysine decarboxylase, ornithine decarboxylase (except S. sonnei is positive), sucrose, adonitol, lactose (2 days), salicin, inositol, KCN, malonate, citrate, and Voges-Proskauer. Most Shigella strains are positive for mannitol (except S. dysenteriae is negative) and methyl red reactions. See Table 3.

F. SEROLOGICAL CHARACTERIZATION

Suspend growth from 24 hr veal infusion slant in 3 ml 0.85% saline to McFarland standard No. 5. Mark nine 3 X 1 cm rectangles on clear glass Petri dish with wax pencil. Add drops of suspension, antisera, and saline in accordance with protocol in Table 4.

Mix contents of each rectangle with needle, taking care that no mixing between rectangles occurs. Rock Petri dish 3-4 min to accelerate agglutination. Read extent of agglutination as follows: 0 = no agglutination, 1+ = barely detectable agglutination, 2+ = agglutination with 50% clearing, 3+ = agglutination with 75% clearing, or 4+ = visible floc with suspending fluid totally cleared. Re-examine suspension in monovalent sera belonging to each polyvalent serum in which a distinct positive reaction (2+, 3+, 4+) has occurred. In event of a negative reaction, heat suspension in steamer 30 min to hydrolyze interfering capsular antigen. Re-examine in polyvalent antiserum, and, if positive, in corresponding monovalent sera.

G. FLOW SHEET

See Figure 6.

H. RECORD SHEET

See Annex 5.

Detection of Shigella

1. MIX
food sample with enrichment

2. DECANT
supernatant into sterile flask

3. INCUBATE
anaerobically for 20 hr at 44 °C (*S. sonnei*) or at 42 °C (other *Shigella* species)

4. STREAK
plates and incubate at 35°C for 20-48 hr

5. SCREENING
with TSI agar and other biochemicals

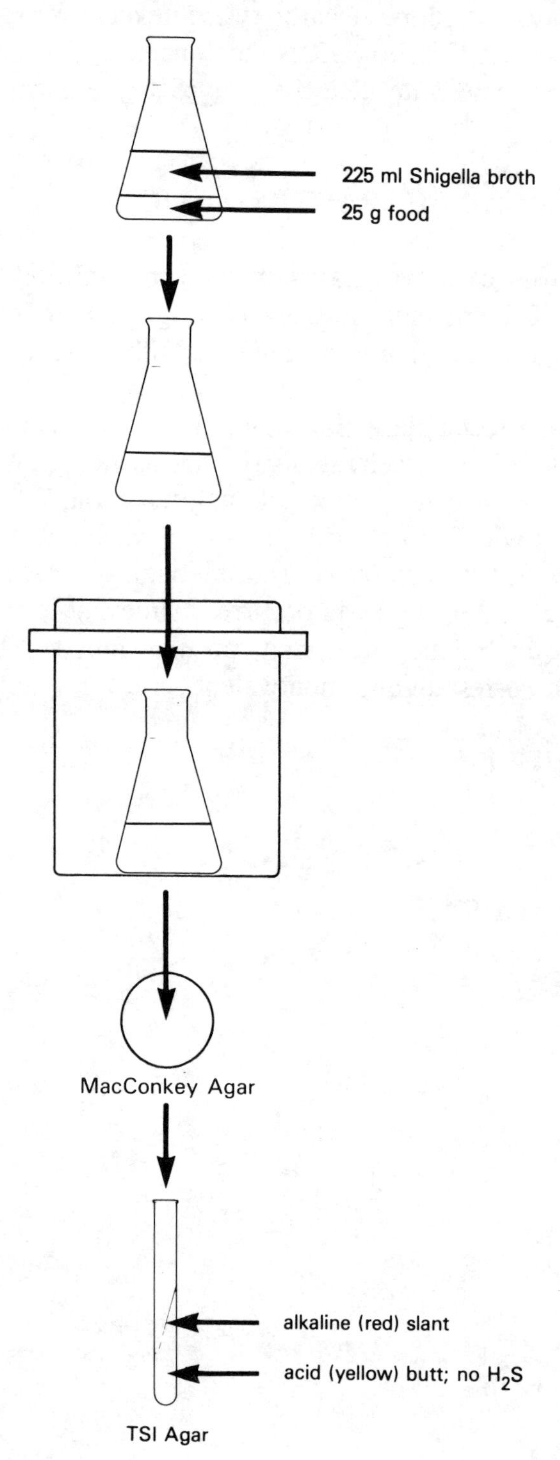

6. SEROLOGICAL CONFIRMATION

Figure 6

I. PRECAUTIONS AND LIMITATIONS OF METHOD

In determining the presence of Shigella in foods, the following points should be observed:

1. Enrichments of Shigella must be incubated in a water bath rather than in an air incubator.

2. Samples should be examined as soon as possible after being collected. Perishable samples should be held under refrigeration at 4°C if they are to be analyzed within 24 hr after collection. If more than 24 hr will elapse between collection and initiation of analysis, then perishable samples must be frozen.

3. Colonial morphology of various strains of Shigella may vary on MacConkey agar. Additionally, many other enteric bacteria may be indistinguishable from Shigella on this selective plating medium. Thus, it is essential to pick several suspect Shigella colonies, if present, from each plate of agar.

4. There are no known methods that are consistently sensitive for Shigella. Recommended methods may not be able to detect Shigella species present in low numbers in certain food samples.

5. Because of the sharing of antigens, there may be some serological cross reactivity between Shigella and other enteric bacteria. Thus, all antisera should be tested with all Shigella species when first used and at periodic intervals thereafter.

Table 3

Biochemical and serological reactions of Shigella

Test or substrate	Positive	Negative	Shigella species reaction[a]
1. Glucose (TSI)	yellow butt	red butt	+
2. Glucose (bromcresol purple broth)	yellow color and/or gas	no gas; no color change	+/-[b]
3. Lactose (TSI)	yellow slant[c]	red slant	-
4. Lactose (bromcresol purple broth)	yellow color and/or gas	no gas; no color change	-
5. Sucrose (TSI)	yellow slant[c]	red slant	-
6. Sucrose (bromcresol purple broth)	yellow color and/or gas	no gas; no color change	-
7. Hydrogen sulfide (TSI)	blackening	no blackening	-
8. Urease	purple-red color	no color change	-
9. Lysine decarboxylase broth	purple color	yellow color	-
10. Ornithine decarboxylase broth	purple color	yellow color	-[d]
11. Indole test	violet color at surface	yellow color at surface	-
12. Motility	diffuse growth	no diffuse growth	-
13. Bromcresol purple broth with adonitol	yellow color and/or gas	no gas; no color change	-
14. Bromcresol purple broth with salicin	yellow color and/or gas	no gas; no color change	-
15. Bromcresol purple broth with inositol	yellow color and/or gas	no gas; no color change	-
16. Bromcresol purple broth with mannitol	yellow color and/or gas	no gas; no color change	+[e]
17. KCN	turbidity; growth	no turbidity; no growth	-
18. Malonate	blue color	no color change	-
19. Citrate	turbidity; growth	no turbidity; no growth	-
20. Methyl red	diffuse red color	diffuse yellow color	+
21. Voges-Proskauer	pink-to-red color	no color change	-
22. Polyvalent somatic test	agglutination	no agglutination	+

[a] +, 90% or more positive in 1 or 2 days; -, 90% or more negative in 1 or 2 days.

[b] Most Shigella strains produce acid, but no gas, from glucose.

[c] Yellow TSI slant indicates lactose and/or sucrose utilized.

[d] Most S. sonnei strains are positive.

[e] Most S. dysenteriae strains are negative.

Table 4

Determination of somatic serological group of Shigella

Rectangle	Suspension	Polyvalent antiserum								Saline
		A	A_1	B	C	C_1	C_2	D	A-D	
1	+	+								
2	+		+							
3	+			+						
4	+				+					
5	+					+				
6	+						+			
7	+							+		
8	+								+	
9	+									+

CHAPTER 6

VIBRIO CHOLERAE

Vibrio cholerae is the causative agent of cholera. Cholera patients excrete this pathogen in great numbers, and the disease is spread by the fecal-oral route. It is also spread indirectly through contaminated water supplies.

V. cholerae differs from the other Vibrio species in that its requirement for the sodium ion can be satisfied by trace amounts present in most media formulations. Strains that are identical or closely related biochemically to the clinical strains but fail to agglutinate in anti-01 V. cholerae serum are referred to as non-agglutinable or non-01 V. cholerae strains. Even though only strains that agglutinate in 01 V. cholerae antiserum are recognized pathogens, there is increasing evidence that non-01 strains are sometimes involved in cholera-like disease.

Foods involved in recent cholera outbreaks have involved raw, undercooked contaminated seafood. Other incriminated foods such as vegetables may have become contaminated by the use of human feces as fertilizer or the use of contaminated water.

A. EQUIPMENT AND MATERIALS

1. Blender, Waring or equivalent, and sterile blender jars
2. Sterile 500 ml Erlenmeyer flasks or sterile containers with large surface area
3. Balance, 2,000 $\pm$ 0.1 g capacity
4. Sterile tools for manipulating food samples: spoons, scissors, forceps, knives, etc.
5. Incubator, 35 $\pm$ 1°C
6. Water bath, covered, 42 $\pm$ 0.2°C
7. Sterile Petri dishes, 15 x 100 mm
8. Sterile pipets, 1, 5, and 10 ml, with 0.1 ml graduations
9. Inoculating needle and loop (about 3 mm)
10. Sterile culture tubes, 13 x 100 mm or 16 x 125 mm, and tube racks
11. Bunsen or Fisher burner
12. pH meter

B. MEDIA AND REAGENTS

(NOTE: Formulations for all biochemical media in this chapter must include at least 1% NaCl)

1. Alkaline peptone water (M3)

2. GPS medium (M32)

3. Thiosulfate-citrate-bile salts-sucrose (TCBS) agar (M98)

4. Gelatin agar medium (M29)

5. Triple sugar iron (TSI) agar (M100) or Kligler iron (KI) agar (M37). KI agar is recommended and may be substituted throughout for TSI agar.

6. Lysine iron (LI) agar (M50)

7. Hugh-Leifson glucose broth (M35)

8. Mannitol, inositol, lactose, sucrose, arabinose, mannose, salicin, and esculin in Andrade's carbohydrate broth (M5)

9. Decarboxylase basal medium (lysine, Falkow) (M25)

10. Decarboxylase basal medium (ornithine) (M25)

1. Arginine dihydrolase (Sutter) (M6)

12. Tryptone (tryptophane) broth, 1% (M111)

13. Mueller-Hinton agar (M62)

14. Trypticase (tryptic) soy agar (TSA) (M102)

15. T_1N_1 medium (M95)

16. Simmons citrate agar (M91)

17. Brain heart infusion (BHI) broth (M12)

18. MR-VP medium (M61)

19. Christensen's urea agar (M23)

20. 1 N hydrochloric acid (R14)

21. 1 N sodium hydroxide solution (R37)

22. Physiological saline solution, 0.85% (sterile) (R28)

23. Oxidase test reagent (R24)

24. Voges-Proskauer test reagents (R40)

25. Gram stain reagents (S6)

26. Group, Inaba, Ogawa, and Hikojima antisera.

27. Mukerjee phage IV

28. 50 Unit polymyxin B antibiotic disk

29. Physiological saline suspension of sheep red blood cells, 5%

30. Physiological saline suspension of chicken red blood cells, 2.5%

C. PREPARATION AND ENRICHMENT OF SAMPLE

1. Oysters

a. Aseptically remove oyster meats and liquor from about 12 shell stock oysters or 12 shucked oysters from container. Aseptically weigh about 200 g of oyster meat and liquor into sterile empty blender jar.

b. Blend at high speed for 1 min.

c. Aseptically weigh 25 g portions into 500 ml flasks containing 225 ml alkaline peptone water. Cover with sterile aluminum foil. Swirl mixture 25 times clockwise and 25 times counterclockwise to suspend oyster homogenate. Incubate 6-8 hr at 42 $\pm$ 0.2°C in water bath.

2. Other foods

a. Weigh 225 g samples into each of two 500 ml tared blender jars. Aseptically cut large samples into smaller pieces before blending.

b. Add 25 ml alkaline peptone water to one jar and 225 ml GPS medium to the other jar.

c. Blend for at least 60 sec and no longer than 2 min at top blender speed.

d. Leave blended suspensions in blender jars or pour into sterile 500 ml Erlenmeyer flasks and incubate 6-8 hr at 35°C. Incubation must not exceed 8 hr.

D. ISOLATION OF V. CHOLERAE

1. After incubation, transfer 5 mm loopfuls of inoculum from pellicle (surface growth) onto 3 plating media: TCBS, gelatin, and GPS agars. Streak from both jars or flasks onto separate plates to yield isolated colonies. Invert inoculated plates and incubate 18-24 hr at 35°C.

2. Examine plates for the following characteristics of colonies.

   a. TCBS agar. Colonies of V. cholerae and related vibrios will typically appear to be large, smooth, yellow or green, and slightly flattened with opaque centers and translucent peripheries.

   b. Gelatin agar. Colonies of V. cholerae and related vibrios may appear transparent or ringed by a cloudy zone. Varying light conditions may make these zones more apparent. When viewed with oblique lighting, colonies of V. cholerae typically appear lightly iridescent, green-to-bronze, and finely granular.

   c. GPS agar. Colonies of V. cholerae and related vibrios will be small, transparent, and ringed by a cloudy zone. Colonies of Vibrio species may be surrounded by small colonies of other bacterial species. The "satellite" phenomenon is due to hydrolysis of gelatin by the vibrios, which allows other nongelatinase-producing bacteria the opportunity to reproduce and grow.

3. Carefully pick at least 2 suspect colonies, if present, and streak for isolation on T_1N_1 agar. Incubate 12-18 hr at 35°C.

E. SEPARATION OF SUSPECT VIBRIOS FROM NON-VIBRIOS

1. **Kligler iron agar, triple sugar iron agar, and lysine iron agar reactions.** Inoculate individual T_1N_1 agar colonies into (1) KI agar or TSI agar and (2) LI agar by streaking slant and stabbing butt. Incubate inoculated slants 18-24 hr at 35°C.

2. **1% Tryptone (tryptophane) broth containing 30 g NaCl/litre.** From cultures giving typical reactions in KI, TSI, and LI agars (Table 5), inoculate 1% tryptone broth, with and without 3% NaCl, and incubate 18-24 hr at 35°C. V. cholerae and related vibrios will grow in 1% tryptone broth containing 3% NaCl. Many other bacterial species that produce reactions similar to those produced by V. cholerae in TSI and LI agar media fail to grow in 1% tryptone medium with 3% NaCl. Use of these media is recommended because the reactions permit early presumptive differentiation between most vibrios, Aeromonas hydrophila, Plesiomonas shigelloides, and other bacteria (Table 5). Cultures that conform to the biochemical patterns listed for V. cholerae in Table 5 and grow in 1% tryptone broth containing 3% NaCl should be considered as possible V. cholerae isolates. Perform biochemical and serological studies on these isolates as described later.

3. **Hugh-Leifson glucose broth.** Inoculate suspect cultures into duplicate tubes of Hugh-Leifson glucose broth. Overlay 1 tube with sterile mineral oil or liquid Vaspar (50% petrolatum, 50% paraffin) to a depth of 1-2 cm. Incubate both tubes for 18-24 hr at 35°C. Members of the genus Vibrio will utilize glucose both oxidatively and fermentatively. Members of the genus Pseudomonas, commonly isolated from seafoods by the enrichment methods used for Vibrio species, will utilize glucose oxidatively only.

f. **BIOCHEMICAL IDENTIFICATION OF V. CHOLERAE**

1. **Cytochrome oxidase test**. Inoculate duplicate TSA slants from suspect culture and incubate 24 hr at 35°C. Use one TSA slant for oxidase test and the other TSA slant for subsequent tests. Let 2-3 drops of oxidase test reagent flow over slant. Rapid development of dark blue color within 2 min is positive reaction. V. metschnikovii is negative by this test; all other halophilic vibrios are positive.

2. **Christensen's urea agar**. Inoculate this agar, using needle containing growth from TSA slant culture. Inoculate by streaking slant and stabbing butt. Incubate 48 ± 2 hr at 35°C. Read results as follows. Positive -- pink or rose color at site of inoculation or throughout medium. Negative -- no color change in medium.

3. **Tests for arginine dihydrolase, lysine, and ornithine decarboxylases**. Inoculate tube of each of 3 decarboxylase media with loopful of TSA culture. After inoculation, add 10 mm thick layer of sterile mineral oil to each tube; include basal medium control. Replace caps loosely and incubate 24 hr at 35°C and examine. Reincubate additional 24 hr and examine. Inoculated media turn yellow as a result of acid production from glucose. When positive reaction occurs, medium becomes alkaline or purple. The basal medium, or control tube, remains acid or yellow. Results of these and other differential tests are shown in Tables 6 and 7.

4. **Fermentation reactions**. Inoculate, from TSA culture, one tube each of Andrade's carbohydrate broth containing glucose, lactose, sucrose, mannose, mannitol, salicin, arabinose, esculin, and inositol; incubate at 35°C for 2 days. Results of these tests are shown in Tables 6 and 7.

5. **Simmons citrate agar**. Inoculate this agar, using needle containing growth from TSA slant culture. Inoculate by streaking slant and stabbing butt. Incubate 48 ± 2 hr at 35°C. Read results as follows. Positive -- presence of growth usually accompanied by color change from green to blue. Negative -- no growth or very little growth and no color change. Refer to Tables 6 and 7 for reactions of V. cholerae.

6. **Gram stain**. Perform Gram stain on each culture (≤ 24 hr old) as described in Chapter 20. V. cholerae cells are Gram-negative, unicellular, nonsporeforming, rod-shaped organisms with a curved or comma-shaped (not spiral) axis.

G. **SEROLOGICAL IDENTIFICATION AND CONFIRMATION OF V. CHOLERAE, V. MIMICUS, AND NON-01 VIBRIOS**

1. **Serological agglutination test**. Use diagnostic antisera of Inaba, Ogawa, Hikojima, and Group for confirming identity and for specific typing of isolates. However, because antigens in different species may be related, biochemical tests must be completed before an isolate is confirmed V. cholerae or Non-01 vibrio. Isolates that give appropriate results with TSI and LI agars are first tested in Group 01 antiserum along with a saline control.

   a. Cultures that agglutinate in Group 01 antiserum and not in plain physiological saline are V. cholerae Group 01 if the biochemical reactions also confirm the isolate as V. cholerae. Cultures that agglutinate in this group-specific antiserum may be further typed with Ogawa, Inaba, and Hikojima antisera.

b. Cultures that do not agglutinate in Group 01 antiserum or in saline are nonagglutinable vibrios if the biochemical reactions confirm them as Vibrio species.

c. Cultures that agglutinate in Group 01 antiserum and in saline cannot be typed. However, use of a richer medium, such as TSA or BHI agar, may discourage self-agglutination, making it possible to type these isolates.

2. **Determination of classical and El Tor biotypes**. Differentiate V. cholerae strains that agglutinate in Hikojima, Inaba, or Ogawa antisera to determine whether they are the classical or El Tor biotype. These 2 biotypes may be distinguished by any of the following methods (polyymyxin B is easiest):

a. Bacteriophage susceptibility--A modification of the method described by Finkelstein and Mukerjee is used (1). Swab surface of Mueller-Hinton agar plate with 4 hr broth culture of strain to be tested in a manner that will result in confluence bacterial growth. Superimpose 1 loopful of appropriate test dilution of Phage IV onto inoculated agar surface with 3 mm platinum loop. Observe plate after overnight incubation at 35°C. Strains belonging to classical biotype are usually sensitive to this bacteriophage and will lyse on plate where phage has been placed (indicated by confluent growth).

b. Polymyxin B sensitivity. This procedure is a modification of the technique described by Han and Khie (2). Swab surface of Mueller-Hinton agar plate with 4 hr broth culture of strain to be tested in a manner that will result in confluent bacteria growth. Let plates absorb inoculum and place 50 unit polymyxin B antibiotics disk on medium surface. Invert plates and incubate for 18-24 hr at 35°C. Classical biotype strains will demonstrate zone of inhibition around disk (generally 12-15 mm). El Tor biotype strains will grow right up to disk or will demonstrate 1-2 mm clearing around disk.

c. Hemolysin test. Mix equal volumes (0.5 or 1 ml) of 24 hr BHI broth culture and 5% saline suspension of sheep red blood cells. Set up similar mixtures with portion of culture that has been heated for 30 min at 56°C and use known hemolytic and nonhemolytic strains of V. cholerae as controls. Incubate mixtures for 2 hr in 35°C water bath; refrigerate overnight at 4-5°C. Examine tubes for hemolysis. Low speed centrifugation may aid in detection of cell lysis. Most El Tor strains will cause lysis of red blood cells. Heated portion of culture should produce no hemolysis because of thermolability of hemolysin. Classical biotypes of V. cholerae and some strains of biotype El Tor will not lyse red blood cells.

d. Chicken red blood cell agglutination. On clean glass slide, mix loopful of washed chicken red blood cells (2.5% in physiological saline) with thick milky suspension of pure culture to be tested. Make suspension in physiological saline from overnight agar plate culture. Visible clumping of red blood cells indicates El Tor biotypes; classical strains usually do not clump red blood cells. Perform positive and negative controls.

e. Voges-Proskauer (V-P) test. Perform this test in MR-VP broth after 18-24 hr incubation at 22°C. El Tor biotypes usually will be positive; classical strains usually will be negative.

H. FLOW SHEET

See Figures 7 and 8

I. RECORD SHEET

See Annex 6.

J. PRECAUTIONS AND LIMITATIONS OF METHOD

In determining the presence of V. cholerae in foods, the following points should be observed:

1. In analyzing oysters and other shellfish, both meat and liquor should be included in the sample.

2. Alkaline peptone enrichment should not exceed 8 hr since competing microflora may overgrowth V. cholerae during longer incubation periods.

3. Incubation of enrichments of oyster samples must be in a water bath thermostatically controlled at 42 $\pm$ 0.2°C.

4. Surface or pellicle growth only must be streaked from alkaline peptone or GPS enrichments onto plating agars.

5. When inoculating KI, TSI, or LI agars, do not repick colony to obtain additional growth and do not heat needle between inoculations of these agars from any one colony.

6. Do not eliminate KI iron agar culture as V. cholerae solely on the basis of acid slant.

7. A negative saline control must be included when performing polyvalent, Inaba, and Ogawa serological tests.

REFERENCES

1. **Finkelstein, R. A., and S. Mukerjee**. 1963. Hemagglutination. A rapid method for differentiating. Vibrio cholerae and El Tor vibrios. Proc. Soc. Exp. Biol. Med. 112:355-359.

2. **Han, G. K. and T. D. Khie**. 1963. A new method for the differentiation of Vibrio comma and Vibrio El Tor. Am. J. Hyg. 77:184-186.

Detection of Vibrio cholerae in Oysters

1. OYSTER HOMOGENATE

~200 g shucked oysters
Blend at high speed for 1 min

2. WEIGH

25 g oyster homogenate into 225 ml alkaline peptone water and swirl

3. INCUBATE

for 6-8 hr at 42 ± 0.2°C in water bath

Water level in water bath

4. PLATING

Incubate 18-24 hr at 35°C

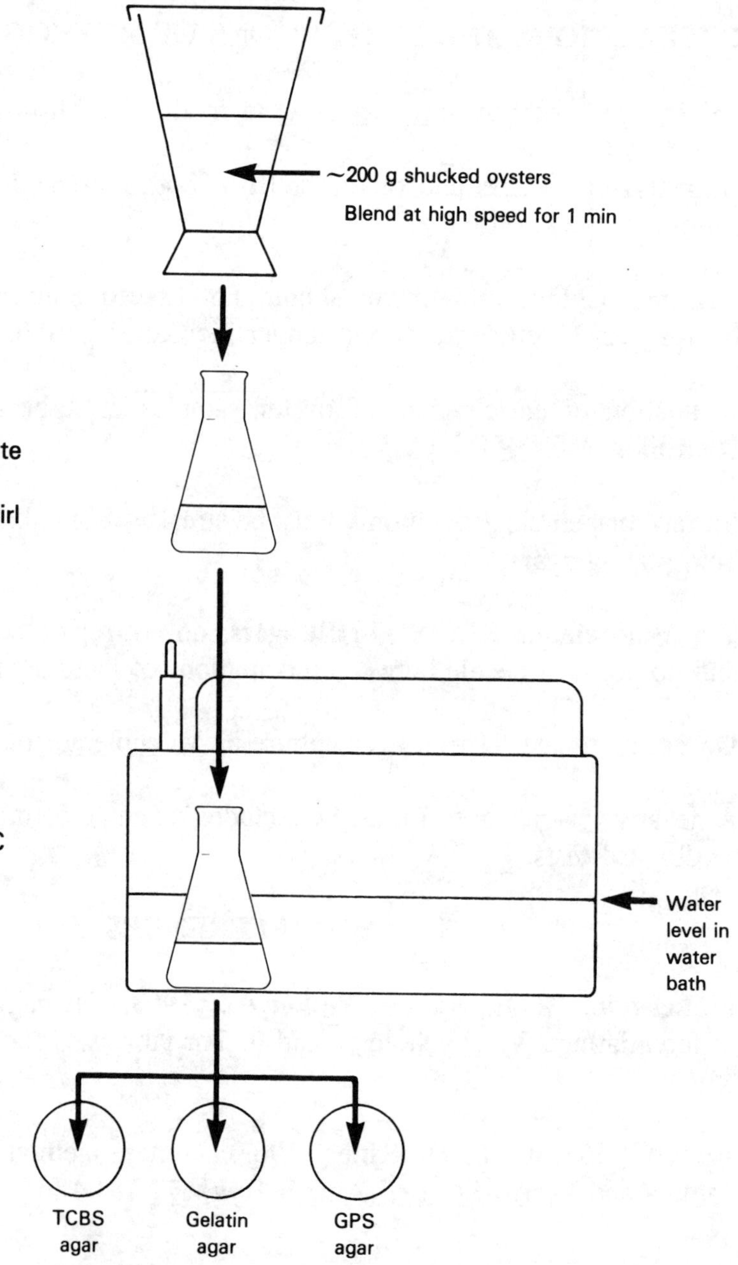

TCBS agar

Gelatin agar

GPS agar

TCBS = Thiosulfate-citrate-bile salts sucrose agar

5. BIOCHEMICAL AND SEROLOGICAL CONFIRMATION

Figure 7

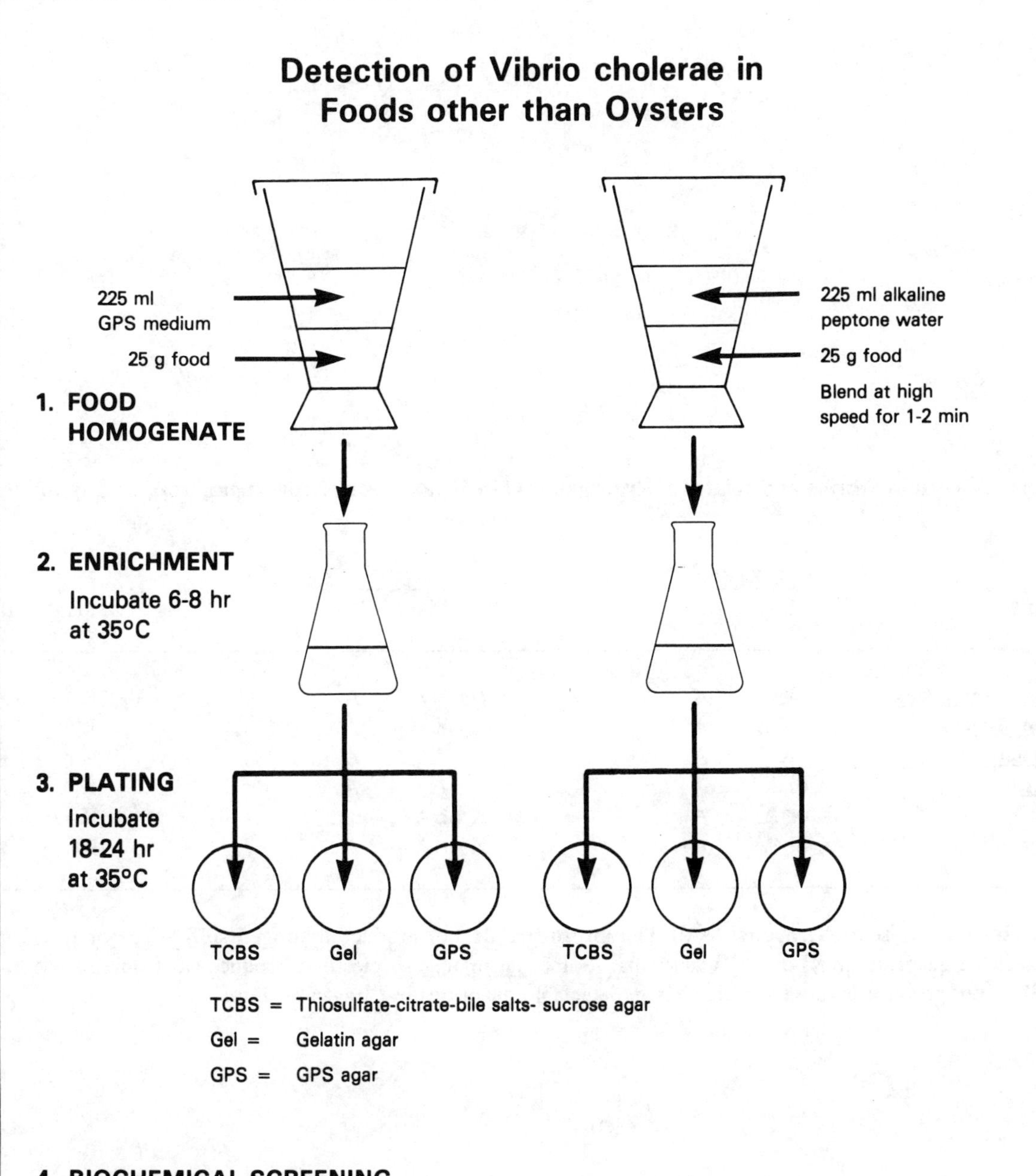

4. BIOCHEMICAL SCREENING

5. SEROLOGICAL CONFIRMATION

Figure 8

Table 5

Reactions[a] of certain vibrios and related microorganisms in Kligler iron, triple sugar iron, and lysine iron agars

Microorganism	KI		TSI		LI	
	Slant	Butt	Slant	Butt	Slant	Butt
V. cholerae, V. mimicus	K	A	A (rarely K)	A	K	K or N
V. parahaemolyticus	K	A	K	A	K	K or N
V. alginolyticus	K	A	A	A	K	K or N
V. vulnificus	A or K	A	A	A	K	K or N
A. hydrophila	K or A	A	K or A	A	K	K or N
P. shigelloides	K or A	A	K or A	A	K	K or N

[a] K, alkaline; A, acid, N. neutral. All microorganisms do not produce hydrogen sulfide gas or gas from glucose in detectable quantities in KI or TSI agar media. Some Aeromonas species may produce gas from glucose in these media. All do not produce hydrogen sulfide gas in detectable quantities in LI agar medium.

Table 6

Comparison of biochemical reactions of V. cholerae, V. parahaemolyticus, V. alginolyticus, and V. vulnificus

Test or substrate[a]	V. cholerae V. mimicus Sign[b]	%	V. parahaemolyticus Sign	%	V. alginolyticus Sign	%	V. vulnificus Sign	%
Voges-Proskauer[c]	+/-	47	-	0	+	100	-	0
Citrate (Simmons)	$+^{w}$	74	+	98	D	20	D	76
Lysine decarboxylase	+	100	+	97	+	100	+	97
Arginine dihydrolase	-	0	-	0	-	0	-	0
Ornithine decarboxylase	+	>99	+	96	+/-	40	+	66
Gas from glucose	-	0	-	0	-	0	-	0
Lactose[c]	-	0	-	0	-	0	+	81
Sucrose[c,d]	+	96	-	5	+	100	-	3
Arabinose[c]	-	10	+	81	-	6	-	0
Mannose	+	99	+	100	+	100	+	100
Mannitol	+	100	+	100	+	100	+	66
Salicin[c]	-	0	-	0	D	13	+	100
Esculin[c]	-	0	+	>99	-	0	NA	NA

[a] Based on incubation for 1-2 days at 35°C. Cultures are oxidase-positive and do not produce hydrogen sulfide in TSI agar.

[b] Sign: ±, 80% or more positive within 1-2 days; -, 80% or more demonstrating no reaction within 1-2 days; +/-, high variability among tested strains; $+^{w}$, weakly positive reaction; NA, data not available; D, differing reactions.

[c] Useful for species differentiation.

[d] Sucrose nonfermenters are identified as V. mimicus.

Table 7

Differentiation of V. cholerae, Aeromonas, and Plesiomonas species.

Test or Substrate[a]	V. cholerae		A. hydrophila		P. shigelloides	
	Sign[b]	Positive (%)	Sign	Positive (%)	Sign	Positive (%)
Voges-Proskauer[c]	+/-	47	+/-	45	-	0
Citrate (Simmons)	+	74[w]	D	54	-	0
Lysine decarboxylase	+	100	-	18[w]	+	98
Arginine dihydrolase	-	0	+	81	+	96
Ornithine decarboxylase	+	96	-	0	+/-	70
Gas from glucose	-	0	+/-	55	-	0
Lactose[c]	-	0	D	13	D	60
Sucrose[c,d]	+	99	D	80	-	1
Arabinose[c]	-	10	D	55	-	0
Mannose	+	99	+	91	D	4
Mannitol	+	100	+	98	-	0
Inositol[c]	-	0	-	0	+	96
Esculin	-	0	+/-	60	-	0

[a] Based on incubation for 1-2 days at 35 $\pm$ 2°C. Cultures are oxidase-positive and do not produce hydrogen sulfide in TSI agar.

[b] Sign: $\pm$, 80% or more positive within 1-2 days; -, 80% or more demonstrating no reaction within 1-2 days; +/-, high variability among tested strains; $+^{w}$, weakly positive reaction; NA, data not available; D, differing reactions.

[c] Useful for species differentiation.

CHAPTER 7

VIBRIO PARAHAEMOLYTICUS

Vibrio parahaemolyticus is the major bacterial cause of epidemic gastroenteritis in Japan, where fish is eaten raw. Foods frequently involved in V. parahaemolyticus outbreaks have been fish, oyster, crab, shrimp, and lobster that had been cooked prior to consumption. Thus, these outbreaks were probably caused by improper refrigeration, insufficient cooking, cross contamination, or recontamination.

V. parahaemolyticus is a halophilic organism. Thus, all media used for the isolation of this pathogen require NaCl at a concentration of 3%. There is increasing evidence that 4 other halophilic Vibrio species are pathogenic for humans: V. alginolyticus, V. fluvialis, V. metschnikovii, and V. vulnificus. All of these halophilic Vibrio species are distinguishable by one or more biochemical tests.

A. EQUIPMENT AND MATERIALS

1. Same as for Salmonella, Chapter 4, A
2. Sterile toothpicks

B. MEDIA AND REAGENTS

(NOTE: All media must contain at least 3% NaCl unless other concentrations are specified)

1. MR-VP medium (M61)
2. Trypticase (tryptic) soy broth (TSB) (M104)
3. Trypticase (tryptic) soy agar (TSA) (M102)
4. Glucose salt Teepol broth (GSTB) (M31)
5. Hugh-Leifson glucose broth (HLGB) (M35)
6. Motility test medium (semi-solid) (M59)
7. Bromcresol purple broth (M14) supplemented individually with the following carbohydrates: mannitol, lactose, sucrose, or arabinose
8. Decarboxylase basal medium (M25) supplemented individually with arginine, lysine, or ornithine
9. Salt trypticase broth (STB) (M86)
10. Thiosulfate-citrate-bile salts-sucrose (TCBS) agar (M98)
11. Triple sugar iron (TSI) agar (M100)

12. Wagatsuma agar (M120)

13. Long-term preservation medium (M47)

14. Oxidase test reagent (R24)

15. Sodium chloride dilution water, 3% (R33)

16. Voges-Proskauer (VP) test reagents (R40)

17. Creatine, crystalline

18. Mineral oil (paraffin oil), sterile (R20)

19. Gram stain reagents (S6)

C. PREPARATION AND ENRICHMENT OF SAMPLE

1. Sample composition

    a. Fish: surface tissues, gut, and gills

    b. Shellfish: entire interior contents of animal; pool 10-12 animals and remove 50 g from composite for test sample

    c. Crustaceans: entire animal, if possible, or central portion of animal, including gills and gut

2. Day 1

    Add 450 ml 3% NaCl dilution water and blend 1 min at 8,000 rpm; this represents 1:10 dilution. Prepare 1:100, 1:1,000, 1:10,000 dilutions and higher, if necessary. Inoculate three 10 ml portions of 1:10 dilution into 3 tubes containing 10 ml double strength GSTB. (This constitutes 1 g portion.) Inoculate three 1 ml portions of 1:10, 1:100, 1:1,000, and 1:10,000 dilutions into single strength GSTB. Incubate broth tubes overnight (18-24 hr) at 35°C.

D. ISOLATION OF V. PARAHAEMOLYTICUS

1. Day 2

    After incubation, streak onto TCBS agar plates 3 mm loopful from top 1 cm of GSTB tubes containing 3 highest dilutions of sample showing growth. Streak known strains of suspect species onto TCBS agar and verify through subsequent tests. Incubate TCBS agar plates 18-24 hr at 35°C.

2. Day 3

Appearance of colonies on TCBS agar plates:

a. V. parahaemolyticus and V. vulnificus colonies are round, 2-3 mm in diameter, with green or blue centers.

b. V. alginolyticus, V. fluvialis, and V. metschnikovii colonies appear larger and yellow.

c. Coliforms, Proteus, and enterococci, if present, appear as small and translucent colonies.

d. When suspect colonies are identified biochemically and/or serologically , apply MPN tables (Chapter 21) for final enumeration of species.

E. BIOCHEMICAL IDENTIFICATION OF V. PARAHAEMOLYTICUS

1. Day 3

Differential Test I. Pick, with needle, 2 or more typical or suspect colonies, if present, from TCBS agar plates onto the following media:

a. TSI agar slant. Streak slant, stab butt, and incubate 18-24 hr at 35°C. V. parahaemolyticus and V. vulnificus produce alkaline slant and acid butt; no gas is produced, and culture growth on TSI agar slant is negative for H_2S. This is typical Shigella-like reaction. V. alginolyticus, V. fluvialis, and V. metschnikovii give acid slant and butt without gas.

b. TSB (3% NaCl) and TSA slant (3% NaCl). Inoculate TSB and TSA; incubate 18-24 hr at 35°C. Use these cultures as source of inoculum for other tests, Gram stains, and microscopic examinations. V. parahaemolyticus and related vibrios are Gram-negative, pleomorphic organisms appearing as curved or straight rods with polar flagella.

c. Motility test medium. Inoculate tube of motility test medium by stabbing center of medium column to bottom of tube. Incubate 18-24 hr at 35°C. Diffuse circular growth from line of stab is positive test. V. parahaemolyticus and related vibrios are motile.

2. Day 4

Differential Test II. Conduct further tests as follows:

a. Hugh-Leifson glucose broth (HLGB). Stab 2 tubes of HLGB medium with inoculating needle from TSA medium. Overlay 1 tube with sterile mineral oil to depth of about 3 cm and incubate 2 days at 35°C, or longer, if necessary. If color of medium in both tubes changes from purple to yellow, organism is carbohydrate fermenter. If color change of medium occurs in open tube, organism is carbohydrate oxidizer. V. parahemolyticus and related vibrios ferment glucose without producing gas.

b. Cytochrome oxidase test. Inoculate TSA slant from control inocula and incubate 24 hr at 35°C. Let 2-3 drops of oxidase test reagent flow over slant or colony on plate. Rapid development of dark blue color within 2 min is positive reaction. V. metschnikovii is negative by this test; all other halophilic vibrios are positive.

c. Test for arginine dihydrolase, lysine, and ornithine decarboxylases. Inoculate tube of each of 3 decarboxylase media with loopful of TSA culture. After inoculation, add 10 mm thick layer of sterile mineral oil to each tube; include basal medium control. Replace caps loosely and incubate 24 hr at 35°C. Examine every 24 hr for 4 days. Inoculated media turn yellow as a result of acid production from glucose. When positive reaction occurs, medium becomes alkaline or purple. The basal medium, or control tube, remains acid or yellow. Results of these and other differential tests are shown in Table 8.

d. Salt trypticase broth (STB), halophilism. Using small loop, inoculate 4 tubes of STB base containing 0, 6, 8, and 10% NaCl, respectively, from TSB-3% NaCl culture. Incubate 24 hr at 35°C. Halophilic Vibrio species that fail to grow in absence of salt grow well in 6% NaCl. Some, including V. parahaemolyticus, V. alginolyticus, V. fluvialis, and occasional strains of V. metschnikovii, grow in 8% NaCl. Only V. alginolyticus grows well in 10% NaCl.

e. Growth at 42°C. Inoculate tube of TSB-2% NaCl with small loopful of 24 hr TSB-3% NaCl culture inoculum and incubate in water bath at 42°C for 24 hr. Consider only profuse growth as positive. V. parahaemolyticus and V. alginolyticus and occasional strains of V. fluvialis and V. metschnikovii are positive in this test.

f. MR-VP broth (Voges-Proskauer test). Inoculate MR-VP medium with loopful of growth from TSA slant culture and incubate 2 days at 35°C. Perform VP test. V. parahaemolyticus, V. vulnificus, and V. fluvialis are VP-negative, whereas V. alginolyticus and V. metschnikovii are positive.

g. Fermentation reactions in bromcresol purple broth. Inoculate one tube each of sucrose, lactose, mannitol, and arabinose broth from TSA slant culture and incubate at 35°C for 4-5 days. Acid reaction appears as change in color of carbohydrate broth from purple to yellow. Results of these fermentation tests are shown in Table 8.

h. Kanagawa phenomenon (Wagatsuma agar). The Kanagawa reaction is a test for the presence of specific heat-stable, direct hemolysin on Wagatsuma agar. A positive reaction has been found to correlate closely with pathogenicity of V. parahaemolyticus isolates. Isolates that have caused illness in humans are usually Kanagawa-positive, whereas those recovered from seafoods are almost always Kanagawa-negative.

To perform the Kanagawa test, spot droplet from 18 hr culture to TSB-3% NaCl with sterile toothpick onto well-dried Wagatsuma blood agar plate. Several spottings may be made in circular pattern on single plate. Always spot known positive and negative cultures on each plate. Incubate at 35°C and observe and record results in exactly 24 hr. Positive test consists of beta-hemolysis, i.e., sharply defined zone of transparent clearing of blood cells around colony. Zone has single, clearly defined edge without multiple concentric rings. Hemolysis is entirely clear without greening. Zones of hemolysis are measured from edge of colony to outer edge of zone. Isolates that produce clear zones of hemolysis $\geq$ 3 mm are considered Kanagawa-positive and are presumed to be pathogenic. Isolates that produce clear zones of hemolysis $<$ 3.0 mm may be weakly pathogenic and should be tested in the rabbit ileal loop assay. **IMPORTANT: No observation beyond 24 hr is valid in this test.**

i. Culture preservation. Inoculate long-term preservation medium by stabbing deeply into semi-solid medium. Leave cap loose until visible growth appears during incubation at 35°C for 24 hr. Tighten caps to prevent dehydration and store at room temperature only. To preserve *V. parahaemolyticus* and *V. vulnificus* for several months at -70°C, place 1 ml of 6-12 hr TSB-3% NaCl culture and 0.09 ml sterile dimethyl sulfoxide (DMSO) into sterile cryotubes; freeze immediately at -70°C.

F. SEROLOGICAL CONFIRMATION OF *V. PARAHAEMOLYTICUS*

The K antigens comprising the various somatic groups are shown in Table 9. Wash slant with 2 ml 3% NaCl solution. (Heavy suspension is best for this microscopic slide agglutination test.) Divide cell suspension into 2 equal volumes in separate tubes. Autoclave tube for 1 hr at 121°C. This autoclaved suspension represents O antigen. Untreated suspension represents K antigen. To perform O agglutination, divide slide into 12 equal compartments, using wax pencil. Place small drop of heavy suspension into each compartment and add 1 drop each of the 11 O-group antisera to each compartment. To 12th compartment containing autoagglutination control, add 1 drop 3% NaCl. Tilt slide gently to mix all components, and rock it back and forth for 1 min. A positive agglutination may be read immediately. The positive O agglutination determines which K type antigens are to be tested. Mark off on a slide the appropriate number of compartments plus control compartment. Place a small drop of heavy, unautoclaved suspension into each compartment and add a drop of appropriate K antisera to each compartment. To the control compartment, add a drop of 3% NaCl. Tilt slide gently to mix components, and rock it back and forth for 1 min. A positive agglutination may be read immediately.

G. FLOW SHEET

See Figure 9.

H. RECORD SHEET

See Annex 7.

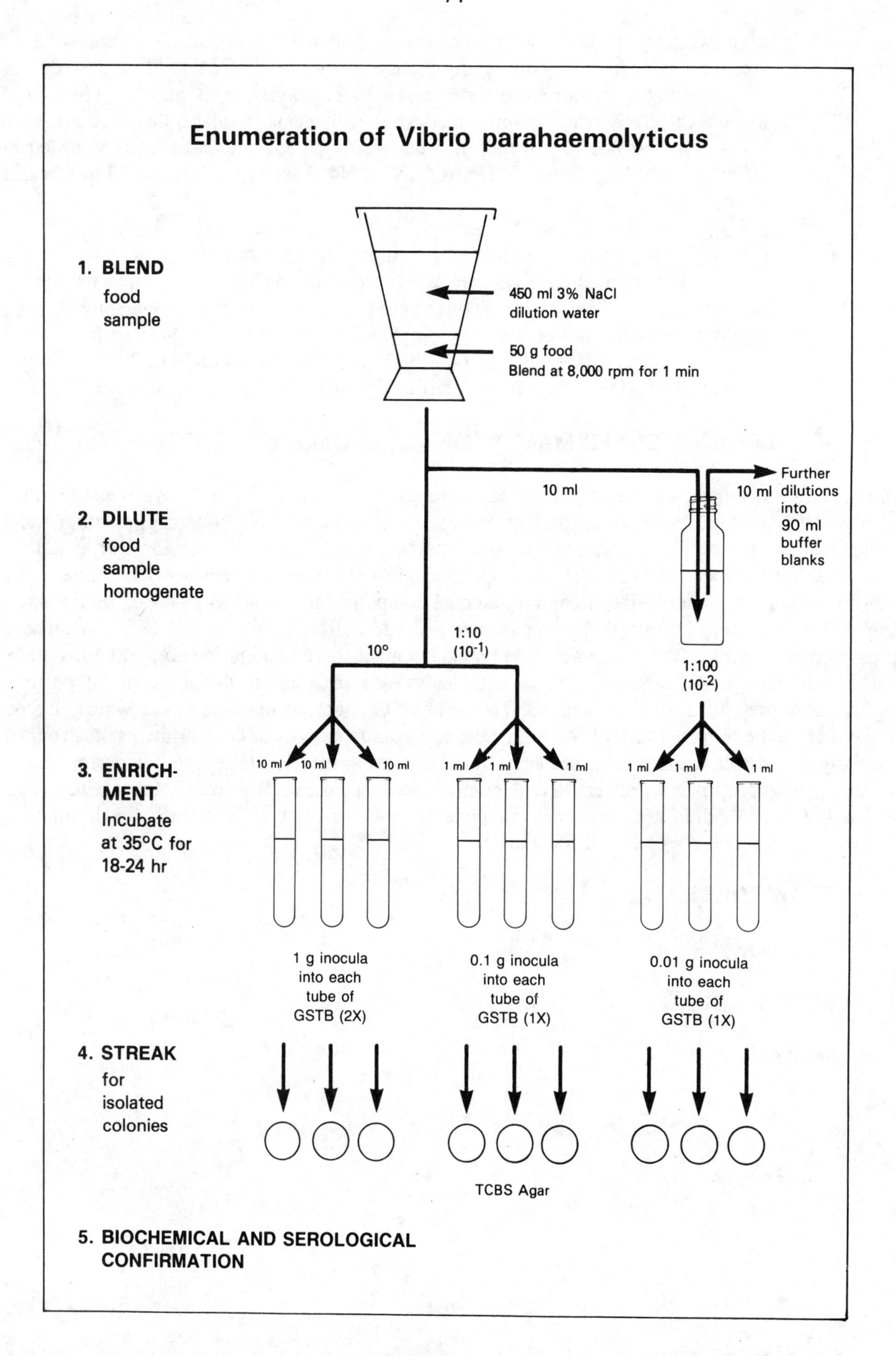
Enumeration of Vibrio parahaemolyticus
1. BLEND
food
sample
450 ml 3% NaCl
dilution water
50 g food
Blend at 8,000 rpm for 1 min
2. DILUTE
food
sample
homogenate
10 ml
10 ml
Further
dilutions
into
90 ml
buffer
blanks
10°
1:10
(10-1)
1:100
(10-2)
3. ENRICH-
MENT
Incubate
at 35°C for
18-24 hr
10 ml
10 ml
10 ml
1 ml
1 ml
1 ml
1 ml
1 ml
1 ml
1 g inocula
into each
tube of
GSTB (2X)
0.1 g inocula
into each
tube of
GSTB (1X)
0.01 g inocula
into each
tube of
GSTB (1X)
4. STREAK
for
isolated
colonies
TCBS Agar
5. BIOCHEMICAL AND SEROLOGICAL
CONFIRMATION

Figure 9

I. PRECAUTIONS AND LIMITATIONS OF METHOD

In determining the presence of V. parahaemolyticus in foods, the following points should be observed:

1. For weighing a 50 g sample of fish, include surface tissue, gut, and gills. For shellfish, include entire interior contents of animal. For crustaceans, include entire animal, if possible, or central portion of animal, including gills and gut.

2. All media for halophilic Vibrio species must include 3% NaCl.

3. In using the MPN procedure for quantitating Vibrio species, double strength glucose salt Teepol broth should be used for the 1 gram inocula portions and single strength medium for all subsequent food sample dilutions.

4. For the cytochrome oxidase test, a fresh (24 hr or less) culture should be used. The oxidase test reagent, N, N, N', N'-tetramethyl-p-phenylenediamine.2HCl, may be pre-prepared fresh, but it should be stored in a dark glass bottle under refrigeration. In performing the actual test, platinum wire should be used for rubbing the test culture on filter paper impregnated with oxidase test reagent. Wire containing iron will give a false positive reaction.

5. For the arginine dihydrolase, lysine decarboxylase, and ornithine decarboxylase tests, a basal medium control should be included.

6. The Kanagawa reaction tests for the presence of specific heat-stable hemolysin on Wagatsuma agar. A positive reaction correlates closely with pathogenicity of V. parahaemolyticus isolates. The presence of beta hemolysis of human or rabbit red blood cells should not be read beyond 24 hr.

7. For long-term preservation of V. parahaemolyticus cultures, stab deeply into semi-solid medium and store at room temperature only.

Table 8. Scheme for differentiating V. parahaemolyticus from related microorganisms

Test	Vibrio species							
	parahaemolyticus	alginolyticus	anguillarum	vulnificus	fluvialis	metschnikovii	Aeromonas hydrophila	Plesioshigelloides
TCBS (green colony)	+	-	-	+	-	-	-	+
Growth at 42°C	+	+	-	ND[a]	Var[b]	Var	-	-
Growth in media with								
0% NaCl	-	-	-	-	-	-	+	+
6% NaCl	+	+	+	Var	+	+	-	-
8% NaCl	+	+	Var	-	Var	Var	-	-
10% NaCl	-	+	-	-	-	-	-	-
Lysine decarboxylase	+	+	-	+	-	Var	-	+
Arginine dihydrolase	-	-	Var	-	+	+	Var	+
Ornithine decarboxylase	+	Var	-	Var	-	-	Var	Var
Sucrose fermentation	-	+	+	-	+	+	Var	-
Lactose fermentation	-	-	-	+	-	Var	Var	Var
Mannitol fermentation	+	+	+	Var	+	+	+	-
Arabinose fermentation	+	-	-	-	+	-	Var	-
Voges-Proskauer	-	+	Var	-	-	+	Var	-

[a] ND, not determined.
[b] Var, variable.

Table 9

V. parahaemolyticus antigen combinations

O group	K type
1	1, 25, 26, 32, 38, 41, 56, 58[a], 64
2	3, 28
3	4[a], 5, 6, 7, 29, 30[a], 31, 33, 37, 43, 45, 48, 54, 57, 58[a], 59
4	4[a], 8, 9, 10, 11, 12, 13, 34, 42 49, 53, 55, 63
5	15, 17, 30[a], 47, 60, 61
6	18, 46
7	19
8	20, 21, 22, 39, 62
9	23, 44
10	19, 24, 52
11	36, 40, 50, 51
Total 11	60 K types/63 serotypes

[a] Occurs in more than one O group.

CHAPTER 8

AEROMONAS HYDROPHILA

The genus Aeromonas consists of 2 well separated groups: (1) the psychrophilic, nonmotile aeromonads known as A. salmonicida and (2) the mesophilic motile aeromonads which include A. hydrophila, A. caviae, and A. sobria.

The aeromonads are found in the aquatic environment, where they are normal inhabitants of the intestines of fish. Some aeromonads are pathogenic for fish. In humans, Aeromonas species were once considered opportunistic pathogens of low virulence. Now, however, they are considered as primary pathogens.

The full potential of Aeromonas in causing food poisoning in humans is not known. It has frequently been reported in the literature that motile Aeromonas is a potential cause of food poisoning and travelers' diarrhea. About 13% of reported incidences of gastroenteritis may be attributed to Aeromonas.

A. EQUIPMENT AND MATERIALS

1. Blender
2. Balance, with weights: 1,200 g capacity; sensitivity 0.1 g
3. Incubator, 35 ± 1°C
4. Sterile spoons or other appropriate instruments for transferring food specimens
5. Sterile Petri dish, 15 x 100 mm, glass or plastic
6. Sterile pipets, 1 ml, with 0.01 ml graduations; 5 and 10 ml, with 0.1 ml graduations
7. Inoculating needle and inoculating loop (about 3 mm id)
8. Sterile wide-mouth, screw-cap blender jars, 500 ml
9. Water bath, 48-50°C
10. pH meter
11. Fisher or Bunsen burner
12. Test or culture tube racks
13. Sterile test or culture tubes, 16 x 150 mm and 20 x 150 mm; serological tubes, 13 x 100 mm

B. **MEDIA AND REAGENTS**

1. Tryptic soy broth with 30 mg ampicillin (TSBA) (M105)
2. MacConkey agar (M52)
3. Peptone-beef extract-glycogen (PBG) agar (M72)
4. Yersinia selective agar (YSA) base (supplemented with cefsulodin and novobiocin) (use commercial formulation from Difco) (M122)
5. Aeromonas hydrophila medium (AHM) (M2)
6. Motility test medium, semi-solid (M59)
7. Triple sugar iron (TSI) agar (M100)
8. Potassium cyanide (KCN) medium (M79)
9. Bacto agar (Difco)
10. Purple carbohydrate fermentation broth (M82), containing arabinose, mannitol, salicin, inositol, glucose, and sucrose
11. Decarboxylase basal medium (ornithine) (M25)
12. Arginine dihydrolase (Sutter) (M6)
13. Esculin agar, modified (M28)
14. MR-VP medium (M61)
15. Trypticase (tryptic) soy broth (M104)
16. Kovacs' reagent (R16)
17. Voges-Proskauer (VP) test reagents (R40)
18. Methyl red indicator (R19)
19. Oxidase test reagent (R24)
20. Sterile water
21. Columbia agar base (use commercial formulation from Oxoid)
22. Defibrinated rabbit blood (commercially available) or washed, citrated rabbit red blood cells

23. NaCl, physiological saline solution, 0.85% (R28)

24. Sodium citrate, 3.8% (R35)

C. PREPARATION AND ENRICHMENT OF SAMPLE

Aseptically weigh 25 g sample into sterile blender jar. Add 225 ml sterile tryptic soy ampicillin broth and blend 2 min at 10,000-12,000 rpm. Decant into 500 ml Erlenmeyer flask and incubate 24 ± 2 hr at 35°C.

D. ISOLATION OF A. HYDROPHILA

1. After incubation, streak 3 mm loopfuls of inoculum onto MacConkey, peptone-beef extract-glycogen (PBG) agar, and Yersinia selective agar (YSA) plating media. PBG agar may be used with or without a 2% non-nutrient agar overlay. With overlay, pour 10 ml 2% non-nutrient agar after streaking. Streak onto separate plates to yield isolated colonies. Invert inoculated plates and incubate 18-24 hr at 35°C.

2. Examine incubated plates for colonies suspected to be Aeromonas:

   a. MacConkey agar. Colonies of Aeromonas typically appear to be large (2-3 mm diameter), circular, colorless (nonlactose fermenters), sometimes with dark centers.

   b. PBG agar. With 2% non-nutrient agar overlay, colonies of Aeromonas typically appear to be small (0.3-0.5 mm diameter) circular, and yellow surrounded by a yellow halo. Without overlay, colonies of Aeromonas typically appear to be large (2-3 mm diameter), circular and yellow-green, with dark centers and translucent peripheries.

   c. YSA supplemented with cefsulodin and novobiocin (YSA-CN). Colonies of Aeromonas typically appear to be large (2-4 mm), raised, irregular with filamentous margins and dark (magenta) centers; light pink edge with translucent peripheries. In some colonies, a thin halo (zone) can be observed between the dark center and peripheries.

E. SEPARATION OF AEROMONAS FROM RELATED SPECIES OF THE FAMILY VIBRIONACEAE

Aeromonas spp. are easily confused with Group F vibrios (V. fluvialis, V. anguillarum). Vibrio and Aeromonas can be differentiated by their phenotypical reactions, e.g., ability to grow in media containing no NaCl, and sensitivity to 0/129 compound. For further ancillary tests, refer to Tables in Chapters 6 and 7.

1. **1% Tryptone broth without NaCl.** Transfer loopful of culture into 1% tryptone without NaCl and incubate 18-24 hr at 35°C. Aeromonas spp. will grow in 1% tryptone without NaCl, whereas Group F vibrios require NaCl for growth.

2. **Vibriostatic compound 0/129**. Prepare TSA plates with 0/129 compound at concentrations of 10 μg/ml and 150 μg/ml. Include a control plate containing no 0/129. Dissolve TSA medium in Erlenmeyer flask and sterilize according to manufacturer's directions. After cooling to 48-50°C, add filter-sterilized 0/129 compound to medium and dispense in Petri plates. Transfer loopful of broth culture (TSB) to plates. Streak about 3 cm inocula on plate (large numbers of isolates can be screened). Aeromonas spp. are resistant to 0/129 compound at these 2 concentrations.

F. IDENTIFICATION OF MOTILE SPECIES OF AEROMONAS

1. **Aeromonas hydrophila medium (AHM)**. Transfer at least 2 colonies per plate to AHM. Inoculate at 35°C for 18-24 hr. If, after incubation, reactions are not definitive, incubate an additional 24 hr.

   The following pattern in AHM indicates potential Aeromonas:

   a. Yellow butt with purple band at or near top of tube (mannitol, +; inositol, -; and ornithine decarboxylase, -); sometimes slight blackening occurs only at or near top of tube (H_2S from cysteine), but not in butt.

   b. Turbidity throughout tube from line of inoculation, indicating motility.

2. **Maintenance medium**. For cultures giving the above-described reactions, inoculate into maintenance medium (motility test medium or TSA slant) for further biochemical testing. Perform Gram stain. Aeromonads are Gram-negative.

3. **Indole production**. After inoculation of maintenance medium, indole production can be detected in AHM, which includes tryptone. Add 0.1 ml Kovacs' reagent to AHM. Development of pink to red color indicates production of indole. Yellow color signifies that indole is not present. Aeromonas is usually (85%) positive for indole production.

4. **Cytochrome oxidase test**. Inoculate from maintenance medium to TSA slant or plate and incubate at 35°C for 18-24 hr. After incubation, add 2-3 drops oxidase test reagent; flow over slant or colony on plate. Rapid development of dark blue color within 2 min is positive reaction. Aeromonas is oxidase-positive. Discard cultures negative for cytochrome oxidase.

5. **TSI agar slant**. Streak slant, stab butt, and incubate 18-24 hr at 35°C. Aeromonas produces alkaline or acid slant and acid butt. Some species may produce gas from glucose. Discard cultures positive for H_2S.

G. CONFIRMATION OF MOTILE SPECIES OF AEROMONAS

1. **API 20E Kit**

   After inoculation, incubate API 20E strip at 35°C for 24 hr. In using the API 20E kit, it should be realized that this system will identify all 3 species of the genus Aeromonas as A. hydrophila. Additional tests are required, as detailed in section H, to differentiate the Aeromonas species.

2. **Conventional biochemical testing**

a. Arginine dihydrolase. Inoculate decarboxylase media with loopful of culture. After inoculation, add 10 mm thick layer of sterile mineral oil to each tube; include basal medium control. Incubate 24 hr at 35°C. Examine every 24 hr for 4 days. Inoculated media turn yellow as a result of acid production from glucose. When positive reactions occurs, medium becomes alkaline, or blue. Basal medium, or control tube, remains acid or yellow.

b. MR-VP medium. Inoculate medium from culture. Incubate 48 ± 2 hr at 35°C. Perform VP and methyl red tests as described in Chapter 4.

c. Growth in 1% tryptone, without NaCl. See section on Separation of Aeromonas from related species of the family Vibrionaceae, 1.

H. DIFFERENTIATION OF MOTILE AEROMONAS SPECIES

Results of these fermentations are shown as Table 10.

1. **Fermentation reactions in bromcresol purple broth**. Inoculate one tube each of salicin, arabinose, and glucose with Durham tubes from maintenance culture, and incubate at 35°C for 4 days. A color change from purple to yellow signifies a positive reaction. NOTE: Incubate salicin at 30°C or at room temperature.

2. **Potassium cyanide (KCN) medium**. Inoculate medium from broth culture. Incubate 48 ± 2 hr at 35°C, but examine after 24 hr. Turbidity indicates growth.

3. **Esculin agar**. Inoculate medium from culture. Incubate at 35°C for 24 ± 2 hr. Esculin hydrolysis is indicated when medium turns black.

4. **Acetoin from glucose(V-P)**. See section G, 2b.

5. **H_2S from cysteine**. See section G, 1a.

6. **Arginine dihydrolase**. See section G, 2a.

Detection of Aeromonas hydrophila

1. BLEND

225 ml tryptic soy ampicillin broth

25 g food

Blend at 10,000-12,000 rpm for 2 min

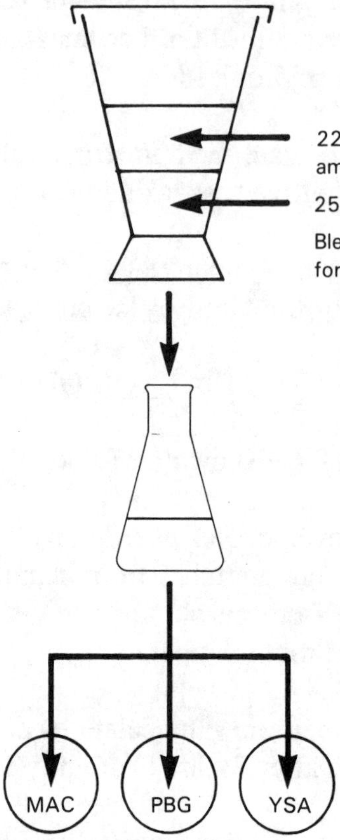

2. ENRICHMENT
Incubate at 35 °C for 24 hr

3. PLATING
Incubate at 35 °C for 18-24 hr

MAC PBG YSA

MAC = MacConkey agar

PBG = Peptone-beef extract-glycogen agar

YSA = Yersinia selective agar

4. BIOCHEMICAL TESTS I –
Identification of motile species in the genus *Aeromonas*

5. BIOCHEMICAL TESTS II –
Confirmation of motile species in the genus *Aeromonas*

6. BIOCHEMICAL TESTS III –
Differentiation between species in the genus *Aeromonas*

Figure 10

I. PROCEDURE FOR IDENTIFICATION OF BETA-HEMOLYTIC ISOLATES OF AEROMONAS

1. **Preparation of blood agar plates.** Add 5-10 ml freshly drawn rabbit blood to 0.6 ml of 3.8% sodium citrate. Mix thoroughly. Centrifuge at 1,500 rpm for 10 min. Wash 3 times with equal volume of 0.85% NaCl. Resuspend in 0.85% NaCl to original volume. Temper red blood cells at 45°C for 1 min.

   Add 5.0 ml red blood cells to 95.0 ml tempered (45°C) Columbia agar base. If commercially available, use defibrinated blood; mix gently, temper blood for 1 min, and add to tempered agar, omitting washing procedure. Pour plates, and let dry 2-4 hr.

2. **Test for beta-hemolytic activity.** Streak loopful of each Aeromonas isolate grown in TSB for 24 hr onto blood agar and incubate 18-24 hr at 35-37°C. Look for clear zone of hemolysis, without greening, around individual colonies.

J. FLOW SHEET

See Figure 10.

K. RECORD SHEET

See Annex 8.

L. PRECAUTIONS AND LIMITATIONS OF METHOD

In determining the presence of A. hydrophila in foods, the following points should be observed:

1. A distinction should be made between the production of H_2S in AHM and in TSI agar. In AHM, A. hydrophila may produce slight blackening only at or near the top of the tube, but not in the butt. In TSI agar, H_2S is not produced by A. hydrophila.

2. If reactions in AHM are not definitive after an incubation period of 18-24 hr, then tubes should be incubated an additional 24 hr before making final readings.

3. In using the API 20E system for biochemical confirmation, the analyst should be aware that this system will identify all 3 species of the genus Aeromonas as A. hydrophila. Additional biochemical tests are needed to differentiate the species.

4. All carbohydrate fermentation tests are incubated at 35°C, with the exception of salicin, which is incubated at 30°C.

5. In determining the growth of the test culture in potassium cyanide broth, the formation of a tight seal between the rim of the tube and the sealing cork must be assured.

6. Only rabbit blood that has been freshly drawn should be used in the test for beta-hemolytic activity.

Table 10

Differentiation between motile Aeromonas species

Characteristic	A. hydrophila	A. caviae	A. sobria
Motility	+[a]	+	+
Esculin hydrolysis	+(100)[b]	+(94)	-(10)
Growth in KCN	+(100)	+(94)	-(20)
L-Arginine utilization	+	+	-
L-Arabinose utilization	+(85)	+(97)	-(8)
Fermentation of salicin	+(90)	+(86)	-(0)
Fermentation of sucrose	+	+	+
Fermentation of mannitol	+	+	+
Breakdown of inositol	-	-	-
Acetoin from glucose (Voges-Proskauer)	+(96)	-(6)	d(93)
Gas from glucose	+(97)	-(0)	+(98)
H_2S from cysteine	+(99)	-(3)	+(100)
Indole production	+	+	+
Oxidase	+	+	+
Beta-hemolysin	+(95)	-(18)	+(93)

[a] +, Typically positive; -, typically negative; d, differs among strains.

[b] (Percent positive). Phenotypic properties of value in identifying isolates of Aeromonas spp. are based on 68 A. hydrophila isolates, 36 A. caviae isolates, and 40 A. sobria isolates.

CHAPTER 9

CAMPYLOBACTER

The genus Campylobacter is composed of organisms once classified as Vibrio spp. The genus contains at least 8 species. The species of greatest interest in foods is C. jejuni, and, to a lesser extent, C. coli. Symptoms of food poisoning from these organisms include abdominal pain or cramps, diarrhea, malaise, headache, and fever. Symptoms last one to 4 days. In more severe cases, bloody stools may occur, and the diarrhea may resemble ulcerative colitis.

Campylobacter spp. are microaerophilic, requiring only about 5% oxygen for growth. About 10% carbon dioxide is required for good growth. The optimal temperature for growth is 42°C. However, Campylobacter grows more slowly than the Enterobacteriaceae. Because of their small size, they can be separated from most other Gram-negative bacteria by the use of a 0.65 um filter.

C. jejuni is not an environmental organism but rather is one that is associated with warm-blooded animals. A large percentage of all major meat animals has been shown to contain these organisms in their feces, with poultry being prominent.

A. EQUIPMENT AND MATERIALS

1. Mechanical blender (see Chapter 1)

2. Blender jar (see Chapter 1)

3. Stomacher, Model 400

4. Sterile Stomacher bags (Tekmar),

    a. 400 ml (10-0049-000)

    b. 400 ml filter bags (10-0441-000)

5. Sterile cheese cloth squares, 13 x 13 cm or cleaned and resterilized filter linings from the Stomacher filter bags

6. Whirl-pak bag racks (Nasco B751) or laboratory baskets

7. Balances

    a. 2,000 g capacity, accurate to 0.1 g

    b. 1 g capacity, accurate to 0.0001 g

8. Centrifuge, refrigerated

9. 250 ml centrifuge bottles (Nalgene 3128, or equivalent) and 50 ml centrifuge tubes, (Nalgene 3119, or equivalent), sterile

10. Sterile tongue depressors (whole or split in half) and swabs

11. Tubes, 13 x 100 mm, sterile

12. Anaerobic jars

    a. Oxoid HP11, with quick-connect valves, and Oxoid BR60 gas-generating envelopes (used when a gas tank is not utilized), or

    b. BBL 60463 and 60607 to be used with Campy pak or pak plus gas generating envelopes, 71034 or 71045

13. Bubbler System (Figure 11)

    a. Water bath (static) capable of being adjusted to 37 ± 1°C and 42 ± 1°C (for example, a coliform bath).

    b. Or, preferably, a shaker bath that can be adjusted from 30 ± 2°C, to 37 ± 1°C, to 42 ± 1°C, with a flat rack to which stainless steel baskets can be fastened (if using Stomacher bags) or a rack with flask clamps (if using Erlenmeyer flasks).

    c. Aquarium gang valves (Penn Plax, Garden City, New York, USA) available at most pet supply stores. Valves are available in sets of 2-5 outlets per set. Six 4-outlet sets will enable analysis of 2 samples, each with 10 analytical units.

    d. Plastic aquarium tubing, available with the valves, approx. 0.3 cm diameter. Approximately 46-61 cm is needed per outlet, approx. 2.5-5.0 cm between valve sets and sufficient tubing to connect the last valve set to the gas tank.

    e. Pinch clamps, Nalgene (6165-0001) and golf tees.

    f. Stomacher bags, 400 ml, plain top (Tekmar 10-0049-000) and filter bags, (10-0441-000), or 250 ml and 500 ml Erlenmeyer flasks.

       If using bags --

       1) 1 ml plastic pipets, sterile

       2) Stainless steel baskets

       3) Long twist ties, approximately 15 cm

       If using flasks --

       1) Parafilm, 5 cm square size pieces with one side cut into the middle.

       2) Weighting rings, 5 cm in diameter

       3) Foam plugs, 27-34 mm

4) 1 ml pipets, sterile

14. Gas Tank System (Figures 11 and 12)

a. Gas tank containing 5% O_2, 10% CO_2 and 85% N_2

b. Brass gas CO_2 regulator, CGA-590, single stage attached to a 3-way connector if one tank is used for bubbling and gassing jars, or a 2-way connector if the tank is used for gassing jars and possibly shaker culture flasks

(1) If using a 3-way connector, also use a graduated 2-way connector (Nalgene T5375) to join the bubbler tubing to a 0.6 cm hose clamped to the 3-way connector.

c. Schrader chuck and clip valves, Oxoid HP20, each clamped to pressure tubing; one is attached to the gas tank and the other to a vacuum source

d. On-off valve, using compatible connections, set between the 2- or 3-way connector and the hose attached to a Schrader valve

e. Ring clamps, to tighten connections

15. Shaking gas flask system (Figure 13)

a. Erlenmeyer vacuum flasks, 250 ml (500 ml optional)

b. Shaker incubator or shaker water bath with platform clamps capable of holding 250 ml flasks (500 ml optional) and of being adjusted to 30 $\pm$ 2, 37 $\pm$ 2, and 42 $\pm$ 1°C.

c. Three-way stopcock, PTFE, Nalgene 87395 series, polypropylene with 4 mm plug bore and tubing to fit

d. Bubble tubing - vacuum grade, 1 cm id

e. Foam plugs (20-26 mm size) cut in eighths

f. Rubber stoppers, Nos. 6 and 7

g. Metal hose-cock clamps

h. Male quick-connect valve (Oxoid HP15) to match female valve attached to the gas tank

16. Incubators

Air, 25 $\pm$ 2°C, 30 $\pm$ 2°C, 35-37 $\pm$ 1°C and 42 $\pm$ 1°C

17. 0.65 μm filtering system (Millipore or Genex) - Dairy samples only

    a. Swinex filtering system (Millipore)

        (1) 0.65 μm filters, 47 mm, sterile (DAWG 047 SO)

        (2) Filter holders, 47 mm (SX00 020 12)

        (3) Syringes 20 ml (XX11 020 12)

    b. Genex Autovial syringe filters, sterile (Genex, Gaithersburg, MD, USA)

18. Long handled forceps or stainless steel kitchen tongs

19. White grease pencils to mark plates

20. Nalgene dilution bottles (2500-0380), or equivalent, sterile

21. Erlenmeyer flasks, 50 ml, sterile with a tin foil top

22. Microscope, phase contrast with a 100X oil immersion objective or dark field with a 63X objective

23. Microscope slides, 1 cm-square cover slips and immersion oil

B. **MEDIA AND REAGENTS**

1. Campylobacter enrichment broth (M19)

    a. Nutrient broth No. 2 (Oxoid M67) with 0.6% yeast extract

    b. FBP concentrate

    c. Defibrinated horse or sheep blood, lysed

    d. Antibiotic supplement

        1) Supplement No. 1 (mg/l): cefoperazone, 30; trimethoprim lactate, 12.5; vancomycin, 10; cycloheximide, 100

        2) Supplement No. 2 (mg/l): cefoperazone, 30; trimethoprim lactate, 12.5; vancomycin, 10; cycloheximide, 100

        3) Supplement No. 3 (mg/l): rifampicin, 10; cefoperazone, 15; trimethoprim lactate, 12.5; cycloheximide, 100

2. Campylobacter isolation agar (M20)

    a. Campylobacter blood-free agar (Oxoid CM739) with 0.2% yeast extract

b. Antibiotic supplement (mg/l): cefoperazone, 30; cycloheximide, 100

3. Campylobacter confirmation broth (antibiotic-free enrichment broth), with and without fetal bovine serum (M18)

4. Heart infusion agar (M33)

5. Peptone, 0.1% (M71)

6. Brucella agar, semi-solid (M16)

7. Triple sugar iron agar (M100)

8. Oxoidative-fermentative test medium (M70)

9. MacConkey agar (M52)

10. Glycine

11. Sodium chloride

12. Cysteine hydrochloride

13. Potassium nitrate

14. Nitrite detection reagents (R23)

15. Sodium hippurate reagent (R36)

16. Ninhydrin reagents (R22)

17. Lead acetate paper strips. Soak filter paper strips in saturated solution of lead acetate and dry.

18. Neutral red solution (R21)

19. Antibiotic discs, nalidixic acid and cephalothin

20. Hydrogen peroxide, 3%

21. Oxidase reagent (R24)

22. Gram stain reagents (S6) with 0.5% carbol fuchsin (S4) as the counterstain

C. PREPARATION OF SAMPLE

1. Food samples that are fragmented; foods with sharp edges such as lobster tails and shrimps with shell intact; semi-solid foods such as tofu or fragile-textured foods such as mushrooms

a. Place a Stomacher filter bag in a Petri dish wire holder (type supplied with anaerobic jars). Place a clip at the lip of the bag to hold the lining adjacent to the plastic bag during filling. Weigh 25 g sample into the bag lining, and add 100 ml enrichment broth. For samples of vegetables, including tofu and bean curd, use Antibiotic Supplement No. 2. For other food types, use Antibiotic Supplement No. 1.

b. Remove the bag from the holder, keeping the clip in place, and wrap a twist wire around the top of the bag. Place the bag(s) in a basket or Whirl-pak bag rack, and gently shake horizontally for 3 min.

c. Open the bag and remove the lining, squeezing lightly as it is withdrawn. If filter bags are not available, rinse in a plain Stomacher bag or sterile quarter jar and decant the rinse into another bag or into a sterile 250 ml Erlenmeyer flask or vacuum flask. See H-1,a, for vacuum flask preparation.

d. Place a 1 ml plastic pipet in the bag and reclose tightly. Be certain that the tip of the pipet does not puncture the bag.

2. Shelled shrimp, frog legs, rabbit pieces, and other solid meat pieces

a. Rinse a 25 g sample with 100 ml enrichment broth with Antibiotic Supplement No. 1 in a plain Stomacher bag for 3 min. Remove the sample using alcohol-flamed forceps or tongs.

b. If the sample is a whole carcass or is bloody or is very fatty, rinse for 3 min with 0.1% peptone at a sample/broth ration of 1:6.

c. Remove the fat by filtering the rinse through sterile cheese cloth into a sterile 250 ml centrifuge container.

d. Centrifuge at 2,000 x g for 10 min at 4°C.

e. Discard the supernatant and resuspend the pellet in 5 ml 0.1% peptone.

f. If the sample weighs less than 500 g, transfer 2 ml to 100 ml enrichment broth containing Antibiotic Supplement No. 1. If the sample weighs more than 500 g (e.g., rabbit carcass), transfer 1 ml to 100 ml enrichment broth containing Antibiotic Supplement No. 1.

g. Make a 1:10 dilution of the 5 ml pellet suspension, above (1 ml pellet and 9 ml 0.1% peptone), and streak both preparations to isolation agar. Incubate 24-48 hr at 42°C under a microaerobic atmosphere. If an incubated plate contains typical Campylobacter colonies (F-1), then the enrichment may be discarded.

3. DAIRY PRODUCTS

a. Milk and ice cream

1) Using pH paper, determine the pH value of milk sample as soon as possible after receipt in the laboratory and adjust, if necessary, to 7.6 with 1 N NaOH. For samples of raw milk, eliminate the preenrichment step. For processed dairy samples, use the 4 hr preenrichment step.

2) Weigh two 25 ml portions into 50 ml centrifuge tubes or centrifuge cans. Centrifuge at 16,000 x g under refrigeration for 40 min.

3) Discard the supernatants. For samples of ice cream, also discard the fat with sterile tongue depressors.

4) Emulsify a large loopful of one of the pellets in 1 ml 0.1% peptone. Pipet 0.1 ml of emulsified culture suspension into a sterile tube and add 0.9 ml 0.1% peptone buffer, and mix gently. Streak both suspensions onto plates of isolation agar. Incubate the plates for 24-48 hr at 42°C under microaerobic conditions. If one of the plates contains typical Campylobacter colonies (F-1), then the enrichments may be discarded.

5) Resuspend one pellet in 100 ml enrichment broth with Antibiotic Supplement No. 1, and the other pellet in 100 ml enrichment with Antibiotic Supplement No. 3.

6) Using pH paper, determine pH value of inoculated broths and adjust, if necessary, to 7.5.

b. Cheese and butter

1) Blend two 25 g samples, each in 100 ml 0.1% peptone, at 10,000-12,000 rpm for 30 s or stomach for 1 min.

2) Strain the samples through sterile cheese cloth into sterile centrifuge cans.

3) Centrifuge under refrigeration at 16,000 x g for 40 min.

4) Remove the fat layers and decant the supernatant.

5) Streak a loopful of each pellet, plus a 1:10 dilution in 0.1% peptone, to isolation agars as described above.

6) Resuspend one undiluted pellet in 100 ml enrichment broth containing Antibiotic Supplement No. 1 and the other undiluted pellet in 100 ml enrichment broth containing Antibiotic Supplement No. 3.

7) Using pH paper, determine pH value of inoculated broths and adjust, if necessary, to 7.5.

c. **Milk "sock" or "strainer"** (gauze piece that is used to filter out solids during milking).

Place a 25 g piece of material into 100 ml each of the 2 enrichment broths, one containing Antibiotic Supplement No. 1 and the other containing Antibiotic Supplement No. 3. Gently shake the broth preparations for 3 min and remove the gauze pieces with flamed forceps, squeezing the liquid from them.

4. **Shellfish**

a. Prepare 2 composites of shucked shellfish, with 5 analytical units per composite, in Stomacher filter bags. Keep the bag lining in place with a clip placed at the top of the bag. The weight of each subsample can range from 25 to 50 g for a total weight of 125 to 250 g per composite. Each shellfish body should be kept intact, if possible. Do not add diluent. If shell fragments are present, place each bag inside another bag.

b. Mash gently, but thoroughly, by hand for 3 min. Withdraw the lining from each bag, squeezing the contents.

c. For each composite, transfer 25 ml of the remaining liquid to 225 ml enrichment broth with Antibiotic No. 2.

d. From the sample enrichments, C-4-c, above, make 1:10 and 1:100 dilutions using enrichment broth with Antibiotic Supplement No.2. The sample enrichments and dilutions of the sample enrichments may be incubated in (1) a microaerobic shaker incubator, (2) an air shaker (split sample enrichments into two 250 ml prepared vacuum flasks), (3) anaerobic jars (split samples enrichments into 2 Stomacher bags or 300 ml Erlenmeyer flasks) or (4) Stomacher bags, if a bubbling system is used.

5. **GROUND MEATS OR ORGAN MEATS**

a. Blend 25 g with 100 ml 0.1% peptone at 10,000-12,000 rpm for 15 s. Centrifuge under refrigeration at 2,000 x g for 2 min.

b. Transfer the supernatant to another sterile centrifuge bottle and centrifuge at 5,000 x g for 30 min. Discard supernatant.

c. Resuspend the pellet in 2 ml 0.1% peptone and make a 1:10 dilution using 0.1 ml suspension and 0.9 ml 0.1% peptone. Streak both suspensions to isolation agar. Incubate the plates at 42°C for 24-48 hr under a microaerobic atmosphere.

d. Place the remaining pellet in 100 ml enrichment broth with Antibiotic Supplement No. 1. If the plates contain typical Campylobacter colonies, the enrichments may be discarded.

6. **Egg yolk and whole egg products**

Weigh 25 g into 100 ml of enrichment broth containing Antibiotic Supplement No. 1. After gently mixing, remove 10 ml and transfer it to another 90 ml enrichment broth containing Antibiotic Supplement No. 1. Analyze both mixtures.

D. PREENRICHMENT AND ENRICHMENT OF SAMPLE

1. If the sample has been refrigerated for 10 days or less, preenrich at 37°C for 4 hr in a microaerobic atmosphere. Then enrich at 42°C for 20 hr in a microaerobic atmosphere. For samples of raw milk, see D-3, below.

2. If the sample has been stored beyond 10 days, has been heated or frozen, or if the product is shellfish, preenrich at 30°C for 3 hr and then at 37°C for 2 hr. The samples are then selectively enriched at 42°C for 19 hr.

3. Raw milk should be analyzed without a preenrichment step. Incubate selective enrichment at 42°C for 24 hr under a microaerobic atmosphere.

4. Use one of the following gassing methods to create a microaerobic atmosphere for optimum growth of Campylobacter:

   a. Bubbling or flowing gas system

      1) If possible, use a shaking water bath that can be set at 3 temperatures (for the 5 hr preenrichment procedure) or 2 temperatures (for the 4 hr preenrichment procedure). If sample enrichments are incubated in Stomacher bags, then the stainless steel baskets that hold the Stomacher bags must be secured to the water bath's rack. If sample enrichments are incubated in flasks, then use a platform with flask clamps. Shaker incubation is preferable because it encourages a more rapid growth rate. Adjust the bath to 30°C and place the bags or flasks in the bath. Connect the pipets to the bubbler lines and begin gassing (H-1). Adjust the shaker speed to 50 rpm. Incubate for 3 hr. Then adjust the temperature to 37°C and incubate for 2 hr. Finally, adjust temperature to 42°C and incubate 19 hr.

      2) If using the 5 hr preenrichment with a static water bath, place the enrichments in a 30°C air incubator for the first step. Incubate 3 hr. Transfer the enrichments to a 37°C water bath and connect the pipets in the bags or flasks to the bubbler lines. Begin gassing (H-1). After 2 hr, change the setting to 42°C. Do not bubble while incubating at 30°C, without shaking, since the broth becomes anaerobic and fosters growth of Clostridium species.

      3) If using the 4 hr preenrichment, place the bags or flasks in a shaker or static water bath set at 37°C. Connect the bubbler lines and begin gassing (H-1). Set shaking bath speed at 50 rpm. After 4 hr, adjust the setting to 42°C and incubate another 20 hr.

NOTE:

Insert golf tees into the ends of unused bubbler lines to prevent gas loss, or the unused valves should be shut off. Using the tees saves the necessity of resetting the valves. After the incubation period is completed, turn off the gas tank and disconnect the tubing from the pipets. Pinch shut the Nalgene clamps on the tubing that connects to the tank. This prevents gas loss through the bubbler when filling the jars with the gas mixture.

b. Shaking system (air incubator)

1) If using the 5 hr preenrichment procedure, place the gassed flasks (H-2) in a 30°C shaker incubator and begin shaking at 200 rpm. After 3 hr, adjust the temperature to 37°C. After 2 hr, adjust to 42°C and incubate 19 hr.

2) If using the 4 hr preenrichment procedure, begin the incubation at 37°C. After 4 hr, adjust the temperature to 42°C and incubate 20 hr.

c. Gassed jar system

1) Gas the enrichments in the jars (H-1-c).

2) If using the 5 hr preenrichment procedure, place the jars in a 30°C incubator for 3 hr. Transfer the jars to a 37°C incubator for 2 hr, then to a 42°C incubator for 19 hr.

3) If using the 4 hr preenrichment procedure, place the jars in a 37°C air incubator for 4 hr, then transfer the jars to a 42°C incubator for 20 hr.

E. ISOLATION OF CAMPYLOBACTER

1. Streak enrichments to isolation agar after 24 hr. Place a small drop of the broth at the edge of the plate using the bubbler pipet (or 1 ml pipet). Then streak agar plate with an inoculating loop.

   When analyzing dairy samples, pipet approximately 3 ml into a 0.65 μm filter. Only 0.25 - 0.5 ml filtrate is required. Streak both the filtered and unfiltered cultures.

2. Fill gassing jar(s) with plates (no more than half full). See H-2 for gassing jar instructions. Incubate the plates at 42°C for 24 hr.

3. If growth on the streaked plates is light or negative after 24 hr, reincubate additional 24 hr. If the plates must be incubated at 37°C, then incubate for 48 hr.

F. IDENTIFICATION OF CAMPYLOBACTER

1. On Campylobacter blood-free agar, typical colonies are round or irregular in shape. They are white to clear and have smooth edges. If plates are not picked within 2 hr of reading the plates, store them under vacuum at 4°C away from light. Pick colonies within 1-2 days.

2. Pick at least 2 colonies per plate and prepare a wet mount slide. Emulsify some of the colony growth in a drop of saline on a slide. Cover with a cover slip and examine without oil under a dark field or with oil under a phase contrast microscope.

3. Cells of Campylobacter are curved, 1.5-5 μm in length, and are often joined in "zigzag" formations. Cells picked from agar usually display only a wiggling motility, whereas cells from a broth swim rapidly in a corkscrew motion. Approximately 20% of C. jejuni isolates are not motile. Older or stressed cells may show decreased motility and a coccoidal form.

4. If the organisms are typical, proceed to the confirmation steps. If organisms do not appear typical, examine up to 5 colonies per plate. Those cultures that contain long rods, short but thick rods, or small cocci, can be discarded. In the event that it is necessary to repick plates, they should be stored at 4°C under a microaerobic atmosphere and protected from light.

5. For samples that have been analyzed using several dilutions and for which more than one dilution is presumptively positive, confirm isolates from only one dilution.

G. CONFIRMATION OF CAMPYLOBACTER

1. Pick a portion of the wet mount-positive colony to a drop of 3% hydrogen peroxide and test for catalase. Presence of bubbles is a positive test.

2. If the culture is positive for catalase, transfer a generous portion of the colony to flasks of Campylobacter confirmation broth (Formulation 1 and Formulation 2) (M18), one with fetal bovine serum and one without fetal bovine serum.

3. Pick 2 colonies per analytical unit to flasks of confirmation broth. When all the flasks are inoculated, incubate at 42°C under microaerobic atmosphere for 24 hr or until broth is visibly turbid.

4. If growth is adequate, use the broth cultures without fetal bovine serum (FBS) to inoculate the biochemicals. Only use the FBS-containing cultures if the plain broth cultures do not grow well.

5. **Antibiotic sensitivity**. Using a sterile swab immersed in the broth culture, inoculate uniformly the surface of a heart infusion agar plate with 5% blood (whole or lysed) and 0.25% FBP. Drop a cephalothin and a nalidixic acid disc on the plate. Incubate the plates at 37°C for 24-48 hr under a microaerobic atmosphere. Read the plates for zones of inhibition around each disc. Any zone indicates sensitivity. Most C. jejuni strains are sensitive to nalidixic acid and resistant to cephalothin.

6. **Gram stain**. Perform Gram stain as directed in S6, Chapter 20, but use 0.5% carbol fuchin (S4) as the counterstain. Campylobacter spp. are Gram-negative.

7. **Hippurate hydrolysis**. Emulsify a generous 2 mm loopful of growth from plate of heart infusion agar (used in antibiotic sensitivity test) with 0.4 ml hippurate solution in a 10 x 75 mm tube. Incubate 2 hr in a 37°C water bath. Add 0.5 ml ninhydrin reagent, agitate, and reincubate 10 min. Read tubes promptly. A distinct violet color (not pale or medium color) is a positive reaction. C. jejuni is positive for hippurate hydrolysis.

8. **TSI agar**. From the broth culture, inoculate TSI agar by streaking the slant and stabbing the butt. Incubate at 35-37°C for 5 days under microaerobic atmosphere. C. jejuni produces an alkaline slant and an alkaline butt with no H_2S production.

9. **Glucose utilization.** Inoculate 2 tubes of oxidative-fermentative test medium by stabbing each tube 3 times with test culture. One tube contains glucose and one tube contains base alone. Incubate at 35-37°C for 4 days under a microaerobic atmosphere. Campylobacter spp. do not utilize glucose or other sugars and will show no change in either tube.

10. **Catalase and oxidase tests**. Inoculate a heart infusion agar slant from the broth culture. Incubate at 35-37°C for 48 hr under microaerobic atmosphere. Remove a loopful of growth from the slant and place it in a drop of 3% hydrogen peroxide.

    Add a drop of oxidase reagent to the heart infusion agar slant culture. If the reagent turns purple, the culture is oxidase positive.

    C. jejuni is both catalase- and oxidase- positive.

11. **Growth temperature tolerance**. Dilute 2 ml of the broth culture with 0.1% peptone to the turbidity of a McFarland Standard No. 1. Using a loopful of diluted culture, streak a line across each of 3 plates of heart infusion-blood-FBP agar. Up to 4 cultures can be inoculated per plate. Incubate one plate at 25°C, one at 35-37°C, and one at 42°C, for 3 days under microaerobic atmosphere. A positive test is indicated by greater turbidity than the initial inoculum. C. jejuni does not grow at 25°C, but does grow at 35-37°C and at 42°C.

12. **Growth on MacConkey agar**. Streak, in a single stroke, the diluted broth culture (G-11) onto MacConkey agar. Up to 4 different cultures may be assayed on a single plate. Incubate at 35-37°C for 3 days under microaerobic conditions. C. jejuni grows on MacConkey agar.

13. **Semi-solid biochemical media**. The following biochemical tests are performed using semi-solid medium. This medium (M17) consists of Brucella broth with 0.18% agar, dispensed 7 ml/tube, in 16 x 125 mm tubes. All the semi-solid media are incubated for 5 days. Drop 0.1 ml of diluted culture (G-11) onto agar surface of semi-solid media.

    a. Growth in 1% glycine. C. jejuni is positive.

    b. Growth in 3.5% NaCl. C. jejuni is negative.

    c. H_2S from cysteine. Inoculate the cysteine medium and hang a lead acetate strip from the top, keeping the cap loose. Do not let the strip touch the medium. Blackening of the strip, even slightly, is a positive reaction. C. jejuni is positive.

    d. Nitrate reduction. After 5 days, add nitrite reagents A and B (R23). The presence of a red color is a positive reaction. C. jejuni is positive.

14. Consult Table 11 for biochemical reactions of C. jejuni and other Campylobacter spp.

H. ASSEMBLY AND OPERATION OF EQUIPMENT

1. Gassing systems

a. Bubbler system

1) Assembly

Remove the side plugs from all but one valve set. Link the sets beginning with the end, plugged set, with 3-5 cm tubing pieces. Join the other end set to the gas tank with a long piece of tubing (See Figures 11 and 12). Place 2 Nalgene pinch clamps on that line. Keep them closed whenever the bubbler is not being used. Attach 45-60 cm tubing pieces to the valve outlets on the front of the valve sets. Attach or hang the bubbler array to a board set behind the water bath (see Figure 11).

2) Operation

If using flasks, place weighing rings over the flasks that contain the sample-enrichment preparations. Rings are not needed if using a shaking bath with platform clamps. Each flask should be sterilized with a foam plug in the top and covered by foil before use. For each flask containing the enrichment preparation, remove the foam plug, insert a 1 ml pipet, and replace the plug. Wrap a 5 cm square piece of parafilm that has been cut 1/2 way up one side around the pipet and over the top of the flask, fitting the slit around the pipet. Place the flasks in the water bath and insert the pipet tops into the tubing ends. Each pipet should rest approximately 0.5 cm above the bottom of each flask.

If using bags, place a plastic pipet into each bag containing the sample preparation. Twist a twist-tie tightly around the top of the bag. Make sure the pipet is not puncturing the bottom of the bag. Place the bags in a stainless steel basket and place in the water bath (or into baskets that are fastened to the platform of a shaker bath). Insert the pipets into the tubing ends. Adjust the pipet ends in the broth so they are in the mid-portion of the broth. If a basket is not filled, place enough water-filled dilution bottles at the end of the basket to secure the bags in an upright position.

When all the flasks or bags are ready, open the gas tank main valve and adjust the regulator to read 5 psi. Use the bubbler valves to adjust the gas flow to the broths. For 100-175 ml volumes, adjust to 1 bubble/sec. For 175-250 ml, adjust to 2 bubbles/sec. Turn off the unused valves or insert golf tees in the valve's tubing.

If using a shaker bath, set the speed to gently shake the enrichments (50 rpm).

Bubbler System for Delivering Modified Atmosphere to Campylobacter Enrichments (Representative of a 12-Unit Array)

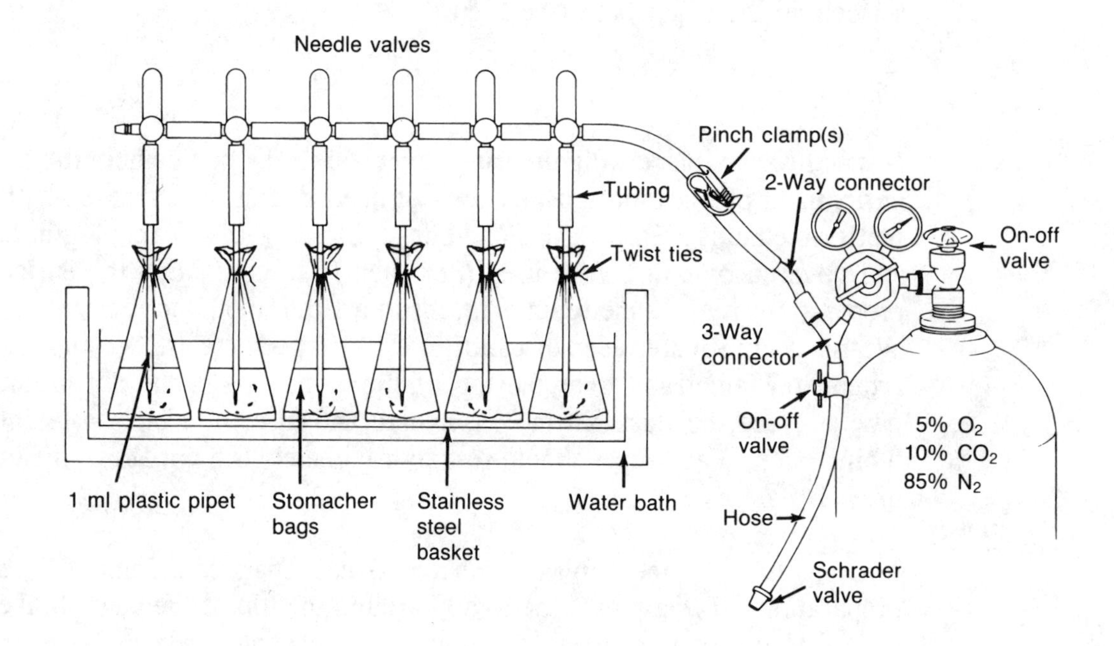

Figure 11

Bubbler System for Delivering Modified Atmosphere to Campylobacter Enrichments (Close-up of Regulator Gauge Area)

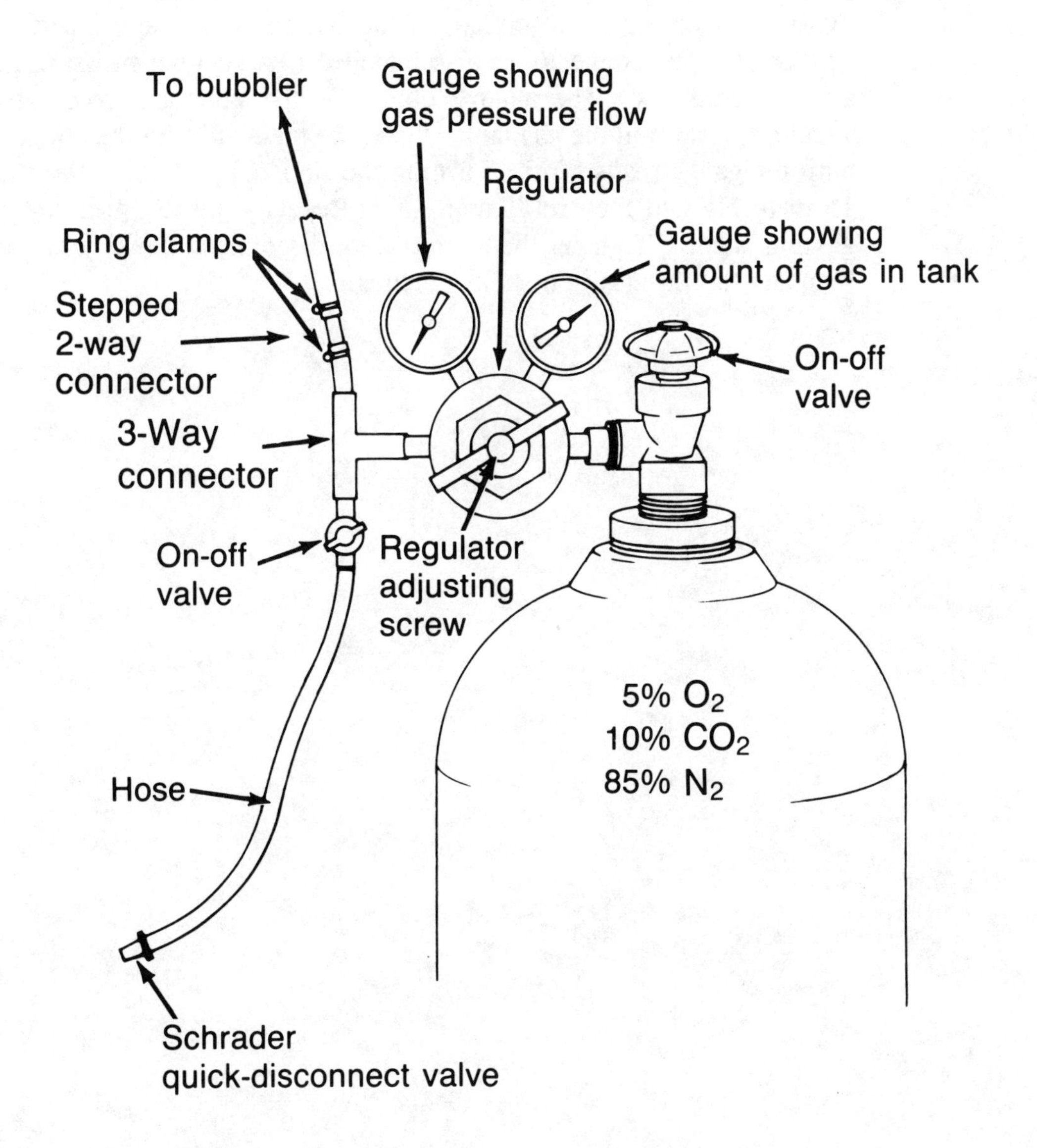

Figure 12

b. Shaker system

1) Assembly

See Figure 13. Autoclave the vacuum flasks with rubber stoppers in the tops, held in place with foil. Bubble tubing (with foam plug inside) should be inserted on the side arm and wrapped in foil.

2) Operation

After placing the enrichment culture in the flask, tape the stopper down tightly. Place the metal clamp on the bubble tubing and insert the 3-way connector. The connector is also attached to a vacuum-pressure gauge and a 3-way stopcock. The other 2 outlets on the stopcock are connected to a vacuum source and the gas tank. Open the tank and turn the adjusting screw until the gauge reads 5 psi. Turning the stopcock, evacuate the flask to 10-15 in of Hg and then refill with gas. Repeat 4 more times ending with a gassing step. Tighten the clamp and disconnect the 3-way connector. Incubate the flask(s) in a shaker incubator.

Vacuum Flask System for Delivering Modified Atmosphere to Campylobacter Enrichments

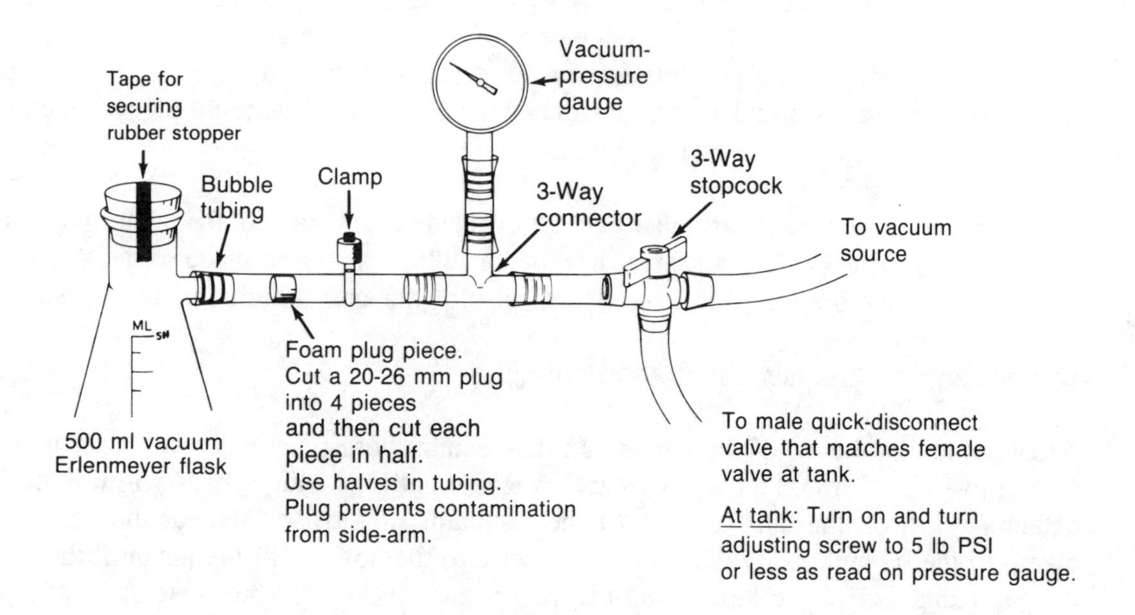

To use:

A. Autoclave flask (with or without broth) with bubble tubing on side arm (no clamp) and with rubber stopper held in place with foil.

B. After sample is prepared, tape rubber stopper firmly in mouth of flask.

C. Put clamp on tubing and insert connector to gassing apparatus.

D. Evacuate and gas 5 times by turning stopcock. End with gassing.

E. Tighten clamp.

F. Use with shaker bath or shaker incubator.

Figure 13

c. Gassed jar system

1) Enrichment broth containers can be incubated in gassed jars (BBL or Oxoid). If using Stomacher bags, place 4 bags per large jar or 2 per small BBL or Oxoid jar. The tops of the bags should be closed with twist-ties loosely. The large jars can hold two 500 ml flasks or four 250 ml flasks. The smaller one only holds one flask per jar. The flasks should have loose fitting foil tops.

2) Use 3 BBL Campy Pak gas generating envelopes with or without catalyst in each large jar and 1 envelope for each BBL small jar. Use Oxoid gas paks with Oxoid jars because of the greater volume in those jars. Because both brands of paks without catalyst produce a small amount of hydrogen, do not open the jars near a flame.

3) The Oxoid jars also can be evacuated and gassed from the gas tank (see below). Use a jar that has the gas intake tubing in place so the gas will enter the bottom of the jar instead of blowing into the tops of the flasks or bags.

2. Evacuating and gassing Oxoid anaerobic jars

After tightening the lid of the jar, attach the vacuum hose's Schrader valve to the jar lid's vacuum valve. Turn on the vacuum and evacuate until the gauge reads 15-20 in of Hg and detach. Turn on the tank and adjust the regulator to 5 psi. Connect the gas tank hose clamp to the lid and open the on-off valve set into the hose. Fill the jar until the gauge on the jar reads 2-3 psi, close the on-off valve, and disconnect the hose from the jar lid. Repeat twice, ending with the gassing step.

NOTE:

Before opening a jar, reduce the positive pressure by depressing a valve stem with a loop handle until the gauge reads 0 psi.

3. Assembly and use of the 65 μm filter units

a. Swinnex filters

1) Autoclave the filter holders and syringes. Separate a holder top and base, laying each in a sterile Petri dish. Place the filter onto the base and center carefully. If the gasket is loose in the holder top, press the gasket into the gasket ring using flamed tweezers run in a circular motion. Screw the top onto the base, twisting tightly.

2) Place the filter tip over a sterile 13 x 100 mm tube and insert the syringe barrel into the top of the filter holder. Pipet 3-4 ml of the enrichment culture into the barrel and insert the plunger. The filtrate should not be turbid. If filtrate is turbid, repeat with another filter. Only 0.25-0.5 ml filtrate is required.

3) After use and autoclaving, separate the holders and syringes. Take apart all components and soak in a mild detergent overnight. Wash by hand, although the tops may need to be washed in a dishwasher. If using a dishwasher, remove the gaskets and wash them by hand.

b. Autovial filters

Filter-syringe units are only available pre-sterilized. To use, place the filter tip into a sterile 13 x 100 mm tube and pipet approximately 2 ml of enrichment culture into the syringe barrel. Insert the plunger and filter 0.25 ml or more into the tube. Discard the filter unit.

I. FLOW SHEET

See Figure 14.

J. RECORD SHEET

See Annex 9.

Detection of Campylobacter

1. SAMPLE PREPARATION

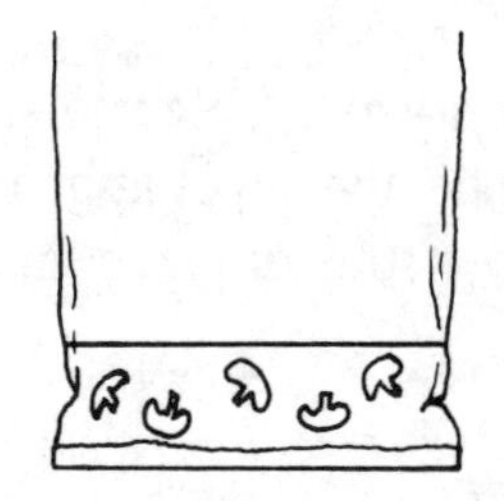

Rinse sample with recommended diluent in Stomacher bag

OR (as directed by method):
1. Rinse, filter, and centrifuge, or
2. Centrifuge, or
3. Blend, or
4. Blend, filter and centrifuge

2. PREENRICHMENT

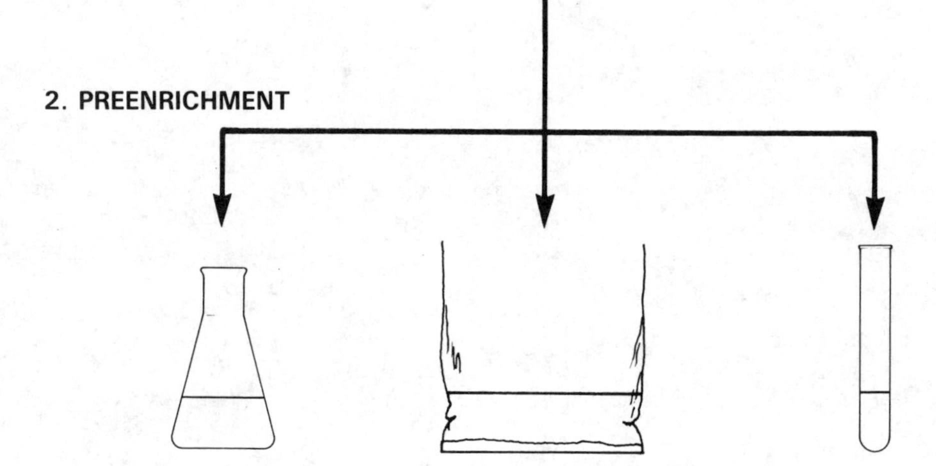

Samples refrigerated ≤ 10 days
Incubate at 37 °C for 4 hr in microaerobic atmosphere

Samples stored > 10 days; heated; frozen; shellfish
Incubate at 30 °C for 3 hr and at 37 °C for 2 hr, both in microaerobic atmosphere

3. ENRICHMENT Raw milk: 42 °C for 24 hr (no preenrichment); microaerobic
Other foods: 42 °C for 19-20 hr; microaerobic

4. ISOLATION AND CONFIRMATION

Figure 14

K. PRECAUTIONS AND LIMITATIONS OF METHOD

In determining the presence of Campylobacter in foods, the following points should be observed:

1. After the sample unit packaging has been opened, initiate sample analysis for Campylobacter before that of any other organisms, since Campylobacter species are sensitive to air and drying.

2. Campylobacter organisms are sensitive to the enzymes in the foods. Thus, samples should either be rinsed and removed from the preenrichment broth, or the sample should be centrifuged to remove sample material.

3. Oxygen-quenching enzymes, added to the Campylobacter media, form toxic radicals when exposed to bright light. Thus, keep the media protected from light during storage and use.

4. Blood hemoglobin readily absorbs oxygen. Use fresh, or frozen fresh, blood and lyse before use. Store lysed blood in frozen condition to inhibit absorption of oxygen.

5. Media, both dehydrated and prepared, absorb oxygen which will form peroxide, a microbial inhibitor. Use only freshly prepared media. Prepared media more than one month old should not be used.

6. For easier viewing, the wet mount of suspect cultures should be examined with an oil immersion objective of a phase contrast microscope rather than with a compound microscope.

7. The semi-solid medium for determining biochemical reactions is inoculated by dropping 0.1 ml of inoculum onto agar surface. Do not stab medium.

8. In determining the growth of cultures at various temperatures, observe the tubes for growth just below the surface. Generalized growth is not due to Campylobacter species.

9. In determining the production of hydrogen sulfide from cysteine, the filter paper strips saturated with lead acetate must not touch the medium.

10. The method described has been shown to be effective for the isolation of Campylobacter from the foods indicated. It may not be used for the analysis of other types of foods.

Table 11

Differentiation of Campylobacter species

Characteristic	C. jejuni	C. coli	C. laridis	C. fetus fetus	C. cinaedi	C. fennelliae	C. cyraerophila	C. hyointestinalis	C. upsaliensis
Growth at									
25°C	-	-	-	+	-	-	+	d	-
35-37°	+	+	+	+	+	+	+	+	+
42°C	+	+	+	d	d	d	-	+	+
Nitrate reduction	+	+	+	+	+	-	+	+	+
3.5% NaCl	-	-	-	-	-	-	-	-	-
H_2S, lead acetate strip	+	+	+	+	+	+	+	+	+
H_2S, TSI	-	d	-	-	-	-	-	+	-
Catalase	+	+	+	+	+	+	+	+	-
Oxidase	+	+	+	+	+	+	+	+	+
MacConkey	+	+	+	+	-	-	-	+	-
Motility (wet mount)	+	+	+	+	+	+	+	+	+
Growth in 1% glycine	+	+	+	+	+	+	-	+	+
Glucose utilization	-	-	-	-	-	-	-	-	-
Hippurate hydrolysis	+	-	-	-	-	-	-	-	-
Resistance to naladixic acid	S	S	R	R	S	S	S	R	S
Resistance to cephalothin	R	R	R	S	S	S	S	S	S

[a] +, 90% or more of strains are positive; -, 90% or more of strains are negative; d, 11-89% of strains are positive; R, resistant; S, sensitive.

CHAPTER 10

YERSINIA ENTEROCOLITICA

Yersinia enterocolitica and Y. pseudotuberculosis are well documented etiological agents of human illness. The most common symptoms are gastroenteritis and ileitis. Only Y. enterocolitica, however, has been isolated from a wide variety of foods, especially those that require refrigeration.

This ability to survive, and even multiply, at refrigeration temperatures, forms the basis for the methodology used for detecting Y. enterocolitica. Primary enrichment is at 10°C for 10 days. This cold enrichment also results in the multiplication of other competing psychrotrophic organisms. The incubated cold enrichment is treated with potassium hydroxide which destroys other Gram-negative bacteria that are more sensitive to alkaline conditions than are the Y. enterocolitica organisms. Thus, this alkaline treatment reduces the competitive background microflora and facilitates the detection of Y. enterocolitica on isolation agars.

A. EQUIPMENT AND MATERIALS

1. Incubators, maintained at 10 ± 1°C and 35-37°C
2. Blender, Waring or equivalent , 8000 rpm, with 500 ml to 1 liter jar
3. Petri dishes, 15 x 100, glass or plastic, sterile
4. Microscope, light 900X and illuminator
5. Syringes, 1 ml; 26-27 gauge needle, sterile
6. Disposable borosilicate tubes, 10 x 75 mm, 13 x 100 mm
7. Wire racks to accommodate 13 x 100 mm tubes
8. Vortex mixer

B. MEDIA AND REAGENTS

1. Peptone sorbitol bile broth (PSBB) (M73)
2. MacConkey agar (M52) (use mixed bile salts)
3. CIN agar (M21)
4. Bromcresol purple broth (M14) supplemented individually with the following carbohydrates, each at 0.5%: mannitol, sorbitol, cellobiose, adonitol, inositol, sucrose, rhamnose, raffinose, melibiose, salicin, xylose, and trehalose; and alpha methyl-D-glucoside at 1%.
5. Christensen's urea agar (M23) (dispense as plated medium or as slants)
6. Phenylalanine deaminase agar (M76) (dispense as plated medium or as slants)

7. Motility test medium (M59). Add 5 ml of 1% 2,3,5-triphenyl tetrazolium chloride

8. Tryptone broth, 1% (M111)

9. MR-VP medium (M61)

10. Simmons' citrate agar (M91)

11. Veal infusion broth (M118)

12. Bile esculin agar (M8)

13. Anaerobic egg yolk agar (M4)

14. Tryptic soy agar (TSA) (M102)

15. Lysine arginine iron agar (LAIA) (M48)

16. Decarboxylase basal medium (Falkow) (M25) supplemented individually with 0.5% arginine, 0.5% lysine, or 0.5% ornithine

17. MOX agar (M60)

18. Blood agar base (BAB) (M10)

19. Pyrazinamidase agar slants (M83)

20. PMP broth (M78)

21. Gram stain reagents (S6)

22. Voges-Proskauer (VP) test reagents (R40)

23. Ferric chloride, 10% in distilled water (R7)

24. Ferrous ammonium sulfate, 1% (R8)

25. Oxidase test reagent (R24)

26. Saline, 0.5% (sterile) (R31)

27. Kovacs' reagent (R16)

28. 0.5% Potassium hydroxide in 0.5% sodium chloride, freshly prepared (R29)

29. Mineral oil, heavy grade, sterile (R20)

30. API 20E system

C. ENRICHMENT OF YERSINIA

1. Analyze samples promptly after receipt, or refrigerate at 4°C. Freezing of samples before analysis is not recommended. However, Yersinia organisms have been recovered from frozen products. Aseptically weigh 25 g sample into 225 ml peptone sorbitol bile broth (PSBB). Homogenize 30 s and incubate at 10°C for 10 days.

2. If high levels of Yersinia are suspected in product, spread plate 0.1 ml on MacConkey agar and 0.1 ml on CIN agar before incubating broth. Also transfer 1 ml homogenate to 9 ml 0.5% KOH in 0.5% saline, mix for several s, and spread plate 0.1 ml on MacConkey and CIN agars. Incubate 48 hr at room temperature (RT) (22-26°C). If high levels of Yersinia contamination are not suspected, omit this step.

3. On day 10, remove enrichment broth from incubator and mix well. Transfer 0.1 ml enrichment to 1 ml 0.5% KOH in 0.5% saline and mix 5-10 s. Successively streak one loopful to MacConkey agar plate and one loopful to cefsulodin-irgasan novobiocin (CIN) agar plate. Transfer additional 0.1 ml enrichment to 1 ml 0.5% saline and mix 5-10 s before streaking, as above. Incubate agars at RT.

D. ISOLATION OF YERSINIA

Examine MacConkey agar plates after 48 hr incubation. Reject red or mucoid colonies. Select small (1-2 mm diameter) flat, colorless, or pale pink colonies. Examine CIN plates after 48 hr incubation. Select small (1-2 mm diameter) colonies having deep red center with sharp border surrounded by clear, colorless zone with entire edge. Inoculate each selected colony into lysine arginine iron agar (LAIA) slant, Christensen's urea agar plate or slant, and bile esculin agar plate or slant by stabbing with inoculation needle. Incubate 48 hr at RT. Isolates giving alkaline slant and acid butt, no gas and no H_2S reactions in LAIA, which are also urease-positive, are presumptive Yersinia organisms. Discard cultures that produce H_2S and/or any gas in LAIA or are urease-negative. Preference should be given to typical isolates that fail to hydrolyze (blacken) esculin.

E. IDENTIFICATION OF YERSINIA

Using growth from LAIA slant, streak culture to one plate of anaerobic egg yolk (AEY) agar and incubate at RT. Use growth on AEY to check culture purity, lipase reactions (at 2-5 days), oxidase test, Gram stain, and inoculum for biochemical tests. From colonies on AEY agar, inoculate the following biochemical test media and incubate all media at RT for 3 days (except one motility test medium and one MR-VP medium, which are incubated at 35-37°C for 24 hr):

1. Decarboxylase basal medium, supplemented with each of 0.5% lysine, arginine, or ornithine; overlay with sterile mineral oil.

2. Phenylalanine deaminase agar. Add 2-3 drops 10% ferric chloride solution to growth on agar slant. Development of a green color is a positive test.

3. Motility test medium, semi-solid, 22-26°C.

4. Motility test medium, semi-solid, 35-37°C.

5. Tryptone broth, 1%. Perform indole test by adding 0.2-0.3 ml Kovacs' reagent to incubated broth culture. Development of deep red color on surface of broth is positive test.

6. MR-VP medium. Use room temperature for autoagglutination test (see F-2, below), followed by V-P test (48 hr). For performance of V-P test, add 0.6 ml alpha-naphthol and shake well. Add 0.2 ml 40% KOH solution with creatine and shake. Read results after 4 hr. Development of pink to ruby red color in medium is positive test.

7. MR-VP medium, 35-37°C for autoagglutination test (see F-2).

8. Bromcresol purple broth with 0.5% of the following filter-sterilized carbohydrates: mannitol, sorbitol, cellobiose, adonitol, inositol, sucrose, rhamnose, raffinose, melibiose, salicin, trehalose, and xylose; and alpha methyl-D-glu-coside at 1%.

9. Simmons citrate agar

10. Veal infusion broth

11. Pyrazinamidase agar slant. After growth of culture appears on slanted pyrazinamidase agar incubated at room temperature, flood 1 ml of 1% freshly prepared ferrous ammonium sulfate over slant. Development of pink color within 15 min is positive test, indicating presence of pyrazinoic acid formed by pyrazinamidase enzyme.

12. Use API 20E system for biochemical identification of Yersinia. Perform tests following manufacturer's instructions.

F. INTERPRETATION

Yersinia are oxidase-negative, Gram-negative rods. Use Tables 1 and 2 to identify species and biotype of Yersinia isolates. Currently, only strains of Y. enterocolitica biotypes 1B, 2, 3, 4, and 5 are known to be pathogenic. These pathogenic biotypes, as well as Y. enterocolitica biotype 6 and Y. kristensenii, fail to hydrolyze esculin or ferment salicin (Tables 12 and 13). However, Y. enterocolitica biotype 6 and Y. kristensenii are relatively rare and can be distinguished by failure to ferment sucrose.

1. **Lipase test**. A positive reaction is indicated by an oily, iridescent, pearly-like colony surrounded by a precipitation ring and an outer clearing zone.

2. **Autoagglutination test**. The MR-VP tube incubated at RT for 24 hr should show some turbidity from bacterial growth. The 35-37°C MR-VP should show agglutination (clumping) of bacteria along walls and/or bottom of tube with clear supernatant fluid. Isolates giving a positive result are presumptive positive for virulence plasmid. Any other pattern for autoagglutination at these 2 temperatures is considered negative.

G. PATHOGENICITY TESTING

1. **Freezing cultures**. Plasmids that determine traits related to pathogenicity of Yersinia can be spontaneously lost during exposure of culture to a temperature above 30°C or during lengthy maintenance of culture and passage of culture below 30°C in the laboratory. It is important, therefore, to immediately freeze presumptive positive cultures to protect plasmid content. Inoculate to veal infusion broth and incubate 48 hr at RT. Add 10% sterile glycerol (e.g., 0.3 ml culture in 3 ml veal infusion broth) and freeze immediately. Storage at -70°C is recommended.

2. **Calcium dependency test**. Inoculate test organism into veal infusion broth. Incubate overnight at 25-27°C. Make decimal dilutions in physiological saline through 10^{-4} dilution. Spread plate 0.1 ml of each dilution in duplicate to blood agar base (BAB) and MOX agars. Incubate one MOX and one BAB agar plate of each dilution at 37°C. Incubate corresponding duplicate plates at 26°C. Examine plates at 24 and 48 hr.

   Growth of Y. enterocolitica strains that contain the virulence plasmid is inhibited on MOX agar at 37°C but not at 26°C. At 37°C, inhibited colonies will be pinpoint-sized at 24 hr and 1 mm at 48 hr on MOX agar, vs 1 and 2-3 mm, respectively, on BAB agar. No inhibition should be seen at 26°C. Inhibited colonies are calcium-dependent and presumptively virulent. Do not subculture from MOX agar because these colonies may have reduced virulence.

3. **Crystal violet binding test** is another rapid screening test to differentiate potentially virulent Y. enterocolitica cultures. Grow suspect cultures for 18 hr at 22-26°C in brain heart infusion (BHI) broth with shaking. Dilute each culture to 1,000 cells/ml in physiological saline. Spread plate 0.1 ml of each culture to each of 2 BHI agar plates. Incubate one plate at 25°C and the other at 37°C for 30 hr. Gently flood each plate with 8 ml of 85 μg/ml crystal violet (CV) solution for 2 min and decant the CV. Observe colonies for binding of CV. Plasmid-containing colonies grown at 37°C will bind CV, but not when grown at 25°C. Plasmidless colonies should not bind CV when grown at either temperature.

H. FLOW SHEET

See Figure 15.

I. RECORD SHEET

See Annex 10.

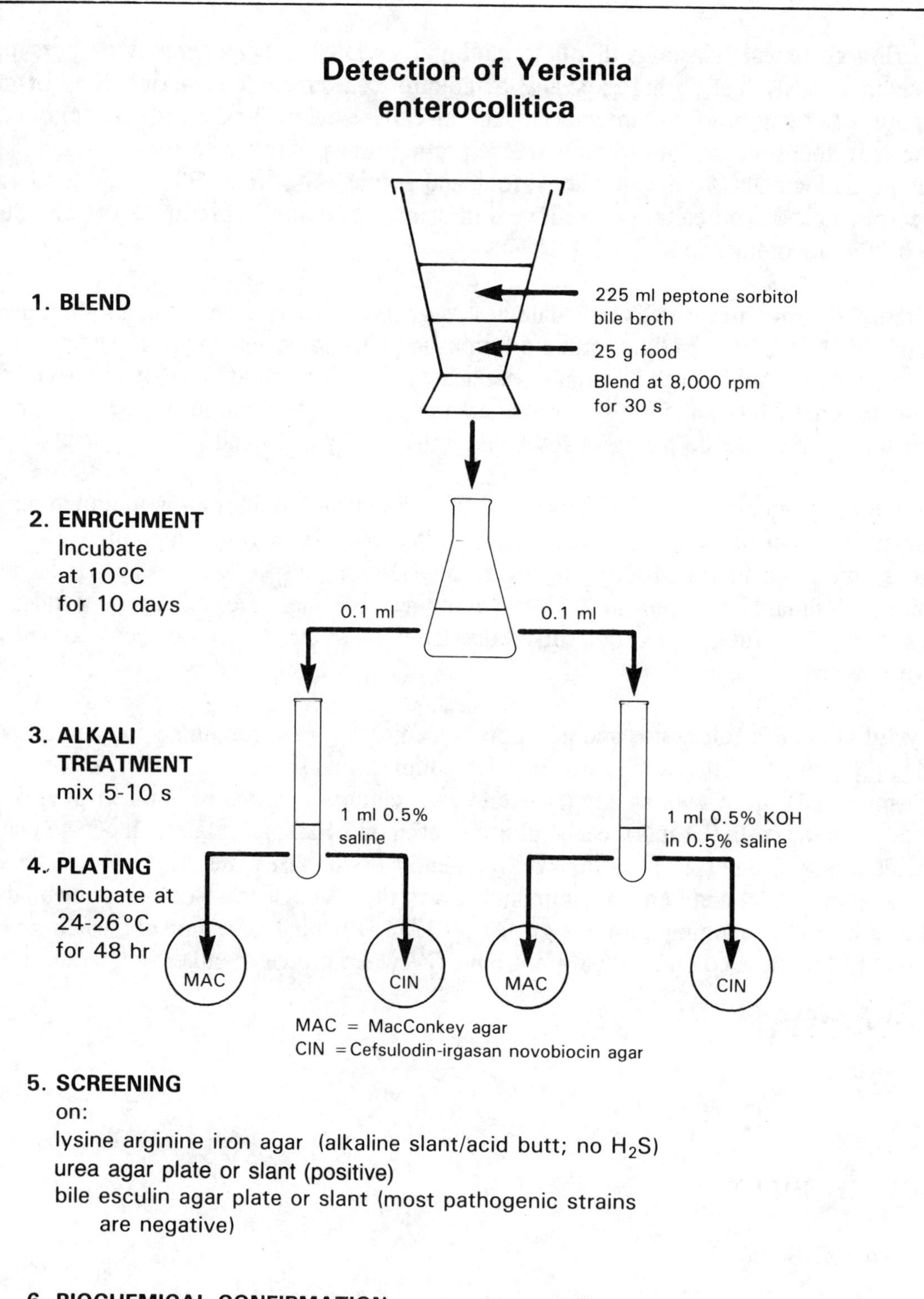
Detection of Yersinia enterocolitica
1. BLEND
225 ml peptone sorbitol bile broth
25 g food
Blend at 8,000 rpm for 30 s
2. ENRICHMENT
Incubate at 10 °C for 10 days
0.1 ml
0.1 ml
3. ALKALI TREATMENT
mix 5-10 s
1 ml 0.5% saline
1 ml 0.5% KOH in 0.5% saline
4. PLATING
Incubate at 24-26 °C for 48 hr
MAC
CIN
MAC
CIN
MAC = MacConkey agar
CIN = Cefsulodin-irgasan novobiocin agar
5. SCREENING
on:
lysine arginine iron agar (alkaline slant/acid butt; no H2S)
urea agar plate or slant (positive)
bile esculin agar plate or slant (most pathogenic strains are negative)
6. BIOCHEMICAL CONFIRMATION
7. PATHOGENICITY TESTING

Figure 15

J. PRECAUTIONS AND LIMITATIONS OF METHOD

In determining the presence of Y. enterocolitica in foods, the following points should be observed:

1. The methodology for Yersinia is still in the investigational stage. Thus, there is no assurance that the recommended method will recover Yersinia from all foods.

2. Y. enterocolitica is especially sensitive to acid conditions. Thus, foods with a low pH value should be analyzed immediately.

3. Growth of other psychrophilic organisms during cold enrichment may interfere with isolation of Yersinia.

4. Incubated sample enrichments in peptone sorbitol bile broth should be not be exposed to alkali more than 10 s.

5. The MR-VP medium tube incubated at 22-26°C for 24 hr should show some turbidity from bacterial growth. The MR-VP medium tube incubated at 35-37°C may show agglutination (clumping) of bacteria along walls and/or bottom of tube with clear supernatant fluid. Any other pattern of agglutination at these temperatures is considered negative.

6. The incubation temperature of 22-26°C for salicin broth is critical.

Table 12. Biochemical characteristics of Yersinia species[a]

Reaction	Yersinia species					
	Y. pestis	Y. pseudo-tuberculosis	Y. entero-colitica	Y. inter-media	Y. frederik-senii	Y. krist-senii
Lysine	-	-	-	-	-	-
Arginine	-	-	-	-	-	-
Ornithine	-	-	+[a]	+	+	+
Motility at						
RT (22-26°C)	-	+	+	+	+	+
35-37°C	-	-	-	-	-	-
Urea	-	+	+	+	+	+
Phenylalanine deaminase	-	-	-	-	-	-
Mannitol	+	+	+	+	+	+
Sorbitol	+/-	-	+	+	+	+
Cellobiose	-	-	+	+	+	+
Adonitol	-	-	-	-	-	-
Inositol	-	-	+/-(+)	+/-(+)	+/-(+)	+/-
Sucrose	-	-	+/-	+	+	-
Rhamnose	-	+	-	+	+	-
Raffinose	-	+/-	-	+	-	-
Melibiose	-	+/-	-	+	-	-
Simmons citrate	-	-	-	+/-	+/-	-
Voges-Proskauer	-	-	+/-(+)	+	+	-
Indole	-	-	+/-	+	+	+/-
Salicin	+/-	+/-	+/-	+	+	-
Esculin	+	+	+/-	+	+	-
Lipase	-	-	+/-	+/-	+/-	+/-
Pyrazinamidase	-	-	+/-	+	+	+

[a]+, Positive after 3 days; (+), positive after 7 days; -, negative.
[b] Some strains of Y. intermedia are negative for either Simmons citrate, rhamnose, and melibiose, or raffinose and Simmons citrate.

Table 13

Biotype schema for Y. enterocolitica

	Reactions for biotypes						
Biochemical test	1A	2B	2	3	4	5	6
Lipase	+	+	-	-	-	-	-
Esculin/salicin (24 hr)	+/-	-	-	-	-	-	-
Indole	+	+	(+)	-	-	-	-
Xylose	+	+	+	+	-	v	+
Trehalose	+	+	+	+	+	-	+
Pyrazinamidase	+	-	-	-	-	-	+
B-D-Glucosidase	+	-	-	-	-	-	-
Voges-Proskauer	+	+	+	+/-	+	(+)	-

[a] ( ), Delayed reaction; v, variable reactions.

CHAPTER 11

LISTERIA

The genus Listeria consists of at least 7 species of Gram-positive, non-sporeforming rods. Only one species, L. monocytogenes, is considered a human pathogen. In human adults, L. monocytogenes is known to cause meningitis, encephalitis, abscesses, and death.

L. monocytogenes is differentiated from the non-pathogenic species by biochemical (Table 1) and virulence testing and by serotyping (Table 2). All virulent strains produce beta-hemolysis on blood agar. The serovars most frequently found in human infections are 1/2 B and 4 B.

The prevalence of L. monocytogenes in foods is not known, and this is probably due to the difficulty of isolating this organism from naturally contaminated samples. Moreover, methodology for detecting this particular pathogen is in a state of flux, with modifications to existing procedures constantly being introduced.

A. EQUIPMENT AND MATERIALS

1. Balance, with weights; 2,000 g capacity, sensitivity of 0.1 g
2. Blender and jars or stomacher and stomacher bags (see Chapter 1)
3. Incubators, 30 and 35 ± 1°C
4. Phase contrast microscope with oil immersion phase objective (100X)
5. Microscope slides
6. Cover slips, glass
7. Immersion oil
8. Erlenmeyer flask, 500 ml
9. Inoculating loops
10. Inoculating needle
11. Petri plates
12. Pipets, 25, 10 and 1 ml
13. Tubes, 16 x 125 mm, screw-cap
14. Beamed white light source (available as Bausch and Lomb Nicholas Illuminator)
15. Dissecting regular or low power microscope with illuminator
16. Fermentation tubes (Durham)

17. Grease pencil or magic marker

18. Needle, 26 gauge, 0.94 cm

19. Syringe, tuberculin, sterile, disposable

B. **MEDIA AND REAGENTS**

1. Listeria enrichment broth (M44)

2. Lithium chloride - phenylethanol - moxalactam (LPM) medium (M45)

3. Oxford medium (M69)

4. Trypticase soy agar with 0.6% yeast extract (TSA-YE) (M103)

5. Trypticase soy broth with 0.6% yeast extract (TSB-YE) (M107)

6. Sheep blood agar (M88)

7. Nitrate reduction medium (M64)

8. SIM motility medium (M90)

9. Purple carbohydrate fermentation broth base (M82), containing 0.5% solutions of glucose, esculin, maltose, rhamnose, mannitol, and xylose

10. Tryptose broth and agar for serology (M113)

11. Acriflavine HC1

12. Agar

13. alpha-Naphthylamine

14. Blood agar base No. 2 (Oxoid)

15. Cycloheximide

16. Sheep blood, defibrinated

17. Glycine anhydride

18. Lithium chloride

19. Moxalactam

20. Nalidixic acid (sodium salt)

21. Nutrient broth (M68)

22. Phenylethanol agar, dehydrated

23. Yeast extract

24. Acetic acid, 5 N (R1)

25. alpha-Naphthol, 5% in absolute ethanol (R40)

26. Hydrogen peroxide solution, 3% (R15)

27. KOH, 40% solution (R30)

28. NaCl, physiological saline solution, 0.85% (R28)

29. Ethanol, absolute

30. Sulfanilic acid (crystal)

31. Fluorescent antibody (FA) buffer (Difco)

32. Listeria-typing sera set

33. Gram stain reagents (S6)

C. ENRICHMENT PROCEDURE

Care should be taken to make the sample representative of the food's outer surface as well as its interior.

Add 25 ml liquid or 25 g cream or solid test material to 225 ml Listeria enrichment broth (EB) in blender or stomacher. Blend at 10,000 - 12,000 rpm for 2 min. Incubate EB culture in blender jar or stomacher bag, or transfer to 500 ml Erlenmeyer flask. Incubate EB culture for 48 hr at 30°C.

D. ISOLATION PROCEDURE

At 24 hr and at 48 hr, streak EB culture onto lithium chloride - phenylethanol - moxalactam (LPM) agar and onto Oxford agar. Incubate plates for 24-48 hr at 30°C (LPM agar) or at 35°C (Oxford agar). Examine LPM plates for suspect colonies by using beamed white light powerful enough to illuminate plate well, striking plate bottom at 45° angle (Henry illumination) (Figure 16).

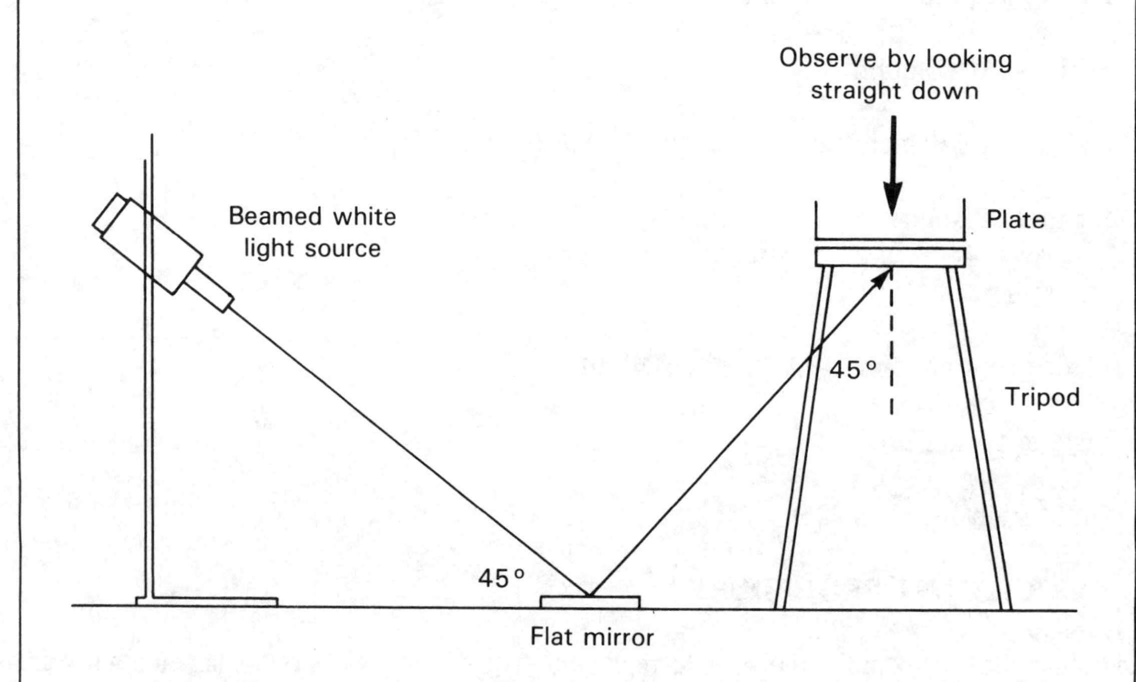
Examination of Plates under Henry Illumination for Suspect Listeria Colonies
Observe by looking straight down
Beamed white light source
Plate
45°
Tripod
45°
Flat mirror

Figure 16

When examined in this oblique-transmitted light from an eye position directly above the plate (i.e., at 90° to the plate) either directly or via a low power microscope or dissecting microscope (with mirrors detached), colonies of Listeria species appear sparkling blue (bluish crushed glass) or white. The use of positive and negative control colonies on LPM agar plates (not the test plates) to attune the observer's eyes is recommended. Although there are variations in the components of the optical systems used, the important points are the 45° angle of incident light and the 90° angle of emergent light. On Oxford agar, colonies are dark brown or black with a brown halo when observed by ordinary illumination. Pick 5 or more typical colonies to trypticase soy agar with 0.6% yeast extract (TSA-YE), streaking for isolated colonies. Incubate TSA-YE plates at 30°C for 24 hr until growth is satisfactory.

E. IDENTIFICATION PROCEDURE

Isolates must be identified by the following classical tests (E-1-13).

1. Examine TSA-YE plates for typical colonies. Using the oblique illumination system (Henry illumination) already described, the colonies will appear blue-gray to blue. Again, the use of known controls on TSA-YE is recommended.

2. Pick a typical colony from culture plate incubated at 30°C or less and examine by wet mount, using 0.85% saline for suspending medium and oil immersion objective of phase-contrast microscope. Be sure to choose a colony with enough growth to make a fairly heavy suspension; emulsify thoroughly. If too little growth is used, the few cells present will stick to the glass slide and appear nonmotile. Listeria species appear as slim, short rods with slight rotating or tumbling motility. Always compare with known culture. Cocci, large rods, or rods with rapid, swimming motility are not Listeria species.

3. Test typical colony for catalase. Listeria species are catalase-positive.

4. Gram stain 16- to 24-hr cultures. All Listeria species are short, Gram-positive rods, but Gram stain reaction can be variable with older cultures. Also, cells in older cultures may appear coccoidal. The cells have a tendency to palisade in thick stained smears. This can falsely lead to rejection as a diphtheroid by those unaware of this property.

5. Pick a typical colony to a tube of trypticase soy broth supplemented with 0.6% yeast extract (TSB-YE) for inoculating biochemicals and fermentations. Incubate at 35°C for 24 hr. This culture may be kept at 4°C for several days and used as inoculum.

6. Inoculate heavily (from a TSA-YE colony) sheep blood agar by stabbing plates that have been poured thick and dried well (check for moisture before using). Draw grid of 20-25 spaces on plate bottom. Stab one culture per grid space. Always stab positive controls (L. ivanovii and L. monocytogenes) and a negative control (L. innocua). Incubate for 48 hr at 35°C.

7. Examine blood agar plates containing culture stabs by using a bright light. L. monocytogenes and L. seeligeri produce a slightly cleared zone around stab. L. innocua shows no zone of hemolysis, whereas L. ivanovii produces a well-defined zone of clearing around stab. Do not try to differentiate species at this point, but note nature of the hemolytic reaction.

8. Using the TSB-YE culture, inoculate urea agar slant by covering the slant well without stabbing the butt. Incubate at 35°C for 5 days. Observe daily for development of a purple color (a positive test). Listeria species do not hydrolyze urea, and no color should develop.

9. Inoculate an MR-VP medium tube, using the TSB-YE culture. Incubate at 35°C for 48 hr. Remove 1 ml to a clean tube and add 0.6 ml alpha-naphthol solution followed by 0.2 ml 40% KOH solution. Shake and examine for a strong red color, a positive test. Incubate remaining culture for additional 2 days. Add methyl red to tube; it should turn red for a positive test. Except for L. denitrificans, other Listeria species are both methyl red and Voges-Proskauer positive.

10. Using the TSB-YE culture, inoculate triple sugar iron agar slant by streaking slant and stabbing butt. Incubate up to 5 days at 35°C. Listeria gives an acid slant and acid butt, with no hydrogen sulfide.

11. Nitrate reduction test. To perform this test, use a TSB-YE culture to inoculate nitrate reduction medium. Incubate at 35°C for 5 days. Add 0.2 ml reagent A, followed by 0.2 ml reagent B (R23). A red color indicates the presence of nitrite, i.e., nitrate has been reduced. If no color develops, add powdered zinc and let stand 1 hr. A developing red color indicates that nitrate is still present and has not been reduced. Only L. murrayi reduces nitrates.

    As an alternative procedure, add 0.2 ml reagent A followed by 0.2 ml reagent C. An orange color indicates reduction of nitrate. If no color develops, add 0.2 ml cadmium reagent. The development of an orange color indicates unreduced nitrate. This procedure eliminates the use of the carcinogen, alpha-naphthylamine.

12. Inoculate SIM motility medium from TSB-YE. Incubate for 7 days at room temperature. Observe daily. Listeria species are motile, giving a typical umbrella-like growth pattern.

13. From TSB-YE culture, inoculate the following carbohydrates set up as 0.5% solutions in purple carbohydrate broth: glucose, esculin, maltose, rhamnose, mannitol, and xylose. Incubate 7 days at 35°C. Listeria species produce acid with no gas or no reaction. Consult Table 14 for xylose-rhamnose reactions of Listeria species. All species should be positive for glucose, esculin, and maltose. All Listeria species, except L. grayi and L. murrayi, should be mannitol-negative.

F. SEROLOGY

This test is used when epidemiological considerations are crucial. Use TSB-YE culture to inoculate tryptose broth. Make 2 successive transfers of cultures incubated in tryptose broth for 24 hr at 35°C. Make a final transfer to 2 tryptose agar slants and incubate 24 hr at 35°C. Wash both slants in a total of 3 ml Difco FA buffer and transfer to sterile 16 x 125 mm screw-cap tube. Heat in water bath at 80°C for 1 hr. Spin at 1,600 x g for 30 min. Remove 2.2-2.3 ml of supernatant fluid and resuspend pellet in remainder of buffer. Follow manufacturer's recommendations for sera dilution and agglutination procedure.

G. MOUSE PATHOGENICITY (OPTIONAL)

Grow isolate for 24 hr at 35°C in TSB-YE. Transfer to 2 tubes of TSB-YE for another 24 hr at 35°C. Place a total of 10 ml culture broth from both tubes into 16 x 125 mm tube and spin at 1,600 x g for 30 min. Discard supernatant and resuspend pellet in 1 ml 0.85% saline diluent. This suspension will contain approximately 10^{10} bacteria/ml and the actual concentration should be determined by a pour or spread plate count. Inject (i.p.) 16- to 18-g Swiss white mice (5 mice/culture) with 0.1 ml of the concentration suspension. Each mouse will receive 10^9 bacterial cells. Observe for death over a 7-day period. Nonpathogenic strains will not kill, but 10^9 pathogenic cells will kill, usually within 5 days. This test should be controlled with known pathogenic and nonpathogenic strains.

H. CAMP TEST

CAMP test cultures are available from several culture collections, including the American Type Culture Collection (ATCC).

Streak weakly beta-hemolytic S. aureus strains (CIP 5710 or NCTC 7428) (ATCC 25923) and Rhodococcus equi (ATCC 6939; NCTC 1621) vertically on sheep blood agar. Separate vertical streaks so that test strains may be streaked horizontally between them without quite touching them. After 24- and 48-hr incubation at 35°C, examine plates for hemolysis in the zone of influence of the vertical streaks. L. monocytogenes hemolysis is enhanced in the vicinity of the S. aureus streak; L invanovii hemolysis is enhanced in the vicinity of the R. equi streak; and hemolysis of L. seeligeri is enhanced near the S. aureus streak. The other species remain nonhemolytic in this test. The CAMP test differentiates L. ivanovii from L. seeligeri and can differentiate a weakly hemolytic L. seeligeri (that may have been read as nonhemolytic) from L. welshimeri. Isolates giving reactions typical for L. monocytogenes, except for the hemolysin production, should be CAMP-tested before they are identified as nonhemolytic L. innocua.

I. INTERPRETATION OF ANALYSES DATA FOR SPECIATION

All Listeria spp. are small Gram-positive rods that demonstrate motility in wet mount and in SIM. They are catalase-positive, do not hydrolyze urea, and produce acid slant and acid butt in TSI without production of H_2S. They utilize dextrose, esculin, and maltose, and some species utilize mannitol, rhamnose, and xylose with production of acid. All species give +/+ reactions in MR-VP broth. An isolate utilizing mannitol with acid production is L. grayi or L. murrayi. Nitrate reduction differentiates between the 2 because only L. murrayi reduces nitrate. L. monocytogenes, L. ivanovii, and L. seeligeri produce hemolysis in sheep blood stabs and are consequently CAMP test-positive. Of the 3, only L. monocytogenes fails to utilize xylose and is positive for rhamnose utilization. The difficulty in differentiating L. ivanovii from L. seeligeri can be resolved by the CAMP test. L. seeligeri shows enhanced hemolysis at the S. aureus streak. L. ivanovii shows enhanced hemolysis at the R. equi streak. Of the nonhemolytic species, L. innocua may give the same rhamnose-xylose reactions as L. monocytogenes but is negative in the CAMP test. L. innocua is the only species that sometimes gives negative results for utilization of both rhamnose and xylose. L. welshimeri that is rhamnose-negative may be confused with a weakly hemolytic L. seeligeri unless resolved by the CAMP test.

After all other results are available, the serotyping of Listeria isolates becomes meaningful. Biochemical, serological, and pathogenicity data are summarized in Tables 14-16. All data collection must be completed before speciation.

J. FLOW SHEET

See Figure 17.

K. RECORD SHEET

See Annex 11.

L. PRECAUTIONS AND LIMITATIONS OF METHOD

In determining the presence of Listeria in foods, the following points should be observed:

1. In weighing the 25 g food sample for analysis, care should be taken to sample the food's outer surface as well as its interior.

2. Sample should not be subjected to blending or stomaching conditions which would result in injury or death to L. monocytogenes organisms.

3. In using Henry illumination, there must be a 45 degree angle of incident light and a 90 degree angle of emergent light.

4. In examining a colony by wet mount, a heavy suspension should be used. If too little growth is used, the few cells present will stick to the glass slide and appear nonmotile.

5. In performing the Gram stain, the culture should not be more than 24 hr old. Gram stain reaction can be variable with older cultures.

6. The sheep blood agar plates must be completely dry so as to avoid confluence of reactions from different cultures.

7. The mouse pathogenicity test should be controlled with known pathogenic and nonpathogenic strains so as to avoid misinterpretation of results.

8. L. welshimeri that is rhamnose-negative may be confused with a weakly hemolytic L. seeligeri unless resolved by the CAMP test.

Detection of Listeria

1. FOOD HOMOGENATE

225 ml Listeria enrichment broth

25 g food

Blend at 10,000-12,000 rpm or stomach for 2 min

2. ENRICHMENT
Incubate at 30°C for 24-48 hr; streak plates at 24 hr and after 48 hr

3. PLATING
Incubate 24-48 hr at 30°C (LPM) or 35°C (OX)

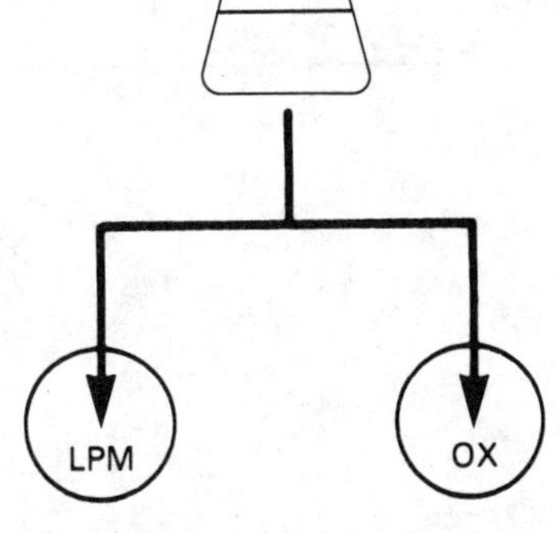

LPM = Lithium chloride-phenylethanol-moxalactam agar

OX = Oxford agar

4. HENRY ILLUMINATION
for LPM plates only; ordinary illumination for OX plates

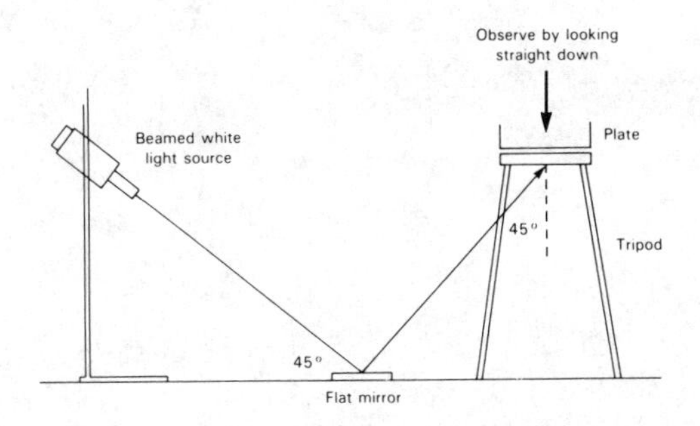

5. BIOCHEMICAL AND SEROLOGICAL CONFIRMATION

Figure 17

Table 14

Differentiation of Listeria species

Species	Hemolytic (Beta)	Nitrate reduction	Utilization Mannitol	Utilization Rhamnose	Utilization Xylose	Sheep blood stab	Virulence (mouse test)
L. monocytogenes	+	-	-	+	-	+	+
L. ivanovii	+	-	-	-	+	+	+
L. innocua	-	-	-	v[a]	-	-	-
L. welshimeri	-	-	-	v[a]	+	-	-
L. seeligeri	+	-	-	-	+	+	-
L. grayi	-	-	+	-	-	-	-
L. murrayi	-	+	+	v[a]	-	-	-

v[a], Variable.

Table 15

Serology of Listeria species

Species	Serotype
L. monocytogenes	1/2A, 1/2B, 1/2C, 3A, 3B, 3C, 4A, 4AB, 4B, 4C, 4D, 4E, "7"
L. ivanovii	5
L. innocua	4AB, 6A, 6B, Un[a]
L. welshimeri	6A, 6B
L. seeligeri	1/2B, 4C, 4D 6B, Un

[a]Un, undefined.

Table 16

CAMP test reactions of Listeria species

Species	Hemolytic reaction	
	Staphylococcus aureus	Rhodococcus equi
L. monocytogenes	+	-
L. ivanovii	-	+
L. innocua	-	-
L. welshimeri	-	-
L. seeligeri	+	-

CHAPTER 12

STAPHYLOCOCCUS AUREUS

In processed foods Staphylococcus aureus is readily destroyed by heating, drying, and other processing conditions to which the food is subjected. Thus, the presence of S. aureus indicates contamination from the skin, mouth, or nose of food handlers. Contamination of processed foods may also occur when contaminated food collects on processing surfaces to which food products are exposed. Large numbers of S. aureus cells in processed foods indicate that sanitation, temperature control, or both, were inadequate. This finding, however, is not sufficient evidence to incriminate a food as the cause of food poisoning. The isolated S. aureus organisms must be shown to produce enterotoxins. Conversely, small staphylococcal populations at the time of testing may exist as remnants of large populations that produced enterotoxins in sufficient quantity to cause food poisoning.

DIRECT PLATE COUNT METHOD

This method is suitable for the analysis of foods in which more than 100 S. aureus cells/g may be expected.

A. EQUIPMENT AND MATERIALS

1. Same basic equipment as for conventional plate count (Chapter 1)

2. Drying cabinet or incubator for drying surfaces of agar plates

3. Sterile bent glass streaking rods, hockey stick or hoe-shaped, with fire-polished ends, 3-4 mm diameter, 15-20 cm long, with an angled spreading surface 45-55 mm long

B. MEDIA AND REAGENTS

1. Baird-Parker medium (M7)

2. Trypticase (tryptic) soy broth with 10% sodium chloride and 1% sodium pyruvate (M106)

3. Trypticase (tryptic) soy agar (TSA) (M102)

4. Brain heart infusion (BHI) broth (M12)

5. Coagulase plasma (rabbit) with EDTA

6. Toluidine blue-DNA agar (M99)

7. Lysostaphin

8. Tryptone yeast extract agar (M112)

9. Paraffin oil, sterile

10. 0.02 M Phosphate saline buffer (R27), containing 1% NaCl

11. Catalase test reagents (R4)

C. PREPARATION OF SAMPLE

See Chapter 1.

D. ISOLATION AND ENUMERATION OF S. AUREUS

1. **For each dilution to be plated**, aseptically transfer 1 ml sample suspension to 3 plates of Baird-Parker agar medium, distributing 1 ml of inoculum equitably to 3 plates (e.g., 0.4 ml, 0.3 ml, and 0.3 ml). Spread inoculum over surface of agar plate, using sterile bent glass streaking rod. Retain plates in upright position until inoculum is absorbed by agar medium (about 10 min on properly dried plates). If inoculum is not readily absorbed by medium, place plates upright in incubator for about 1 hr. Invert plates and incubate 45-48 hr at 35°C. Select plates containing 20-200 colonies, unless only plates at lower dilutions (>200 colonies) have colonies with typical appearance of S. aureus. Colonies of S. aureus are typically circular, smooth, convex, moist, 2-3 mm in diameter on uncrowded plates, gray to jet-black, frequently with light-colored (off-white) margin, surrounded by opaque zone and frequently with an outer clear zone; colonies have buttery to gummy consistency when touched with inoculating needle. Occasionally, nonlipolytic strains of similar appearance may be encountered, except that surrounding opaque and clear zones are absent. Strains isolated from frozen or desiccated foods that have been stored for extended periods frequently develop less black coloration than typical colonies and may have rough appearance and dry texture.

2. **Count and record colonies.** If several types of colonies that appear to be S. aureus are present, count the number of colonies of each type. Use plates of the lowest dilution plated, which contain <20 colonies. If plates containing >200 colonies of typical S. aureus appearance and plates of higher dilution have no typical colonies, use these plates to enumerate S. aureus. Select more than 1 colony of each type counted and test for coagulase production. Add number of colonies on set of triplicate plates represented by colonies giving positive coagulase test and multiply by sample dilution factor. Report this value as number of S. aureus/g of product tested.

E. COAGULASE TEST

Transfer suspect S. aureus colonies into small tubes containing 0.2-0.3 ml BHI broth and emulsify thoroughly. Inoculate agar slant of suitable maintenance medium, e.g., TSA, with loopful of BHI suspension. Incubate BHI culture suspension and slants 18-24 hr at 35°C. Retain slant cultures at room temperature for ancillary or repeat tests in case coagulase test results are questionable. Add 0.5 ml reconstituted coagulase plasma with EDTA (B-5, above) to the BHI culture and mix thoroughly. Incubate at 35°C and examine periodically over 6 hr period for clot formation. Only firm and complete clot that stays in place when tube is tilted or inverted is considered positive for S. aureus. Partial clotting, i.e., 2+ and 3+ coagulase reactions, must be tested further. Test known positive and negative cultures simultaneously with suspect cultures of unknown coagulase activity. Make Gram strains of all suspect cultures and observe microscopically.

F. **ANCILLARY TESTS**

1. **Catalase test.** Use growth from TSA slant for conducting catalase test on glass slide or spot plate properly illuminated to observe production of gas bubbles.

2. **Anaerobic utilization of glucose.** Inoculate tube of carbohydrate fermentation medium containing glucose (0.5%). Immediately inoculate each tube heavily with wire loop. Make certain inoculum reaches bottom of tube. Cover surface of agar with layer of sterile paraffin oil at least 25 mm thick. Incubate 5 days at 37°C. Acid is produced anaerobically if indicator changes to yellow throughout tube, indicating presence of S. aureus. Run controls simultaneously (positive and negative cultures and medium controls).

3. **Anaerobic utilization of mannitol.** Repeat F-2, above, using mannitol as carbohydrate in medium. S. aureus is usually positive but some strains are negative. Run controls simultaneously.

4. **Lysostaphin sensitivity.** Transfer isolated colony from agar plate with inoculating loop to 0.2 ml phosphate saline buffer, and emulsify. Transfer half of suspended cells to another tube (13 x 100 mm) and mix with 0.1 ml phosphate saline buffer as control. Add 0.1 ml lysostaphin (dissolved in 0.02 M phosphate saline buffer containing 1% NaCl) to original tube for concentration of 25 ug lysostaphin/ml. Incubate both tubes at 35°C for not more than 2 hr. If turbidity clears in test mixture, test is considered positive. If clearing has not occurred in 2 hr, test is negative. S. aureus is generally positive.

5. **Thermostable nuclease production.** This test is claimed to be as specific as the coagulase test. It is also less subjective than the coagulase test since it is a color change from blue to bright pink. However, it is not intended as a substitute for the coagulase test but rather as a supportive test, particularly for 2+ coagulase reactions. Prepare microslides by spreading 3 ml toluidine blue-deoxyribonucleic acid agar on surface of each microscope slide. When agar has solidified, cut 2 mm diameter wells (10-12 per slide) in agar and remove agar plug by aspiration. Add about 0.01 ml of heated sample (15 min in boiling water bath) of broth cultures used for coagulase test to well on prepared slide. Incubate slides in moist chamber 4 hr at 35°C. A positive reaction is development of bright pink halo extending at least 1 mm from periphery of well.

G. **SOME TYPICAL CHARACTERISTICS OF TWO SPECIES OF STAPHYLOCOCCI AND THE MICROCOCCI**, which may be helpful in their identification, are shown in Table 17.

H. **FLOW SHEET**

See Figure 18.

I. **RECORD SHEET**

See Annex 12.

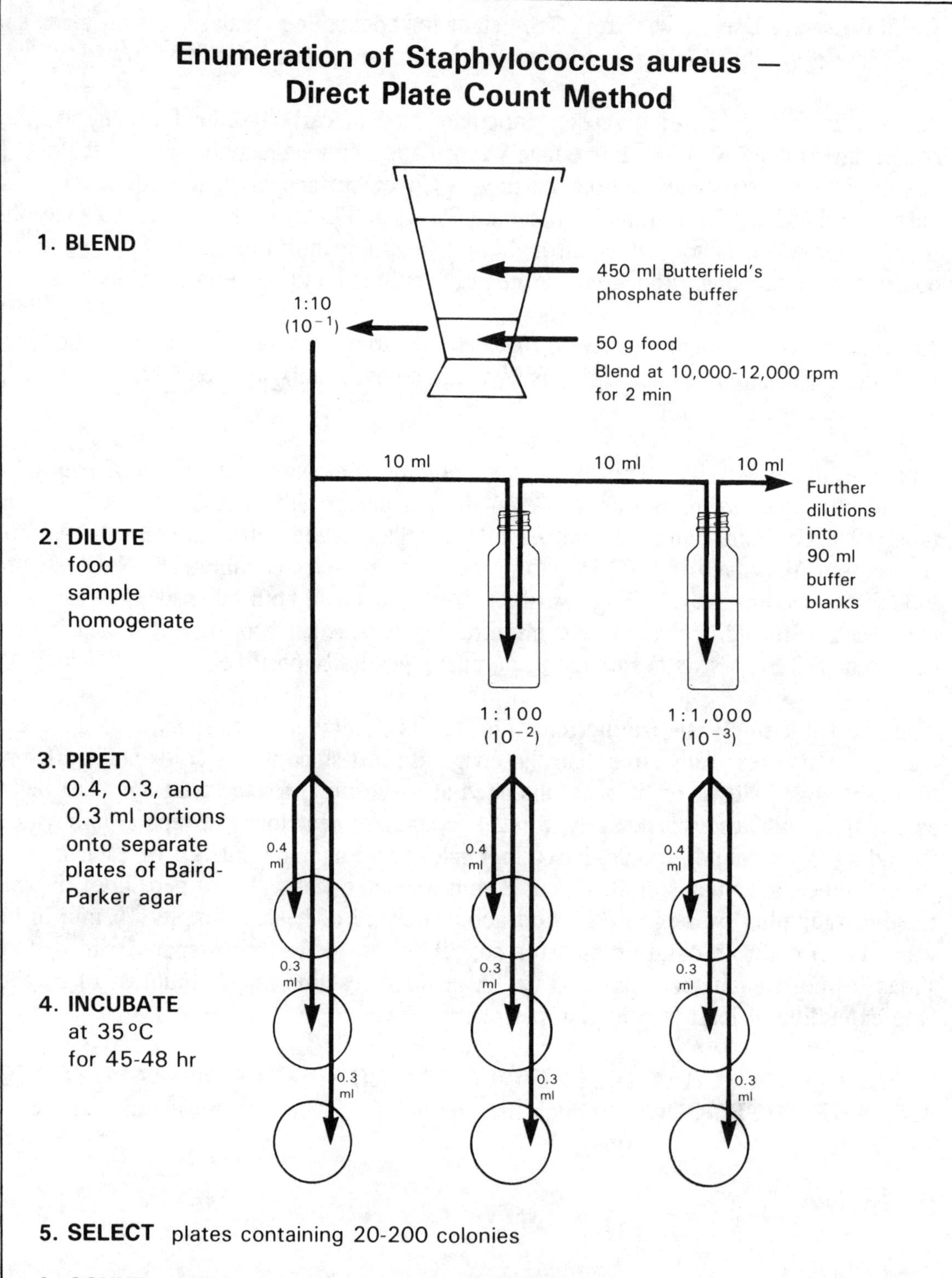

5. SELECT plates containing 20-200 colonies

6. COUNT number of colonies of each morphological type appearing to be *S. aureus;* test > 1 colony of each type for coagulase; add number of colonies on set of triplicate plates giving positive coagulase test and multiply by dilution factor; express this value as number of *S. aureus* cells per g

Figure 18

J. PRECAUTIONS AND LIMITATIONS OF METHOD

In enumerating S. aureus in foods by the direct plate count method, the following points should be noted:

1. Many of the S. aureus cells surviving in processed foods may be expected to be physiologically damaged. Thus, the selective agents used in Baird-Parker agar may have an adverse effect on S. aureus cells as well as on the cells of other genera and species. Moreover, metabolically impaired cells surviving the toxic chemicals of selective media may fail to show a typical form and appearance on Baird-Parker agar.

2. Even though Baird-Parker agar is generally regarded as the most productive of the selective plating agars, it will not always prevent the growth of all competing species.

3. Not all strains of S. aureus have the capacity to hydrolyze egg yolk. Thus, these strains will not demonstrate a typical morphology on Baird-Parker agar.

4. In testing for coagulase, only a firm and completed clot of the coagulase plasma with ethylenediamine tetra-acetic acid that stays in place when the tube is tilted or inverted should be considered positive for S. aureus. Cultures giving partial clotting reactions must be tested further (Gram stain, catalase activity, thermonuclease production, lysostaphin sensitivity, and anaerobic utilization of glucose and mannitol).

5. Instability of biochemical traits has been observed in strains of S. aureus. Thus, one should be aware that the usual biochemical characteristics of the species may not always be evident.

MOST PROBABLE NUMBER METHOD

This method is recommended for use in routine surveillance of products for sanitary quality in which small numbers of S. aureus are expected and in foods expected to contain a large population of competing species.

A. EQUIPMENT AND MATERIALS

Same as for Direct Plate Count Method, above.

B. MEDIA AND REAGENTS

Same as for Direct Plate Count Method, above. In addition: Trypticase (tryptic) soy broth containing 10% NaCl and 1% sodium pyruvate (M106).

C. PREPARATION OF SAMPLE

Same as for Direct Plate Count Method, above.

D. DETERMINATION OF MOST PROBABLE NUMBER (MPN)

Inoculate 3 tubes of Trypticase (tryptic) soy broth containing 10% NaCl and 1% sodium pyruvate (B, above) with 1 ml portions of decimal dilutions of each sample. Highest dilution must give negative end point. Incubate tubes 48 $\pm$ 2 hr at 35°C. Using 3 mm loop, transfer 1 loopful from each tube showing growth (turbidity) to plate of Baird-Parker medium with properly dried surface. Vortex-mix tubes before streaking if growth is visible only on bottom or sides of tubes. Streak inoculum to obtain isolated colonies. Incubate plates 48 hr at 35°C. From each plate showing growth, pick 1 or more colonies, suspected of being S. aureus, to BHI broth as described in D and E of the Direct Plate Count Method, above. Continue procedure for identification and confirmation of S. aureus (E and F, Direct Plate Count Method, above). Report S. aureus/g as most probable number/g, according to tables in Chapter 21, Most Probable Number Determination.

E. FLOW SHEET

See Figure 19.

F. RECORD SHEET

See Annex 13.

G. PRECAUTIONS AND LIMITATIONS OF METHOD

See J, Direct Plate Count Method.

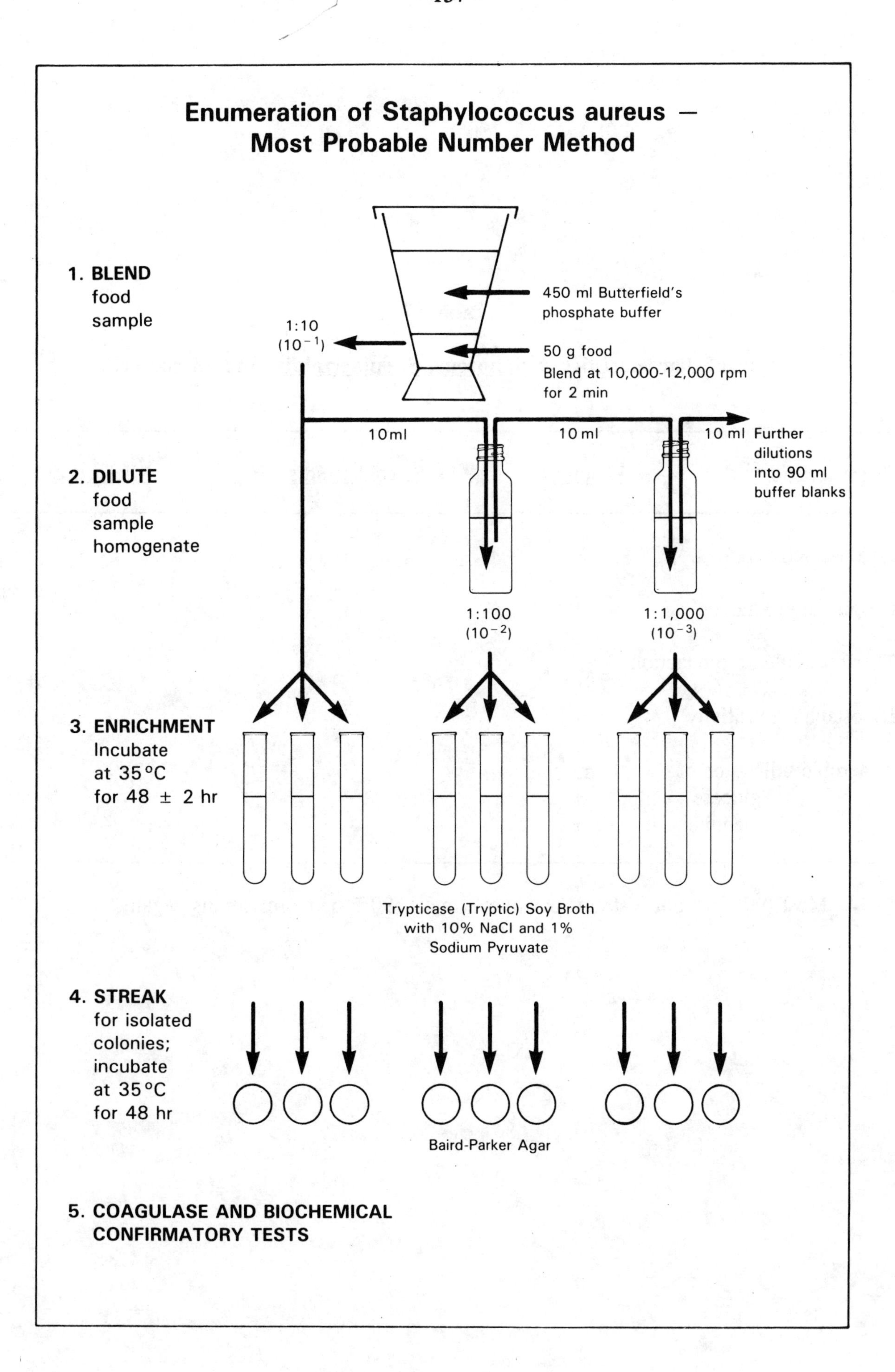
Enumeration of Staphylococcus aureus — Most Probable Number Method
1. BLEND food sample
450 ml Butterfield's phosphate buffer
1:10 (10^{-1})
50 g food
Blend at 10,000-12,000 rpm for 2 min
2. DILUTE food sample homogenate
10 ml
10 ml
10 ml
Further dilutions into 90 ml buffer blanks
1:100 (10^{-2})
1:1,000 (10^{-3})
3. ENRICHMENT Incubate at 35°C for 48 ± 2 hr
Trypticase (Tryptic) Soy Broth with 10% NaCl and 1% Sodium Pyruvate
4. STREAK for isolated colonies; incubate at 35°C for 48 hr
Baird-Parker Agar
5. COAGULASE AND BIOCHEMICAL CONFIRMATORY TESTS

Figure 19

Table 17

Typical characteristics of S. aureus, S. epidermidis, and micrococci[a]

Characteristic	S. aureus	S. epidermidis	Micrococci
Catalase activity	+	+	+
Coagulase production	+	-	-
Thermonoculease production	+	-	-
Lysostaphin sensitivity	+	+	-
Anaerobic utilization of			
glucose	+	+	-
mannitol	+	-	-

[a]+, Most (90% or more) strains positive; -, most (90% or more) strains negative.

CHAPTER 13

STAPHYLOCOCCAL ENTEROTOXINS

Some strains of staphylococci are capable of producing enterotoxins. If these strains are allowed to reach large populations in foods, then the ingested enterotoxin is capable of producing illness in humans. The most common symptoms of staphylococcal enterotoxin food poisoning are nausea, vomiting, and diarrhea which usually develop 2 to 6 hr after ingestion of the contaminated food. The illness is normally mild, with complete recovery occurring after one day. More severe cases, however, may require hospitalization.

These are several antigenically different enterotoxins (A-E). Methods for the detection of the enterotoxins involve the use of specific antibodies to each of them. Several different methods employing specific antibodies have been developed for the detection and measurement of enterotoxins. One of the most useful has been the microslide gel double diffusion test. The minimum detectable level of enterotoxin with this method is 0.1 - 0.25 μg of enterotoxin per 100 g of food. This level of sensitivity appears adequate since information from food poisoning outbreaks indicates that those made ill probably consumed less than 1 μg total enterotoxin.

A. EQUIPMENT AND MATERIALS

1. Blender and sterile blender jars (<u>see</u> Chapter 1)
2. Incubator, 35 $\pm$ 1°C
3. Centrifuge, high speed
4. Centrifuge tubes, 50 ml
5. Sterilizer (Arnold), flowing steam
6. Hot plate, electric
7. Sterile bent glass spreaders (<u>see</u> Chapter 12)
8. Templates, plastic (<u>see</u> Figure 20)
9. Electrical tape, 0.25 mm thick, 19.1 mm wide
10. Silicone grease, high vacuum
11. Sponges, synthetic
12. Test tubes, 25 x 100 mm and 20 x 150 mm
13. Petri dishes, 15 x 100 mm and 20 x 150 mm, sterile
14. Bottles, prescription, 200 ml
15. Microscope slides, plain glass, pre-cleaned, 7.6 x 2.5 cm

16. Pipets, sterile, 1, 5, and 10 ml, graduated

17. Wood applicator sticks

18. Glass tubing, 7 mm, for capillary pipets and de-bubblers

19. Pasteur pipets or disposable 30 or 40 μl pipets

20. Staining jars (Coplin or Wheaton)

21. Desk lamp

22. Timer, interval

B. MEDIA AND REAGENTS

1. Brain heart infusion (BHI) agar, 0.7% (M11)

2. Agar, bacteriological grade, 0.2%

3. Gel diffusion agar, 1.2% (R10)

4. Staphylococcus agar No. 110 (M94)

5. Nutrient agar, slants (M66)

6. Distilled water, sterile

7. Butterfield's phosphate-buffered dilution water (R3)

8. 0.2 M sodium chloride solution, sterile (R34)

9. Physiological saline solution, sterile (antisera diluent) (R28)

10. Slide preserving solution (R32)

11. No. 1 McFarland standard (R17)

12. Antisera and reference enterotoxins

13. Thiazine red R stain (S8)

C. PREPARATION OF MATERIALS AND MEDIA

1. **BHI agar, 0.7%.** Adjust BHI broth to pH 5.3; add agar to prepare 0.7% concentration, and dissolve by minimal boiling. Dispense 25 ml portions into 25 x 200 mm test tubes and autoclave 10 min at 121°C. Just before use, aseptically pour sterile medium into standard Petri dishes.

2. **No. 1 McFarland standard.** Prepare turbidity standard No. 1 of McFarland nephelometer scale (1). Mix 1 part 1% $BaCl_2$ with 99 parts 1% H_2SO_4 in distilled water.

3. **1.2% Gel diffusion agar for gel diffusion slides.** Prepare fluid base for agar in distilled water as follows: NaCl 0.85%; sodium barbital 0.8%; merthiolate 1:10,000 (crystalline). Adjust pH to 7.4. Prepare agar by adding 1.2% Noble Special Agar (Difco). Melt agar mixture in Arnold sterilizer (steamer) and filter while hot, in steamer, through 2 layers of filter paper; dispense in small portions (15-25 ml) in 200 ml prescription bottles. (Remelting more than twice may break down purified agar.)

4. **Thiazine red R stain.** Prepare 0.1% solution of Thiazine red R stain in 1.0% acetic acid.

5. **Preparation of slides.** Wrap double layer of electrician's plastic insulating tape around both sides of glass slide, leaving 2.0 cm space in center. Apply tape as follows: Start piece of tape (about 9.5-10 cm long) 0.5 cm from edge of undersurface of slide and wrap tightly around slide twice. Wipe area between tapes with cheesecloth soaked with 95% ethanol, and dry with dry cheesecloth. Coat upper surface area between tapes with 0.2% agar in distilled water as follows: Melt 0.2% bacteriological grade agar, and maintain at 55°C or higher in screw-cap bottle. Hold slide over beaker placed on hot plate adjusted to 65-85°C and pour or brush 0.2% agar over slide between 2 pieces of tape. Let excess agar drain into beaker. Return agar collected in beaker to original container for reuse. Wipe undersurface of slide. Place slide on tray and dry in dust-free atmosphere (e.g., incubator). NOTE: If slides are not clean, agar will roll off slide without coating it uniformly.

6. **Preparation of slide assembly.** Prepare plastic templates as described by Casman et al. (2). (See Figure 20 for specifications.) Spread thin film of silicone grease on side of template that will be placed next to agar, i.e., side with smaller holes. Place 0.4 ml melted and cooled (55-60°C) 1.2% gel diffusion agar between tapes. Lay silicone-coated template immediately on melted agar and edges of bordering tapes. Place one edge of template on one of the tapes and bring opposite edge to rest gently on other tape. Place slide in prepared Petri dish (C-7, below) soon after agar solidifies and label slide with number, date, or other information.

Microslide Assembly for Preparation and Specifications for Plastic Template

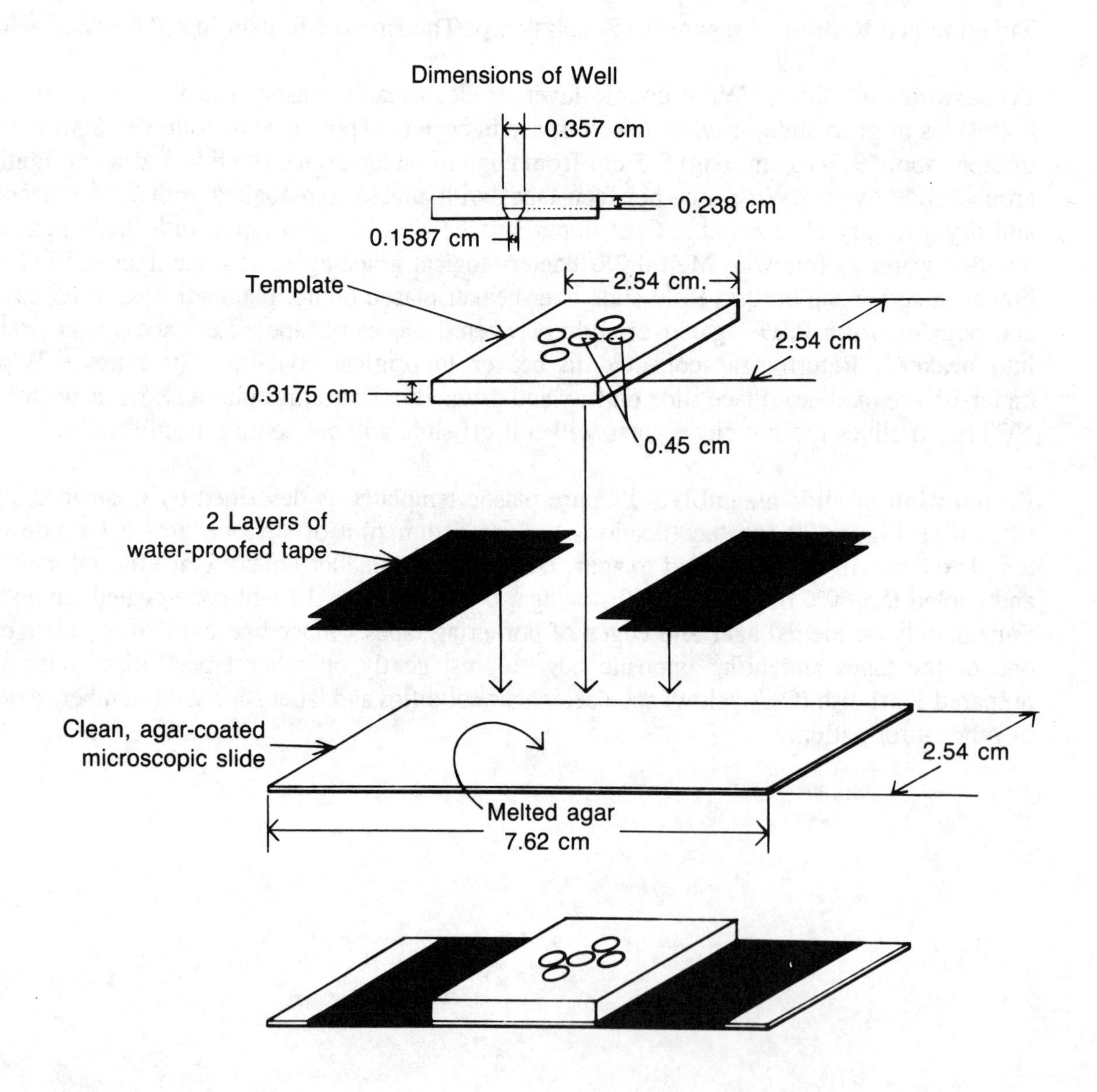

Figure 20

7. **Preparation of Petri dishes for slide assemblies.** Maintain necessary high humidity by saturating 2 strips of synthetic sponge about 1.2 cm (1/2 inch) wide x 1.2 cm (1/2 inch) deep x 6.4 cm (2 1/2 inches) long with distilled water and placing them in each 20 x 150 mm Petri dish. From 2 to 4 slide assemblies can be placed in each dish.

8. **Recovery of used slides and templates.** Clean slides without removing tape; rinse with tap water, brush to remove agar gel, boil in detergent solution 15-20 min, rinse about 5 min in hot running water, and boil in distilled water. Place slides on end, using test tube rack or equivalent, and place in incubator to dry. If slides cannot be uniformly coated with hot 0.2% agar, they are not clean enough and must be washed again. Avoid exposure to excessive heat or plastic solvents when cleaning plastic templates. Place templates in a pan, pour hot detergent solution over them, and let soak 10-15 min. Brush with soft nylon brush to remove residual silicone grease. Rinse sequentially with tap water, distilled water, and 95% ethanol. Spread templates on towel to dry.

9. **Directions for dissolving reagents used in slide gel.** The reagents are supplied as lyophilized preparations of enterotoxins and their antisera. Rehydrate antisera in physiological saline. Rehydrate reference enterotoxins in physiological saline containing 0.3% proteose peptone, pH 7.0, or physiological saline containing 0.37% dehydrated BHI broth medium, pH 7.0. These preparations should produce faint but distinct reference lines in the slide gel diffusion test. The lines may be enhanced (E-3, below).

D. PROCEDURE FOR ENUMERATION AND SELECTION OF COLONIES OF STAPHYLOCOCCI

In examining food products, use procedures described for detecting coagulase-positive staphylococci (<u>see</u> Chapter 12). Test resulting isolates for enterotoxigenicity as described in E, below. For examining food in a suspected staphylococcal food poisoning outbreak, however, the following method is recommended:

1. **Enumeration of total organisms/g.** Blend food with sterile 0.2 M NaCl solution for 3 min at high speed (20 g food in 80 ml of 0.2 M NaCl or 100 g in 400 ml, or whatever amount gives a 1:5 dilution). Prepare decimal dilutions as follows: 10^{-1} = 1 part 1:5 dilution plus 1 part Butterfield's buffer; prepare dilutions 10^{-2}, 10^{-3}, 10^{-4}, 10^{-5}, 10^{-6}, etc., as directed in Chapter 1. Place 0.1 ml portion of each dilution onto prepared Staphylococcus 110 agar and spread with sterile bent glass rod. Incubate plates at 35°C for 48 $\pm$ 2 hr. Count plates at dilution having 30-300 well distributed colonies. Calculate staphylococci organisms/g as follows: Total count x dilution factor of slurry x 10.

2. **Enumeration of enterotoxigenic staphylococci/g.** Note any variation in type or amount of pigment or other morphological characteristics produced by colonies. Count number of colonies in each group type, and record. Pick minimum of 2 colonies from each type and transfer to nutrient agar slants or comparable medium. Test for enterotoxigenicity as described in E, below. Calculate enterotoxigenic staphylococci/g as follows: Number of enterotoxigenic staphylococcal colonies x dilution factor of slurry x 10.

3. **Production of enterotoxin.** Of the methods described by Casman and Bennett (3) for the production of enterotoxin, cultivation of staphylococci on semi-solid BHI agar (pH 5.3) is simple and does not require the use of special apparatus. Add loopful of growth from nutrient agar slants to 3-5 ml sterile distilled water or saline. Inoculate surface of semi-solid

BHI agar with 4 drops of this aqueous suspension, which should contain about 300 million organisms per ml. Turbidity of suspension should be equivalent to No. 1 on McFarland nephelometer scale. Deliver suspension with sterile 1.0 ml pipet. Spread drops of aqueous culture suspension over entire surface with sterile spreader and incubate plates at 35°C. Good surface growth is obtained after 48 hr incubation, at which time pH of culture should have risen to 8.0 or higher. Transfer contents of Petri dish to 50 ml centrifuge tube with wooden applicator stick or equivalent. Remove agar and organisms by high speed centrifugation (e.g., 10 min at 33,000 x g). Examine supernatant for presence of enterotoxin by filling depots in the slide gel diffusion assembly, as directed in E, below.

E. SLIDE GEL DIFFUSION TEST

To prepare record sheet, draw hole pattern of template on record sheet, indicate contents of each well, and give each pattern a number to correspond with number on slide.

1. **Addition of reagents** (see Figure 21). Place suitable dilution of anti-enterotoxin (antiserum) in central well and place homologous reference enterotoxin in upper peripheral well (if diamond pattern in used); place material under examination in well adjacent to well containing reference enterotoxin(s). If bivalent system is used, place other reference toxin in lower well. Use reference toxins and antitoxins (antiserum), previously balanced, in concentrations that give line of precipitation about halfway between their respective wells. Adjust dilutions of reagents to give distinct but faint lines of precipitation for maximal sensitivity. (See C-9 for directions for dissolving reagents.) Prepare control slide with only reference toxin and antitoxin. Fill wells to convexity with reagents, using Pasteur pipet (prepared by drawing out glass tubing of about 7 mm od) or a disposable 30 or 40 ul pipet. Remove bubbles from all wells by probing with fine glass rod. Make rods by pulling glass tubing very fine, as in making capillary pipets; break it into 6.4 cm (2 1/2 inch) lengths and melt ends in flame. It is best to fill wells and remove bubbles against a dark background. Insert rods into all wells to remove trapped air bubbles that may not be visible. Let slides remain at room temperature in covered Petri dishes containing moist sponge strips for 48-72 hr before examination or 24 hr at 37°C.

2. **Reading the slide**. Remove template by sliding it to one side. If necessary, clean slide by dipping it momentarily in water and wiping bottom on slides; then stain slides as described below. To examine slide, hold over source of light and against dark background. Identify lines of precipitation through their coalescence with reference line of precipitation (see Figure 22). If concentration of enterotoxin in test material is excessive, formation of reference line will be inhibited, and test material must then be diluted and retested. Figure 23, diagram A, shows typical precipitate line inhibition due to enterotoxin excess by test preparation reactant arrangement in Figure 21. Figure 24 shows typical line formation of diluted preparation. Occasionally, atypical precipitate patterns may be difficult for inexperienced analyst to interpret. One of the most common atypical reactions is formation of lines not related to toxin but caused by other antigens in test material. Examples of such patterns are given in Figure 25.

Arrangement of Antiserum (a) and Homologous Reference Enterotoxin(s) for Bivalent and Monovalent Systems

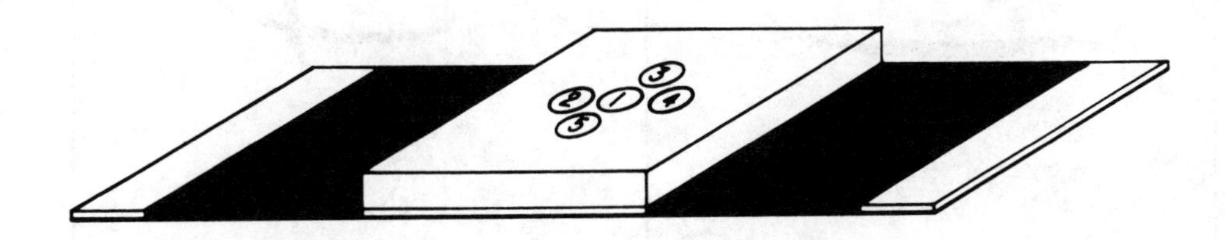

(1) Bivalent

1. Combination antisera (e.g., Anti A and Anti B)
2. Preparation under test
3. Reference enterotoxin (e.g., Anti A)
4. Preparation under test
5. Reference enterotoxin (e.g., Anti B)

(2) Monovalent

1. Antiserum (e.g., Anti A)
2. Dilutions of preparation under test
3. Reference enterotoxin (e.g., type A)
4. Dilutions of preparation under test
5. Dilutions of preparation under test

Figure 21

Microslide Gel Double Diffusion Test as Bivalent System

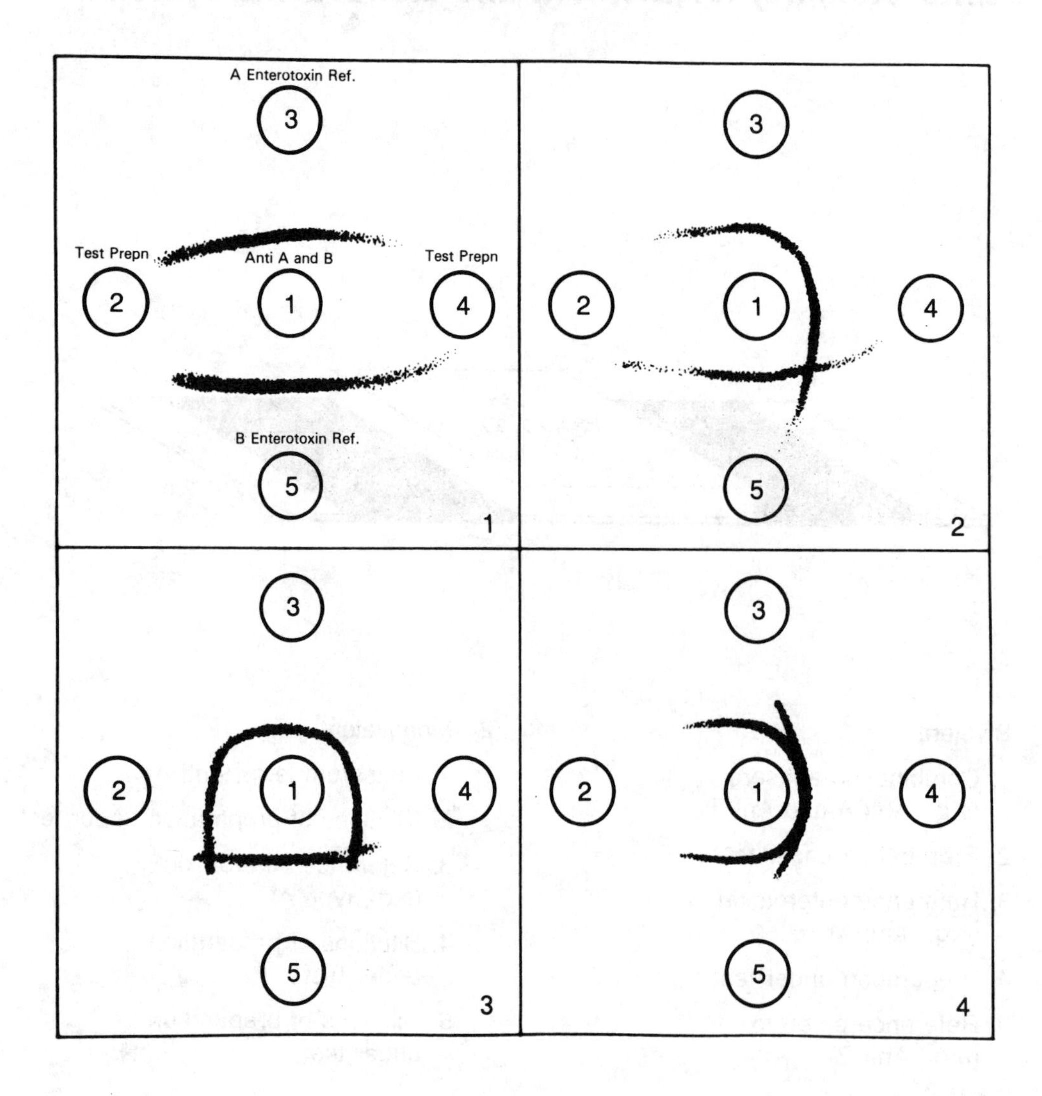

Antisera to staphylococcal enterotoxins A and B are in well 1; known reference enterotoxins A and B are in wells 3 and 5, respectively, to produce reference lines of A and B; preparations under test are in wells 2 and 4. Interpret 4 reactions as follows: (1) No line development between test preparations and antisera—absence of enterotoxins A and B; (2) coalescence of test preparation line from well 4 with enterotoxin A reference line (intersection of test preparation line with enterotoxin B reference line)—absence of enterotoxins A and B in well 2, presence of enterotoxin A and absence of enterotoxin B in well 4; (3) presence of enterotoxin A and absence of enterotoxin B in both test preparations; and (4) absence of enterotoxins A and B in test preparation in well 2, presence of enterotoxins A and B in well 4.

Figure 22

Effect of Amount of Staphylococcal Enterotoxin on Reference Line of Precipitation

A	B	C	D	E	F
(1) Antiserum	(1) Antiserum	(1) Antiserum	(1) Antiserum	(1) Antiserum	(1) Antiserum
(2) 10 µg/ml Enterotoxin	(2) 4 µg/ml Enterotoxin	(2) 2 µg/ml Enterotoxin	(2) 0.5 µg/ml Enterotoxin	(2) 0.125 µg/ml Enterotoxin	
(3) Enterotoxin Ref.	(3) Enterotoxin Ref.	(3) Enterotoxin Ref.	(3) Enterotoxin Ref.	(3) Enterotoxin Ref.	(3) Enterotoxin Ref.
(4) 5 µg/ml Enterotoxin	(4) 3 µg/ml Enterotoxin	(4) 1 µg/ml Enterotoxin	(4) 0.25 µg/ml Enterotoxin	(4) 0.0625 µg/ml Enterotoxin	

Diagram A demonstrates inhibition (suppression) of reference line when 10 and 5µg enterotoxin/ml, respectively, are used. Diagrams B-E show precipitate patterns when successively less enterotoxin (test preparation) is used. Diagram F shows typical formation of reference line of precipitation observed in slide test control system.

Figure 23

Microslide Gel Double Diffusion Test as Monovalent System

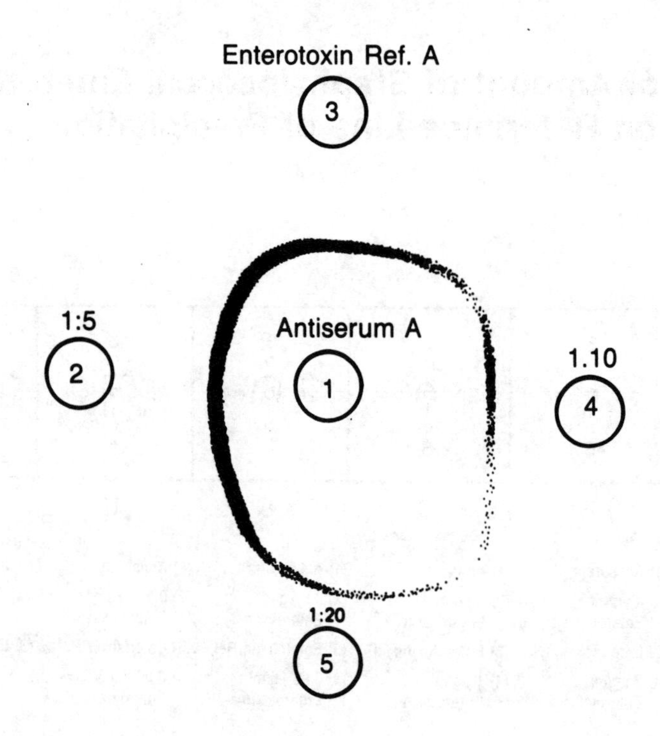

Varying dilutions of test preparation are assayed for presence of staphylococcal enterotoxin.

Figure 24

Atypical Reactions in Microslide Gel Double Diffusion Test

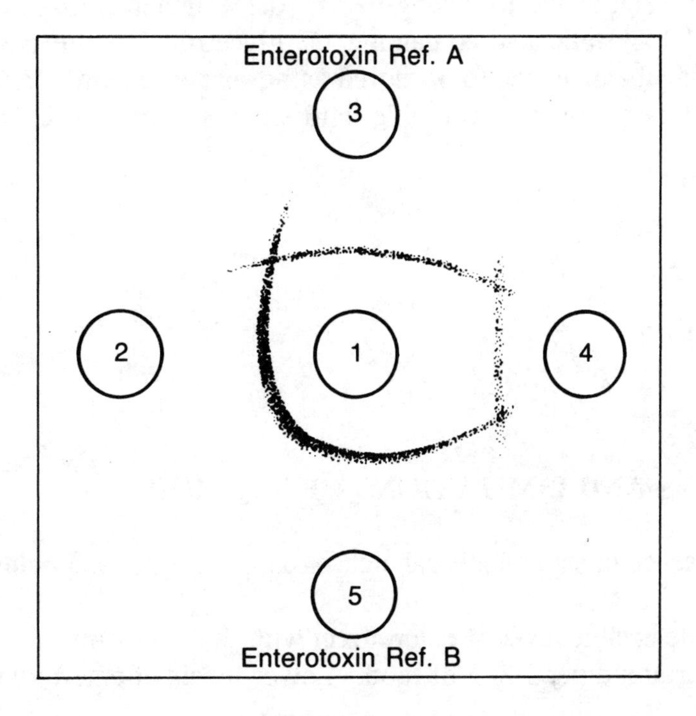

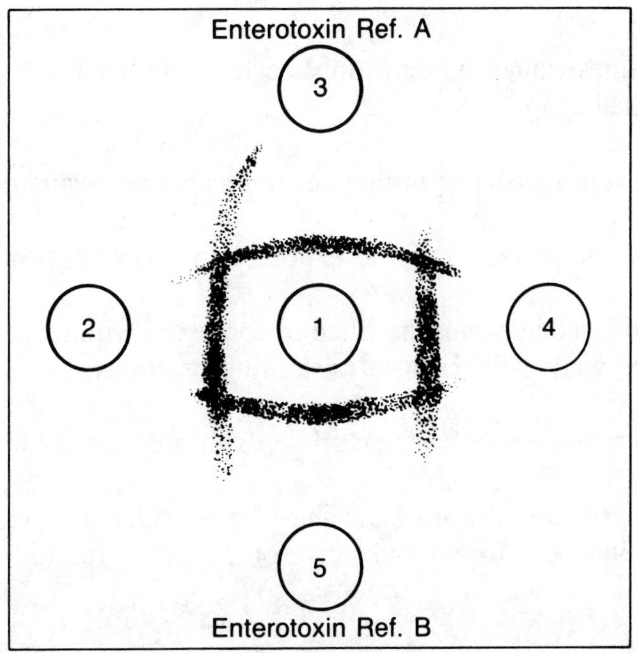

In pattern 1, test preparation in well 4 produces atypical reaction indicated by nonspecific line of precipitation (lines of nonidentity with enterotoxin references A and B) which intersects with enterotoxin reference lines. In pattern 2, both test preparations (wells 2 and 4) are negative for enterotoxins A and B but produce nonspecific lines of precipitation which intersect enterotoxin A and B reference lines of precipitation.

Figure 25

3. **Staining of slides.** Enhance lines of precipitation by immersing slides for 5-10 min in thiázine red R stain, and then examine. Such enhancement is necessary when reagents have been adjusted to give lines of precipitation that are only faintly visible. Use staining procedure described by Crowle (4), modified slightly, when slide is to be preserved. Rinse away any remaining reactant liquid by dipping slide momentarily in water and immersing it for 10 min in each of the following baths: 0.1% thiazine red R in 1% acetic acid; 1% acetic acid; and 1% acetic acid containing 1% glycerol. Drain excess fluid from slide and dry it in 35°C incubator if it is to be stored as permanent record. After prolonged storage, lines of precipitation may not be visible until slide is immersed in water.

F. FLOW SHEET

See Figure 26.

G. RECORD SHEET

See Annex 14.

H. PRECAUTIONS AND LIMITATIONS OF METHOD

In determining the presence of staphylococcal enterotoxin, the following points should be observed:

1. The minimum detectable level of enterotoxin with the microslide gel diffusion test is 0.1 to 0.25 μg of enterotoxin per 100 g of food. Lower levels of enterotoxin will not be detected by this method.

2. If gel diffusion slides cannot be uniformly coated with hot 0.2% agar, they are not clean and must be washed again.

3. Sponges must be kept moistened during entire incubation period of slides.

4. When cleaning plastic templates, avoid exposure to excessive heat or plastic solvents.

5. Wells on gel diffusion slides must be filled to convexity with reagents, using Pasteur pipet. Inadequately filled wells will result in misleading reactions.

6. It is necessary to remove bubbles from all wells in order to obtain accurate reactions.

7. Only lines of precipitation that coalesce must be considered a positive reaction. Lines of precipitation that intersect are not indicative of a positive reaction.

Detection of Staphylococcal Enterotoxin

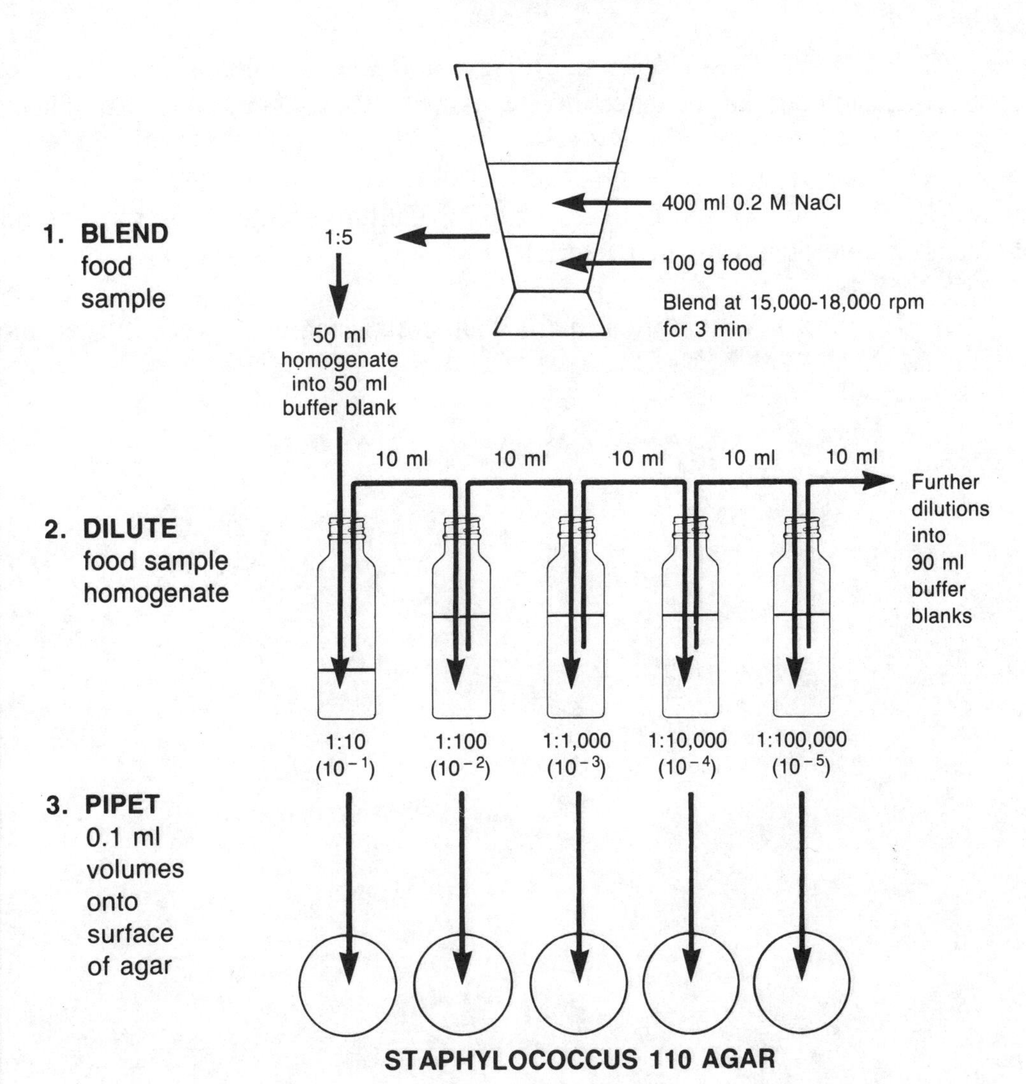

4. **INCUBATE** at 35°C for 48 hr
5. **PICK** to nutrient agar slant and incubate at 35°C for 24 hr
6. **EMULSIFY** loopful of nutrient agar culture with 3-5 ml sterile distilled water; turbidity should be equivalent to McFarland Standard No. 1
7. **INOCULATE** 4 drops of suspension onto surface of semi-solid BHI agar plate and incubate at 35°C for 48 hr
8. **TRANSFER** Petri dish contents to centrifuge tube and centrifuge for 10 min at 33,000 × *g*
9. **ASSAY** supernatant for staphylococcal enterotoxin

Figure 26

REFERENCES

1. **McFarland, J**. 1907. The nephelometer: An instrument for estimating the number of bacteria in suspensions used for calculating the opsonic index and for vaccines. J. Am. Med. Assoc. 49:1176.

2. **Casman, E. P., R. W. Bennett, A. E. Dorsey, and J. E. Stone**. 1969. The micro-slide gel double diffusion test for the detection and assay of staphylococcal enterotoxins. Health Lab. Sci. 6:185-198.

3. **Casman, E. P., and R. W. Bennet**. 1963. Culture medium for the production of staphylococcal enterotoxin A. J. Bacteriol. 86:18-23.

4. **Crowle, A. J.** 1958. A simplified micro double-diffusion agar precipitin technique. J. Lab. Clin. Med. 52:784.

CHAPTER 14

CANNED FOODS

Canned foods are those foods which have been preserved by heat in hermetically sealed containers or in containers which prohibit the passage of gas or microorganisms. If processing of these containers has been conducted properly, then the contents of these cans should remain unspoiled indefinitely. When spoilage does occur, however, certain changes occur in the product, e.g., gas production, swelling of the can, alteration of the pH, or increase in the microorganisms of the product.

The type(s) of microorganisms present in the product may give an indication of the cause of spoilage. A mixed microflora consisting of rods and cocci is indicative of leakage, but this initial diagnosis must be confirmed by an analysis for container integrity. Leakage may occur because of can defects or punctures. Bacteria in the contaminated cooling water will gain access to the canned food through these leakage areas and the food becomes spoiled.

When normal cans contain only mesophilic, Gram-positive, sporeforming rods, they should be considered underprocessed. Underprocessing may be caused by any of several factors: undercooking, faulty retort operations, bypassing of the retort altogether and changes in the handling of the product which results in tighter packing in the can with the consequent lengthening of the heat penetration time.

The single most important factor for determining the degree of thermal processing needed to achieve product stability is pH. Moreover, the pH of the canned food forms the basis for deciding how the food is to be analyzed microbiologically. A pH value of 4.6 is used to distinguish low-acid canned foods (pH above 4.6) from acid canned foods (pH of 4.6 or lower). Thus, two separate methods of canned food analysis are used.

The preferred type of opener is a bacteriological can opener consisting of a puncturing device at the end of a metal rod mounted with a sliding triangular blade with a set screw holding it in place. The advantage is that it does no damage to the double seam and therefore will cause no interference with subsequent seam examination of the can.

A large enough number of cans should be examined bacteriologically to obtain reliable results. Samples for laboratory examination should consist of all abnormal cans and 6 normal containers. The sample will include as many abnormal cans as were found (e.g., if 21 abnormal containers are observed during the inspector's examination, then the sample will consist of 21 abnormal and 6 normal containers). Do not collect leakers, but do report the number of leakers noted. Randomly select one normal container from each of 6 cases, if available, in the lot. The 6 cases are to be randomly selected from the entire lot. A lot is defined as one production code.

A. EQUIPMENT AND MATERIALS

1. Incubators, thermostatically controlled at 30, 35, and 55°C.

2. pH meter, potentiometer

3. Microscope, slides, and coverslips

4. Can opener, bacteriological can opener, and can punch, all sterile

5. Petri dishes, sterile

6. Test tubes, sterile

7. Serological pipets, cotton-plugged, sterile

8. Nontapered pipets, cotton-plugged (8 mm tubing), sterile

9. Soap, water, brush, and towels, sterile and nonsterile

10. Indelible ink marking pen

11. Diamond point pen for marking cans

12. Examination pans (Pyrex or enamel baking pans)

B. **MEDIA AND REAGENTS**

1. Bromcresol purple dextrose broth (BCP) (M15)

2. Chopped liver broth, (M22), or cooked meat medium (CMM) (M24)

3. Malt extract broth (M54)

4. Liver-veal agar (without egg yolk) (LVA) (M46)

5. Acid broth (M1)

6. Nutrient agar (NA) (M66)

7. Sabouraud's dextrose agar (SAB) (M85)

8. 4% Iodine in 70% ethanol (R5)

9. Methylene blue stain (S7), crystal violet (S5), or Gram stain (S6)

C. **CAN PREPARATION**

Remove labels. With marking pen, transfer subnumbers to side of can to aid in correlating findings with code. Mark labels so that they may be replaced in their original position on the can to help locate defects indicated by stains on label. Separate all cans by code numbers and record size of container, code, product, conditions, evidence of leakage, pinholes or rusting, dents, buckling or other abnormality, and all identifying marks on label. Classify each can according to definitions below. Before observing cans for classification, make sure cans are at room temperature.

Flat - a can with both ends concave, and remaining in this condition even when the can is brought down sharply on its end on a solid, flat surface.

Flipper - a can that normally appears flat, but when brought down sharply on its end on a flat surface, one end flips out. When pressure is applied to this end, it flips in again and can appears flat.

Springer - a can with one end permanently bulged. When sufficient pressure is applied to this end, it will flip in, but the other end will flip out.

Soft swell - a can bulged at both ends, but not so tightly that the ends cannot be pushed in somewhat with thumb pressure.

Hard swell - a can bulged at both ends, and so tightly that no indentation can be made with thumb pressure.

A hard swell will generally "buckle" before the can bursts. Bursting usually occurs at the double seam over the side seam lap, or in the middle of the side seam.

D. EXAMINATION OF CAN AND CAN CONTENTS

After classification of cans, the can itself and its contents may be described according to terms in Table 18.

1. Sampling can contents

a. Swollen cans. Immediately analyze springers, swells, and a representative number (at least 6, if available) of flat and flipper cans. Retain examples of each, if available, when reserve portion must be held. Place remaining flat and flipper cans (excluding those held in reserve) in incubator at 35°C. Examine at frequent intervals for 14 days. When abnormal can or one becoming increasingly swollen is found, make note of it. When can becomes a hard swell or when swelling no longer progresses, culture sampled contents, examine for preformed toxin of C. botulinum (Chapter 18), and perform remaining steps of canned food examination. After 14 days, remove flat and flipper cans from incubator. If no swelling has occurred and cans are otherwise normal, culturing is not necessary.

b. Flat and flipper cans. Place cans (excluding those held in reserve) in incubator at 35°C. Observe cans for progressive swelling at frequent intervals for 14 days. When swelling occurs, follow directions in above paragraph. After 14 days, remove flat and flipper cans, and test at least 6, if available. (It is not necessary to analyze all normal cans.) Do not incubate cans at temperature above 35°C. After incubation, bring cans back to room temperature before classifying them.

2. Opening the can.

Open can in an environment that is as aseptic as possible. Use of vertical laminar flow hood is recommended.

a. Hard swells, soft swells, and springers. Chill hard swells in refrigerator before opening. Scrub entire uncoded end and adjacent sides of can using abrasive cleanser, cold water, and a brush, steel wool, or abrasive pad. Rinse and dry with clean sterile towel. Sanitize can end to be opened with 4% iodine in 70% ethanol for 30 min and wipe off with sterile towel. DO NOT FLAME. Badly swollen cans may spray out a portion of the contents, which may be toxic. Take some precaution to guard against this hazard, e.g., cover can with sterile towel to invert sterile funnel over can. Sterilize can opener by flaming until it is almost red, or use separate presterilized can openers for each can. At the time a swollen can is punctured, test for headspace gas by holding mouth of sterile test tube at puncture site to capture some escaping gas, or by using can-puncturing press to capture some escaping gas in a syringe. Flip mouth of tube to flame of Bunsen burner. A slight explosion indicates presence of hydrogen. Immediately turn tube upright and pour in a small amount of lime water. A white precipitate indicates presence of CO_2. Make opening in sterilized end of can large enough to permit removal of sample.

b. Flipper and flat cans. Scrub entire uncoded end and adjacent sides of can using abrasive cleanser, warm water, and a brush, steel wool, or abrasive pad. Rinse and dry with clean sterile towel. Gently shake cans to mix contents before sanitizing. Flood end of can with iodine-ethanol solution and let stand at least 15 min. Wipe off iodine mixture with clean sterile towel. Ensure sterility of can end by flaming with burner in a hood until iodine-ethanol solution is burned off, end of can becomes discolored from flame, and heat causes metal to expand. Be careful not to inhale iodine fumes while burning off can end. Sterilize can opener by flaming until it is almost red, or use separate presterilized can openers for each can. Make opening in sterilized end of can large enough to permit removal of sample.

3. **Removal of material for testing**

Remove large enough portions from center of can to inoculate required culture media. Use sterile pipets, either regular or wide-mouthed. Transfer solid pieces with sterile spatulas or other sterile devices. Always use safety devices for pipetting. After removal of inocula, aseptically transfer at least 30 ml or, if less is available, all remaining contents of cans to sterile closed containers, and refrigerate at about 4°C. Use this material for repeat examination, if needed, and for possible toxicity tests. This is the reserve sample. Unless circumstances dictate otherwise, analyze normal cans submitted with sample organoleptically and physically (see 5-b, below), including pH determination and seam teardown and evaluation. Simply and completely describe product appearance, consistency, and odor. If analyst is not familiar with decomposition odors of canned food, another analyst, preferably one familiar with decomposition odors, should confirm this organoleptic evaluation. In describing the product in the can, include such things as low liquid level (state how low), evidence of compaction, if apparent, and any other characteristics that do not appear normal. Describe internal and external condition of can, including evidence of leakage, etching, corrosion, etc.

4. **Physical examination**

Perform net weight determinations on a representative number of cans examined (normal and abnormal). Determine drained weight, vacuum, and headspace on a representative numbers of normal-appearing and abnormal cans, (see Official Methods of Analysis reference). Examine metal container integrity of a representative number of normal and all abnormal cans that are not too badly buckled for this purpose (see Chapter 15). **CAUTION: Always use care when handling the product, even apparently normal cans, since there is a possibility that botulinum toxin may be present.**

5. **Cultural examination of low-acid food (pH greater than 4.6)**

If there is any question as to product pH range, determine pH of a representative number of normal cans before proceeding. From each container, inoculate 4 tubes of chopped liver broth or cooked meat medium previously heated to 100°C (boiling) and rapidly cooled to room temperature; also inoculate 4 tubes of bromcresol purple dextrose broth. Inoculate each tube with 1-2 ml of product liquid or product-water mixture, or 1-2 g of solid material. Incubate as shown in Table 19.

After culturing and removing reserve sample, test material from cans (other than those classified as flat) whose pH is 4.6 or higher for preformed toxins of C. botulinum when appropriate, as described in Chapter 18.

a. Microscopic examination. Prepare direct smear from contents of each can after culturing. Dry, fix, and stain with methylene blue, crystal violet, or Gram stain. If product is oily, add xylene to a warm, fixed film, using a dropper; rinse and stain. If product washes off slide during preparation, examine contents as wet mount or hanging drop, or prepare suspension of test material in drop of chopped liver broth before drying. Check liver broth before use to be sure no bacteria are present to contribute to the smear. Examine under microscope; record types of bacteria seen and estimate total number per field.

b. Physical and organoleptic examination of can contents. After removing reserve sample from can, determine pH on remainder, using pH meter. DO NOT USE pH PAPER. Pour contents of cans into examination pans. Examine for odor, color, consistency, texture, and overall quality. DO NOT TASTE THE PRODUCT. Examine can lining for blackening, detinning, and pitting.

E. CULTURAL FINDINGS

Check incubated medium for growth at frequent intervals up to maximum time of incubation, as listed in Table 19. If there is no growth in either medium, report and discard. At time growth is noted, streak 2 plates of liver-veal agar (without egg yolk) or nutrient agar from each positive tube. Incubate one plate aerobically and one anaerobically, as in schematic diagram (Table 20). Reincubate CMM at 35°C for maximum of 5 days for use in future toxin studies. Pick representatives of all morphologically different types of colonies into CMM and incubate for appropriate time, i.e., when growth is sufficient for subculture. Dispel oxygen from CMM broth to be used for anaerobes but not from those to be used for aerobes. After obtaining pure isolates, store cultures to maintain viability.

1. **If mixed microflora is found only in BCP**, report morphological types. If rods are included among mixed microflora in CMM, test CMM for toxin, as described in Chapter 18. If Gram-positive or Gram-variable rods typical of other Bacillus or Clostridium organisms are found in the absence of other morphological types, search to determine whether spores are present. In some cases, old vegetative cells may appear to be Gram negative and should be treated as if they are Gram positive. Test culture for toxin according to Chapter 18.

2. **Examination of acid foods (pH 4.6 and below) by cultivation**. From each container, inoculate 4 tubes of acid broth and 2 tubes of malt extract broth with 1-2 ml or 1-2 g of product, using the same procedures as for low-acid foods, and incubate as in Table 21.

Record presence or absence of growth in each tube, and from those that show evidence of growth, make smears and stain. Report types of organisms seen. Pure cultures may be isolated as shown in Table 22.

F. **INTERPRETATION OF RESULTS** (see Tables 23-26)

1. The presence of only sporeforming bacteria, which grow at 35°C, in cans with satisfactory seams and no microleaks indicates underprocessing if their heat resistance is equal to or less than that of C. botulinum. Spoilage by thermophilic anaerobes such as C. thermobutylicum may be indicated by gas in cooked meat at 55°C and a cheesy odor. Spoilage by C. botulinum, C. sporogenes, or C. perfringens may be indicated in cooked meat at 35°C by gas and a putrid odor; rods, spores, and clostridial forms may be seen on microscopic examination. Always test supernatants of cultures of this type for botulinum toxin even if no toxin was found in the product itself, since viable spores of this organism in canned foods indicate a potential public health hazard, requiring recall of all cans bearing the same code. Spoilage by mesophilic organisms such as Bacillus thermoacidurans or B. coagulans and/or thermophilic organisms such as B. stearothermophilus, which are flat-sour types, may be indicated by acid production in BCP tubes at 35 and/or 55°C in high-acid or low-acid canned foods. No definitive conclusions may be drawn from inspection of cultures in broth if the food produced an initial turbidity on inoculation. Presence or absence of growth in this case must be determined by subculturing.

2. Spoilage in acid products is usually caused by nonsporeforming lactobacilli and yeasts. Cans of spoiled tomatoes and tomato juice remain flat but the products have an off-odor, with or without lowered pH, due to aerobic, mesophilic, and thermopholic sporeformers. Spoilage of this type is an exception to the general rule that products below pH 4.6 are immune to spoilage by sporeformers. Many canned foods contain thermophiles which do not grow under normal storage conditions, but which grow and cause spoilage when the product is subjected to elevated temperatures (50-55°C). B. thermoacidurans and B. stearothermophilus are thermophiles responsible for flat-sour decomposition in acid and low-acid foods, respectively. Incubation at 55°C will not cause a change in the appearance of the can, but the product has an off-odor with or without a lowered pH. Spoilage encountered in products such as tomatoes, pears, figs, and pineapples is occasionally caused by C. pasteurianum, a sporeforming anaerobe which produces gas and a butyric acid odor. C. thermosaccolyticum is a thermophilic anaerobe which causes swelling of the can and a cheesy odor of the product. Cans which bypass the retort without heat processing usually are contaminated with nonsporeformers as well as sporeformers, a spoilage characteristic similar to that resulting from leakage.

3. A mixed microflora of viable bacterial rods and cocci usually indicates leakage. Can examination may not substantiate the bacteriological findings, but leakage at some time in the past must be presumed. Alternatively, the cans may have missed the retort altogether, in which case a high rate of swells would also be expected.

4. A mixed microflora in the product, as shown by direct smear, in which there are large numbers of bacteria visible but no growth in the cultures, may indicate precanning spoilage. This results from bacterial growth in the product before canning. The product may be abnormal in pH, odor, and appearance.

5. If no evidence of microbial growth can be found in swelled cans, the swelling may be due to development of hydrogen by chemical action of contents on container interiors. The proportion of hydrogen varies with the length and condition of storage. Thermophilic anaerobes produce gas, and since cells disintegrate rapidly after growth, it is possible to confuse thermophilic spoilage with hydrogen swells. Chemical breakdown of the product may result in evolution of carbon dioxide. This is particularly true of concentrated products containing sugar and some acid, such as tomato paste, molasses, mincemeats, and highly sugared fruits. The reaction is accelerated at elevated temperatures.

G. FLOW SHEET

See Figures 27 and 28.

H. RECORD SHEET

See Annexes 15 and 16.

Analysis of Low Acid (pH Above 4.6) Canned Foods

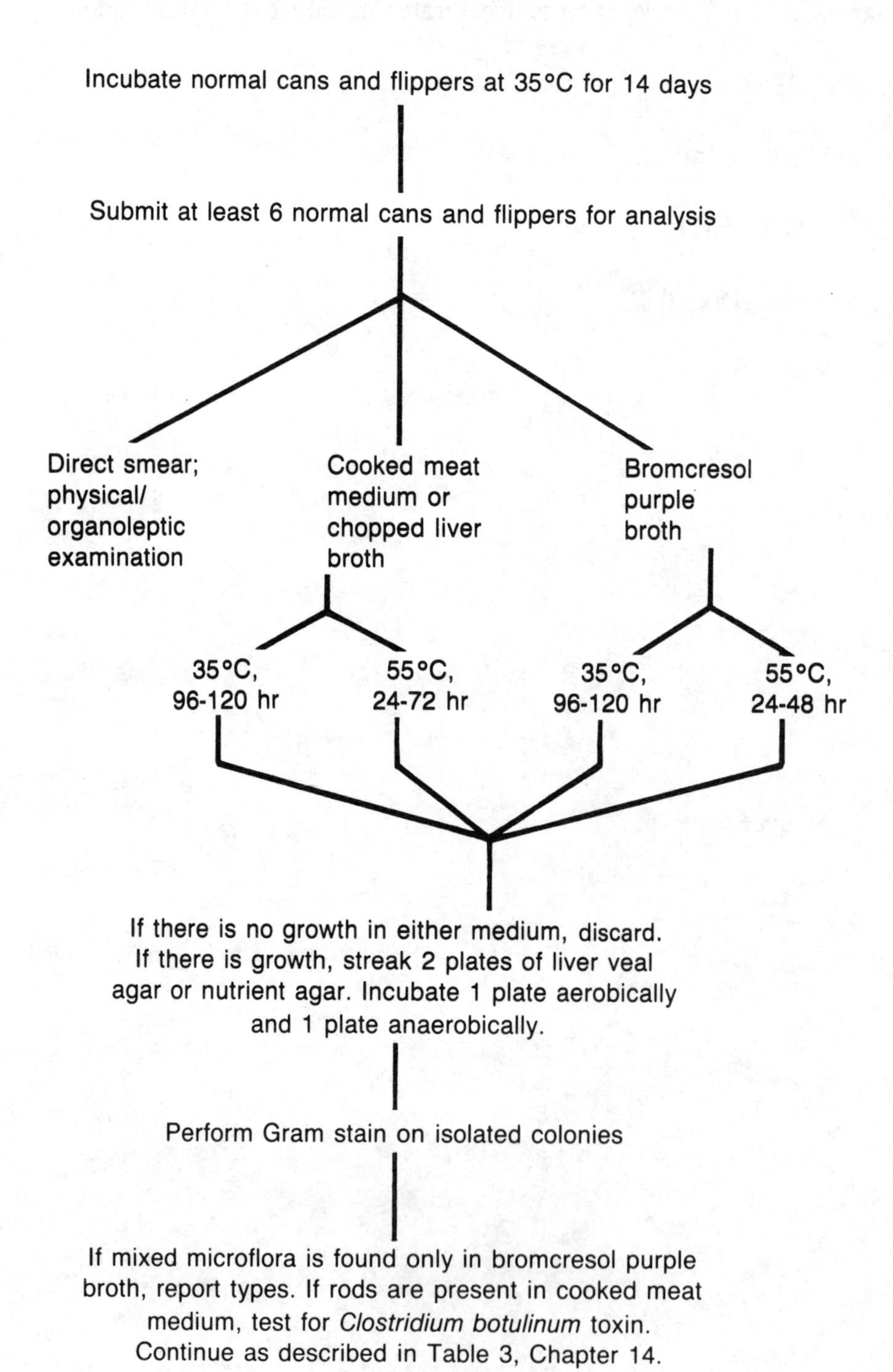

Figure 27

Analysis of Acid (pH 4.6 or Below) Canned Foods

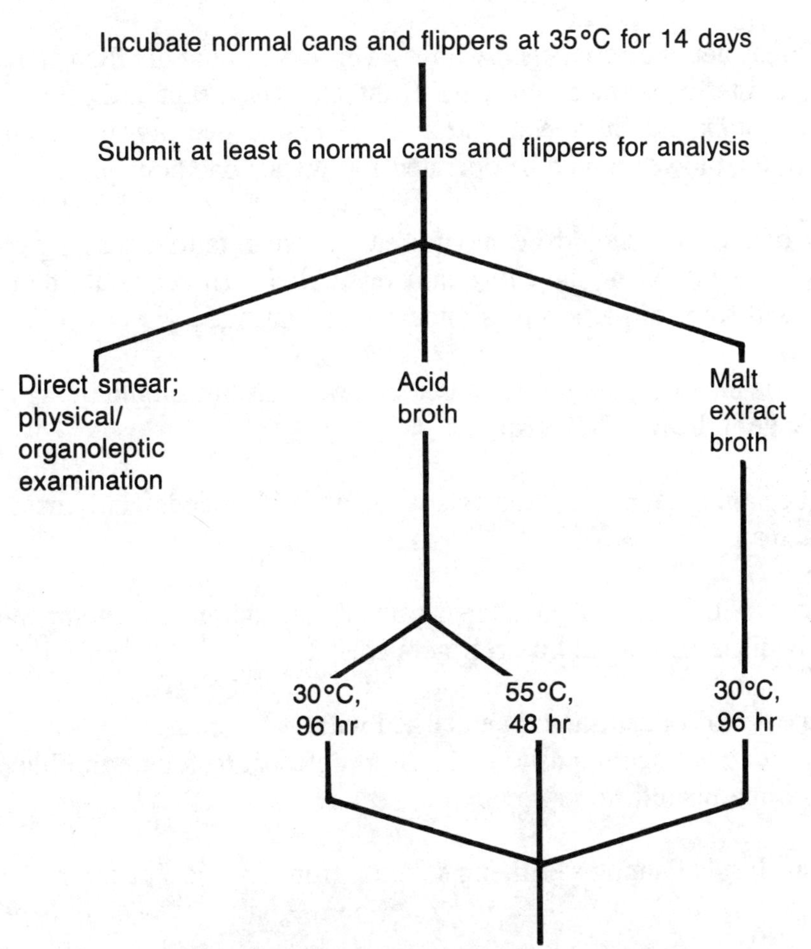

If there is no growth in either medium, discard. If there is growth, streak 2 plates of Sabouraud's dextrose agar or nutrient agar. Incubate 1 plate aerobically and 1 plate anaerobically.

Perform Gram stain on isolated colonies and report types of microorganisms seen. Continue as described in Table 5, Chapter 14.

Figure 28

I. PRECAUTIONS AND LIMITATIONS OF METHOD

For the microbiological analysis of canned foods, the following points should be noted:

1. Prior to microbiological analysis, canned food samples must be incubated under appropriate conditions. Ordinarily, these conditions will be at 35°C for 14 days. However, for low-acid products intended for storage above 40°C, samples should be stored at 55°C for 5-7 days.

2. The work area should be contained within a vertical laminar air flow station. Prior to being used, all the interior surfaces must be disinfected by a standard iodophore or quaternary ammonium compound at a concentration recommended by the manufacturer. After disinfection, the blower should be operated for at least one hour prior to initiating analyses.

3. Efficiency of the unit should be monitored by using fallout plates exposed to the work station environment during the entire transfer period. There should be fallout plates to the left, right, and front of the samples undergoing analysis.

4. Nondisposable plates should not be stored in pipet cans but should be wrapped no more than 5 to a package in heavy Kraft paper.

5. Prior to beginning analysis, analysts should scrub hands and arms thoroughly with germicidal soap.

6. Except for swollen cans, all incubated cans should be at room temperature before being opened. Swollen cans should be refrigerated.

7. Defects on exterior of can should be marked with waterproof ink. Mark "X" on end of can showing defect and open opposite end. If no defects are found on either end of can, save the end with the manufacturer's code.

8. When the analyst is flaming sterilizing solution from lid, the side seam should be away from analyst.

9. A bacteriological can opener should be used for opening all cans. A common type kitchen can opener should not be used since the sample may become contaminated and the double seam may become distorted.

10. Under no circumstances should the canned product be tasted since botulinum toxin may be present.

11. The Gram stain must be performed on 24 hr cultures only. Older cultures may not give typical staining reactions.

REFERENCE

Association of Official Analytical Chemists. 1990. Official methods of analysis, 15th ed. AOAC, Arlington, VA.

Table 18

Useful descriptive terms for canned food analysis

Exterior can condition	Internal can condition
leaker	normal
dented	peeling
rusted	slight, moderate or severe etching
buckled	slight, moderate or severe blackening
paneled	slight, moderate or severe rusting
bulge	mechanical damage

Micro-leak test	Product odor	Product liquor
packer seam	putrid	cloudy
side panel	acidic	clear
side seam	butyric	foreign
cut code	metallic	frothy
pinhole	sour	
	cheesy	
	fermented	
	musty	
	sweet	
	fecal	
	sulfur	
	off odor	

Solid product	Liquid product	Pigment	Consistency
digested	cloudy	darkened	slimy
softened	clear	light	fluid
curdled	foreign	changed	viscous
uncooked	frothy		ropy
undercooked			
overcooked			

Table 19

Incubation times for various media

Medium	No. of tubes	Temp. (°C)	Time of incubation (hr)
Chopped liver (cooked meat)	2	35	96-120
Chopped liver (cooked meat)	2	55	24-72
BCP[a]	2	55	24-48
BCP	2	35	96-120

[a] BCP, Bromcresol purple dextrose broth.

Table 20

Schematic diagram of culture procedure for low-acid canned foods

Original media	Subculture		Pure culture		Characterization
	LVA, NA[a] aerobic incubate 35°C	Growth	Agar slant CMM 35°C		Gram stain LVA, NA anaerobic
CMM and BCP 35°C	LVA, NA anaerobic incubate 35°C	Growth	CMM 35°C	Gram stain	LVA, NA aerobic 35°C
	LVA, NA aerobic 55°C	Growth	Agar slant CMM 55°C		Gram stain LVA, NA anaerobic 55°C
CMM and BCP 55°C	LVA, NA anaerobic 55°C	Growth	CMM 55°C	Gram stain	LVA, NA aerobic 55°C

[a]LVA, liver-veal agar; NA, nutrient agar; CMM, cooked meat medium; BCP, bromcresol purple dextrose broth.

Table 21

Incubation of acid broth and malt extract broth

Medium	No. of tubes	Temp (°C)	Time of incubation (hr)
Acid broth	2	55	48
Acid broth	2	30	96
Malt extract broth	2	30	96

Table 22

Pure culture scheme

Original media	Subculture		Pure culture		Characterization
	NA, SAB[a] aerobic 30°C	Growth	Agar slant Acid broth Malt extract broth 30°C		Gram stain NA, SAB anaerobic 30°C
Acid broth Malt extract broth 30°C	NA, SAB anaerobic 30°C	Growth	Acid broth Malt ex- tract broth 30°C	Gram stain	NA, SAB aerobic 30°C
	NA aerobic 55°C	Growth	Agar slant Acid broth 55°C		Gram stain NA anaerobic 55°C
Acid broth 55°C	NA anaerobic 55°C	Growth	Acid broth 55°C	Gram stain	NA aerobic 55°C

[a]NA, nutrient agar; SAB, Sabouraud's dextrose agar.

Table 23

Spoilage microorganisms that cause high and low acidity in various vegetables and fruits

Spoilage type	pH groups	Examples
Thermophilic		
Flat-sour[a]	≥5.3	Corn, peas
Thermophilic[a]	≥4.8	Spinach, corn
Sulfide spoilage[a]	≥5.3	Corn, peas
Mesophilic		
Putrefactive anaerobes[a]	≥4.8	Corn, asparagus
Butyric anaerobes[a]	≥4.0	Tomatoes, peas
Aciduric flat-sour[a]	≥4.2	Tomato juice
Lactobacilli	4.5 - 3.7	Fruits
Yeasts	≤3.7	Fruits
Moulds	≤3.7	Fruits

[a] The responsible organisms are bacterial sporeformers.

Table 24

Spoilage manifestations in low-acid products

Group of organisms	Classification	Manifestations
Flat-sour	Can flat	Possible loss of vacuum on storage
	Product	Appearance not usually altered; pH markedly lowered, sour; may have slightly abnormal odor; sometimes cloudy liquor
Thermophilic anaerobe	Can swells	May burst
	Product	Fermented, sour, cheesy, or butyric odor
Sulfide spoilage	Can flat	H_2S gas absorbed by product
	Product	Usually blackened, rotten egg odor
Putrefactive anaerobe	Can swells	May burst
	Product	May be partially digested; pH slightly above normal; typical putrid odor
Aerobic spore-formers	Can flat or swollen	Usually no swelling, except in cured meats when nitrate and sugar present; coagulated evaporated milk, black beets

Table 25

Spoilage manifestations in acid products

Type of organism	Classification	Manifestation
Bacillus thermoacidurans (flat, sour tomato juice)	Can flat	Little change in vacuum
	Product	Slight pH change; off-odor
Butyric anaerobes (tomatoes and tomato juice)	Can swells	May burst
	Product	Fermented, butryic odor
Nonsporeformers (mostly lactic types)	Can swells	Usually burst, but swelling may be arrested
	Product	Acid odor

Table 26

Laboratory diagnosis of bacterial spoilage

	Underprocessed	**Leakage**
Can	Flat or swelled; seams generally normal	Swelled, may show defects
Product appearance	Sloppy or fermented	Frothy fermentation; viscous
Odor	Normal, sour or putrid, but generally consistent from can to can	Sour, fecal; generally varying from can to can
pH	Usually fairly constant	Wide variation
Microscopic and cultural	Cultures show sporeforming rods only. Growth at 35 and/or 55°C. May be characteristic on special media, e.g., acid agar for tomato juice. If product misses retort completely, rods, cocci, yeasts or molds, or any combination of these may be present.	Mixed cultures, generally rods and cocci; growth only at usual temperatures
Distribution	Spoilage usually confined to certain portions of pack	Spoilage scattered

CHAPTER 15

CONTAINER INTEGRITY EXAMINATION

The quality of a food container is determined by its ability to protect the product it contains from chemical deterioration or microbiological spoilage. Good double and side seams are essential for an effective hermetic seal. The integrity of the can, however, is also affected by several other factors. Spoilage within the can may be caused by leakage, underprocessing, or elevated storage temperatures. Leaker spoilage occurs mainly from seam defects and mechanical damage. Improper pressure control during retorting and cooling operations may stress the seam, resulting in poor seam integrity and subsequent leaker spoilage. Chemical corrosion that results in hydrogen swells and sulfide stains sometimes occurs. In addition, prolonged storage of cans at elevated temperatures is likely to promote corrosion which may result in perforations. Improper retorting operations, such as rapid pressure changes, may cause deformation of cans and damage the integrity of the seams. Postprocess contamination by nonchlorinated cooling water or excessive buildup of bacteria in can-handling equipment may also be responsible for many spoilage incidents. Abusive container handling may result in leaker spoilage.

A. SAMPLING AND SAMPLE SIZE

1. Sample size required for product analysis and can examination depends on the type of spoilage and the complexity of the problem. When the cause of spoilage is clear-cut, 6 cans may be enough. In more complex cases, it may be necessary to examine 50 cans or more.

2. Sample adequate number of normal (flat) cans from same case or lot for can examination.

B. PRELIMINARY EXAMINATION

1. Either container specialist or microbiologist may remove labels, assign subnumbers, if necessary, and separate code numbers. Use same coding or subnumber system for product examination and container examination.

2. Before any product sample is removed, container specialist should perform complete external can examination, observing such defects as evidence of leakage, pinholes, rusting, dents, and buckling, as well as general exterior conditions.

3. Classify each can as (a) flat, (b) flipper, (c) springer, (d) soft swell, or (e) hard swell according to criteria in Chapter 14. (Leakage tests and external double seam dimensions may not be valid when cans are buckled. However, these cans should be examined and then torn down and re-examined for seam defects that may have existed before buckling.)

4. If possible, set aside reserve cans representing the classifications noted. It may be advisable to refrigerate reserve cans to prevent blown cans.

5. Examine cans classified as springers and soft and hard swells immediately. Do not incubate.

6. The microbiologist should remove sample from uncoded end of can in manner that will not disturb double seam, e.g., with bacteriological can opener. If can end has been punctured as result of gas sampling, bacteriological can opener may be used if puncture is in center of end. If puncture is not in center, remove end with pair of metal cutters.

C. CAN EXAMINATION

Include the following in examination of container: Note condition of cans (exterior and interior) and quality of seams; observe and feel for gross abnormalities, mechanical defects, perforations, rust spots, and dents; perform pressure and/or vacuum tests to detect invisible microleaks either in double seam or side seam areas; measure seam dimensions and perform teardown examination; note condition of double seam formation and construction (by micrometer, seam scope, or seam projector).

1. Visual examination

The double seam (Figure 29) consists of 5 thicknesses of plate (7 thicknesses at the juncture of the end and the side seam for 3-piece cans) interlocked or folded and pressed firmly together, plus a thin layer of sealing compound. It is formed in 2 rolling operations. The side seam consists of 4 thicknesses of metal body plate, except at the lap or cross-over area, which has 2 thicknesses of metal. The side seam is bonded with solder, which is generally applied to the outer surface of the side seam. The side seams of cemented cans have 2 thicknesses of metal body plate and welded cans have 1.3 to 1.4 times the metal plate thickness at the seam. Welded 3-piece cans permit reduction of the side seam thickness and the double seam thickness at the cross-over juncture. Drawn cans eliminate the side seam and the bottom end seam, resulting in fewer areas that affect can integrity. The can ends (see the enlarged profile in Figure 30) are punched from sheets. The edges of the ends are curled and a sealing compound is applied and dried in the lining channel (curl and flat areas) of the can end. Once the lined can end is double seamed onto a can body, the sealing compound in this compressive seal fills the voids (spaces) between the folds of metal in the properly made double seam to form an abuse-resistant hermetic seal.

Double Seam

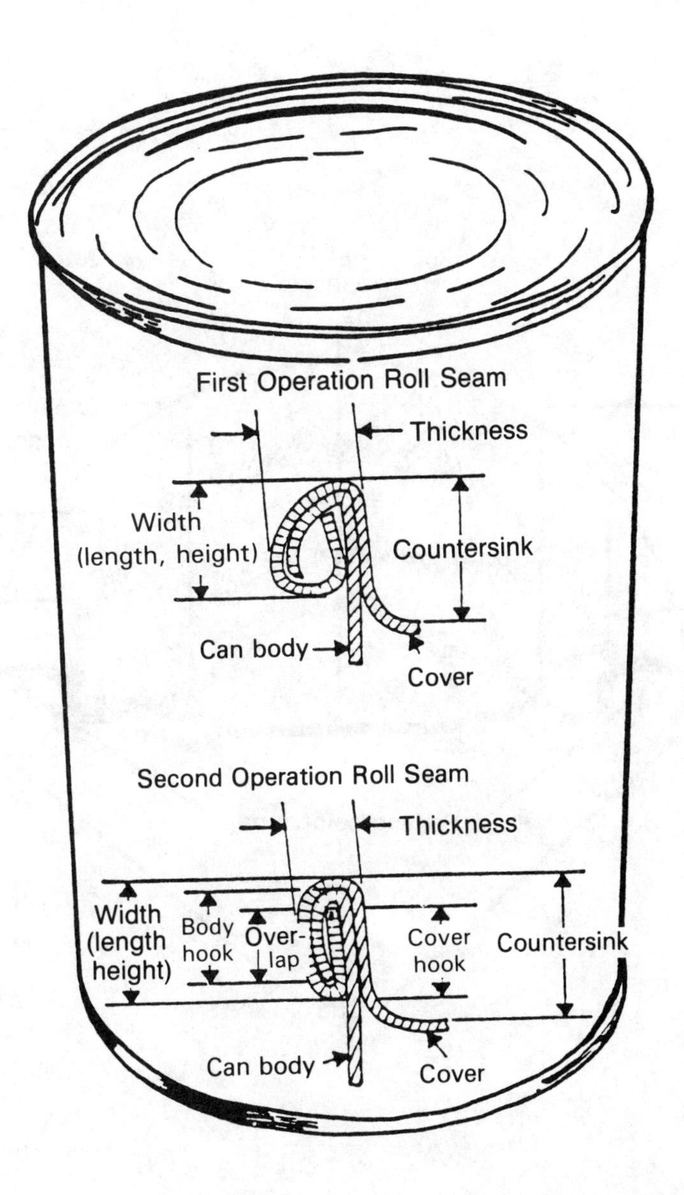

Figure 29

Can End Profile

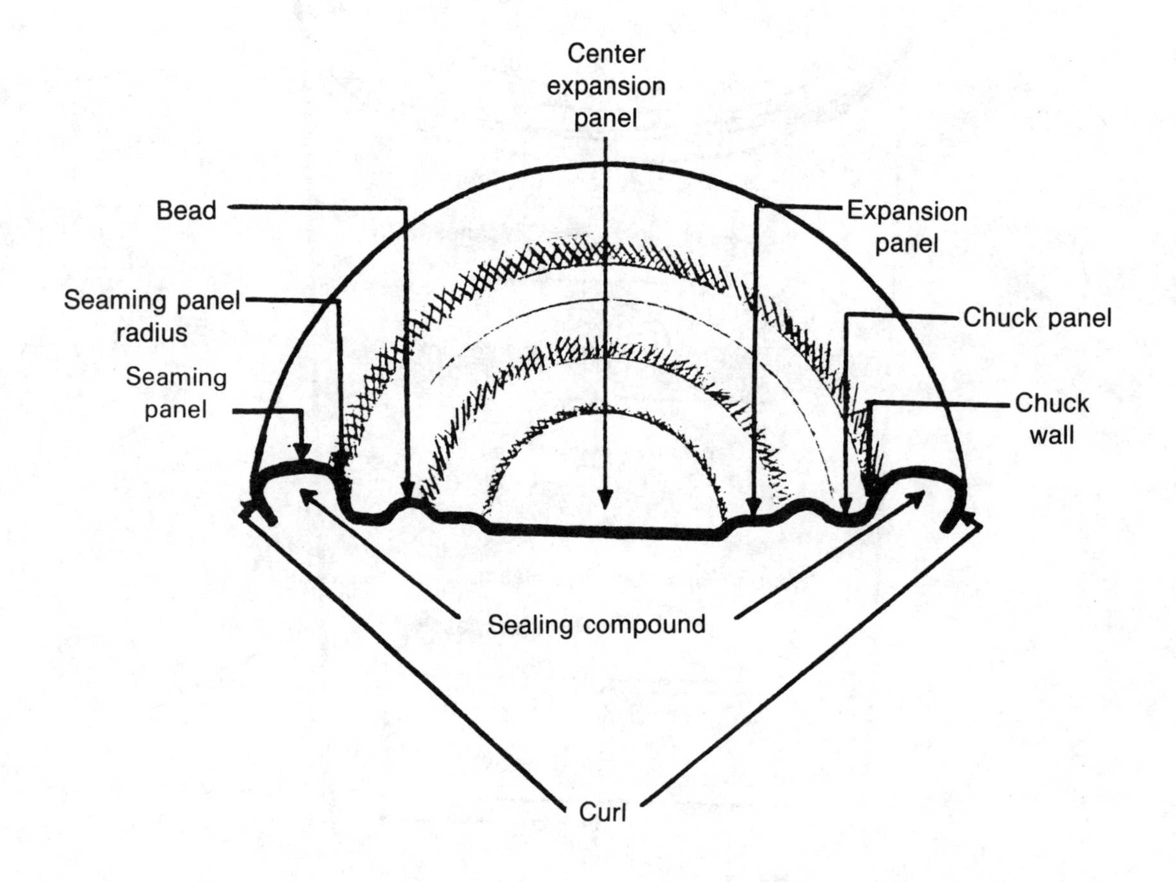

Figure 30

Use hand as well as eye. A magnifying glass with proper illumination is helpful. Run thumb and forefinger around seam on inside (chuck wall) and outside of seam to locate any roughness, unevenness, or sharpness. By visual observation and fingerfeel examination, determine presence of following defects that may result in can leakage (for definition of terms, see glossary at end of this chapter):

- sharp seam
- cutovers or cut-throughs
- dents
- deadheads (incomplete seam)
- code cut
- false seam (although some false seams may not be detected by external examination)
- excessive droop
- excessive scuffing in chuck wall area
- excess solder
- jumpover
- knocked-down flange
- cable cuts on double seam

2. **Microleak detection**

The microleak tests are not listed in their order of sensitivity nor is it necessary to use them all. Each has its advantages and disadvantages depending on the particular set of conditions. In some instances, a test may be chosen as a personal preference. They are all presented to provide the analyst with all the procedures and options open to him.

Make all external measurements of can double seams before any microleak testing. See section on double seam measurements for description of these measurements.

a. Vacuum leak test

This test applies a vacuum to the can, which more closely duplicates the condition existing in the can when it contains product and is sealed. Proponents of this method feel that the use of a vacuum to detect leakage in a can designed to hold a vacuum is more effective than the use of pressure in that it may remove food particles from leakage paths in can seams. Pressure, on the other hand, may force particles more deeply into leakage paths.

1) Materials

a) Bacteriological can opener (Wilkens-Anderson Co., 4515 W. Division St, Chicago, IL 60651 USA, or equivalent)

b) Plexiglass plate

c) Plastic tubing

d) Rubber gasket, to fit container being examined

e) Vacuum source with gauge

f) Nonfoaming wetting agent, e.g., Triton X-100 (R39)

g) Outside light source, such as high intensity lamp

2) Procedure

Remove uncoded end of can with bacteriological can opener adjusted to cut out can end, leaving 0.6 cm border around outer edge. Empty and wash container with water and suitable detergent. Wash to remove food particles lodged in seam area. (Ultrasonic cleaner may be used to remove small food particles lodged in seam areas). Add wetting agent plus water to depth of about 2.5 cm. Place Plexiglass plate with tubing attached and wetted rubber gasket on open end of container. Increase vacuum until gauge indicates vacuum of between 15 and 25 in. Swirl water in container to dissipate small bubbles produced by application of vacuum. Tilt container slowly to immerse all seam surfaces, letting light source focus through Plexiglass into can for better observation. Rotate tilted can so that all surfaces are observed and covered with water. Depending on size of hole, path of leakage, pressure differential, and surface tension of the test water, bubbles will appear smaller or larger and with lesser or greater frequency. Release vacuum by first closing main vacuum petcock and then opening intake petcock.

b. Mead jar test

This is a nondestructive type of examination to determine leakage paths of finished cans that have both ends double seamed. It is primarily used for vacuum- or nonvacuum-packaged dried or semi-dried products. Deaerated water is preferred because, as the vacuum is pulled on the jar, the dissolved air in the water comes out of solution as bubbles and prevents a clear view of the container and of the possible leakage bubbles coming out of the container. This test would rarely be used for items thermally processed in metal cans, but it is described as an example of the other microleak tests available, which could be used by the innovative container expert under a particular set of conditions.

1) Materials

a) Mead jar (battery jar), glass or plastic

b) Mesh protector, to fit around Mead jar

c) Metal top with rubber gasket on bottom side and air inlet, vacuum inlet, and vacuum gauge on top side

d) Vacuum source

e) Rubber tubing

f) Deaerated water (prepare by applying 25-30 inches of vacuum on water for 8 hr or overnight)

g) Vaseline or stopcock grease

h) Device or weight to hold can under water

2) Procedure

Place enough deaerated water into Mead jar to completely submerge test container. Put container in water and, if necessary, place device or weight on it to hold it under water. Place mesh protector around jar, and put Vaseline on rubber gasket on underside of top piece to act as sealant between gasket and lip of Mead jar. Place lid on top of jar and rotate it to improve seal. Attach rubber hose to vacuum line. Turn off air inlet stopcock and open vacuum line stopcock. Turn on vacuum and record vacuum reading on gauge if leakage or air bubbles emanating from container are noted. Note where leakage point is located. Turn up to full vacuum available if no leakage is noted. Turn off vacuum and open air inlet valve to release vacuum. Take off lid, remove container, and mark leakage point.

3) Precautions

Do not use defective or cracked Mead jars because of danger of collapsing them. Always put wire mesh protector in place before turning on vacuum. For a container to be classified as a leaker, a continuous stream of bubbles from a single point is necessary. Unless leakage is obvious, the can should be observed under test for at least 30 seconds. Generally, a few bubbles will be observed when the vacuum is first applied because of air entrapment in the double seams. These should not be confused with leakage point bubbles.

c. Air pressure test

For detection of container leakage caused by minute body pinholes and perforations, and/or defective side seams, air pressure testing is the most convenient and conclusive. It is also helpful in locating position of double seam leaks. During pressure testing, the container is subjected to pressures that may distort double seams, and this may either produce false leakers or seal off minute leakage paths. For this reason, the air pressure testing method should be used in conjunction with the fluorescein test or penetrant dye test to trace actual leakage path through double seams.

1) Materials

a) Single pocket (or multiple pocket) foot tester with attached air and water source

b) Manually operated air pressure control valve and pressure gauge installed on air line

c) Two gate valves installed in air line, one to pressurize and the other to exhaust the can during testing

2) Procedure

a) Preparation of samples

Open filled cans at one end with bacteriological can opener so that double seam remains intact. Container manufacturer's end is usually opened, but if enough samples are available, half the cans may be opened at packer's end. After removing contents, wash cans thoroughly. Wash containers of fatty or oily product with warm detergent and water, or boil them in detergent and water, and use ultrasonic cleaner to remove all fat or oil trapped in double seam. Dry cans at least 8 hr at 38-49°C before pressure testing.

b) Pressure test

Fill foot tester with warm water (38-49°C) to about 5.1 cm from top. Set air pressure control valve at desired pressure level (20 lb for most sanitary cans). Seat open end of test can against rubber base plate, side seam up, and lower pressure bar against can with enough pressure just to hold can in place. Secure pressure bar in this position with 2 nuts so that additional adjustment is not necessary for pressure testing of additional cans of same size. Close exhaust gate valve and completely immerse can in tank by stepping on foot pedal. With can completely immersed, open pressurized gate valve, letting air flow into can. Hold can in this position long enough to detect any leakage points. Leakage may be shown by steady stream of large bubbles or continued intermittent escape of very tiny bubbles. When it is determined whether leakage is present in can being observed, close pressurized gate valve and exhaust can by opening exhaust gate valve. Release foot pressure and lift can out of water. Rotate can 180° against rubber base plate until new area of can is exposed to view, and repeat pressure testing operation. Mark location of any leaks noted during test for use as reference when cans are further examined. Do not exceed 30 psi because can may burst beyond this level.

After pressure testing, use fluorescein test to obtain additional information on presence of any leakage paths, or strip can seams for further examination if point of leakage is conclusively located.

3) Precautions

During pressure testing operation, pay particular attention to cross-over areas and double seams on both ends of can. Also observe side seam area closely and scan bodies for pinholes or perforations. Use warm water during test. During drying, small leakage paths will be opened; these may be reclosed by contraction of can during testing in cold water.

The most important factor in completing air pressure test is that operation must not be rushed. Very small leaks may take several seconds to show up and then the only evidence may be the intermittent flow of very tiny bubbles.

d. <u>Fluorescein dye test</u>

Fluorescein dye testing has been used for many years to detect minute double seam, lap, and side seam leakage paths in all types of containers. The fluorescein test is especially useful for examining sanitary-style containers that are normally packed with some initial vacuum. Experience has shown that, in many cases, fluorescein dye can detect minute leakage paths on suspected cans that do not leak under air pressure test. Fluorescein testing of most types of containers under vacuum simulates actual packed condition, i.e., with ends pulled inward.

1) Materials

a) Vacuum line, with bleed-valve attachment for regulating amount of vacuum to be used

b) Rubber-faced plate, for attaching containers to vacuum line

c) Ultraviolet light (Black Light Lamp)

d) Fluorescein dye solution. Mix 100 ml triethylene glycol, 300 ml water, 15 g glycerine, 3 g wetting agent (such as Triton X-100 (R39)), and 3 g sodium fluorescein, technical grade. (Zyglo dye solution ZL-4B is available from Magnaflux Corporation, Chicago, IL, USA)

2) Procedure

a) Preparation of sample

Open filled cans at one end with bacteriological can opener so that double seam remains intact. Container manufacturer's end is usually opened, but if enough samples are available, half the containers may be opened at packer's end. After removing contents, wash containers thoroughly and dry them. Wash containers of fatty or oily products with hot detergent and water, or boil them in detergent and water to remove all fat or oil trapped in double seams. (Ultrasonic bath may be used to remove small food particles lodged in seam areas.)

b) Vacuum application

Place opened end of emptied and dried container against vertically mounted rubber-faced plate connected to vacuum line. As a general rule, 15-20 in of vacuum is used for most pressure-processed, sanitary-style containers. For other container styles, maximum allowable vacuum to be used depends on panel resistance of bodies.

(i) After container is mounted in place and vacuum drawn, apply fluorescein solution with a small brush or eyedropper to outside double seam and side seam areas. Or:

(ii) Place can in dye bath to cover seam and/or score areas to be examined.

Run test for 30 min to 2 hr, depending on what style of container is being examined. Longer test period may be required to detect microleaks. If any doubt exists as to time needed to demonstrate leakage, remove containers every 30 min and check inside with UV light for presence of fluorescein. Since fluorescein solution runs off mounted container, apply fresh solution at 15 min intervals.

c) Container examination

At end of vacuum test period, thoroughly remove fluorescein remaining on outside of containers with water and wipe containers until dry. Take special care not to splash any fluorescein into opened end of container. After cleaning excess solution from outside, strip double seam and examine it under UV light to detect any evidence of fluorescein on inside. Use very careful technique so that wet solution trapped in outside of double seam does not creep to inside of container, thus giving false positive results. Also, keep tools clean of fluorescein to prevent contamination during stripping examination. Examine containers immediately after vacuum testing, since dried fluorescein solution does not fluoresce.

3. **Double seam measurements**

Take seam width (length, height), thickness, and countersink measurements on both can ends before can is opened, if possible. Seam width and thickness measurements may be taken after opening if proper can opener (bacteriological) has been used. However, take care to prevent distortion of area around seaming panel. Countersink measurements cannot be taken after removing end. Perform teardown examination and record results for seam integrity evaluation. Either seam scope or projector technique, or micrometer measurement system, or all 3 systems for verification may be used to determine adequacy of double seam formation.

a. Teardown and cross-section strip examinations

1) Materials (stripping tools) see Figure 31

2) Procedure

(The seam projector examination is easier to accomplish before the cover hook is removed and after the seam has been cut.)

a) If uncoded end is still on can, remove with bacteriological can opener. Empty contents from can, wash and dry it, and test it for leakage as previously described. After leakage test, cut 2 strips about 1 cm wide through double seam and into body of can about 3.2 cm, leaving bottom of strips attached to can. One strip should be at least 1.3 cm counterclockwise from juncture, and other should be 180° from lap or side seam. This will leave adequate length of cover hook for thorough examination. Remove remaining end metal, beginning at cutout strip closest to side seam, moving clockwise until entire cover hook is removed. Take care to prevent any injury or cut. Partially remove coded end with bacteriological can opener, leaving it attached about 90° counterclockwise from side seam to allow the can to be identified after stripping. Then proceed to cut strips into coded end of can and to tear it down, using procedures previously discussed in this section.

b) Strip off cover hook to point where remaining end metal has not been removed (about 1.3 cm) from second metal strip). Measure body and cover hooks. Grade cover hook for tightness by examining and evaluating wrinkles. Scrutinize cross-over area of 3-piece cans for wrinkles and for crawled laps (body hook of one side of cross-over drops down or crawls below bottom of other side) in soldered cans. Observe double seams for plate fractures. Inspect interior of body wall in double seam area for well-defined continuous impression around circumference of can. This is referred to as pressure ridge and is one indication of seam tightness, although pressure ridge may or may not be present in a good seam. Tape cover hooks and coded end to can so that they may be identified as belonging to can in question. (If cover hooks are still attached to can, bend them inward inside the can body to prevent cut injury.)

Materials for Teardown and Cross-section Strip Examination

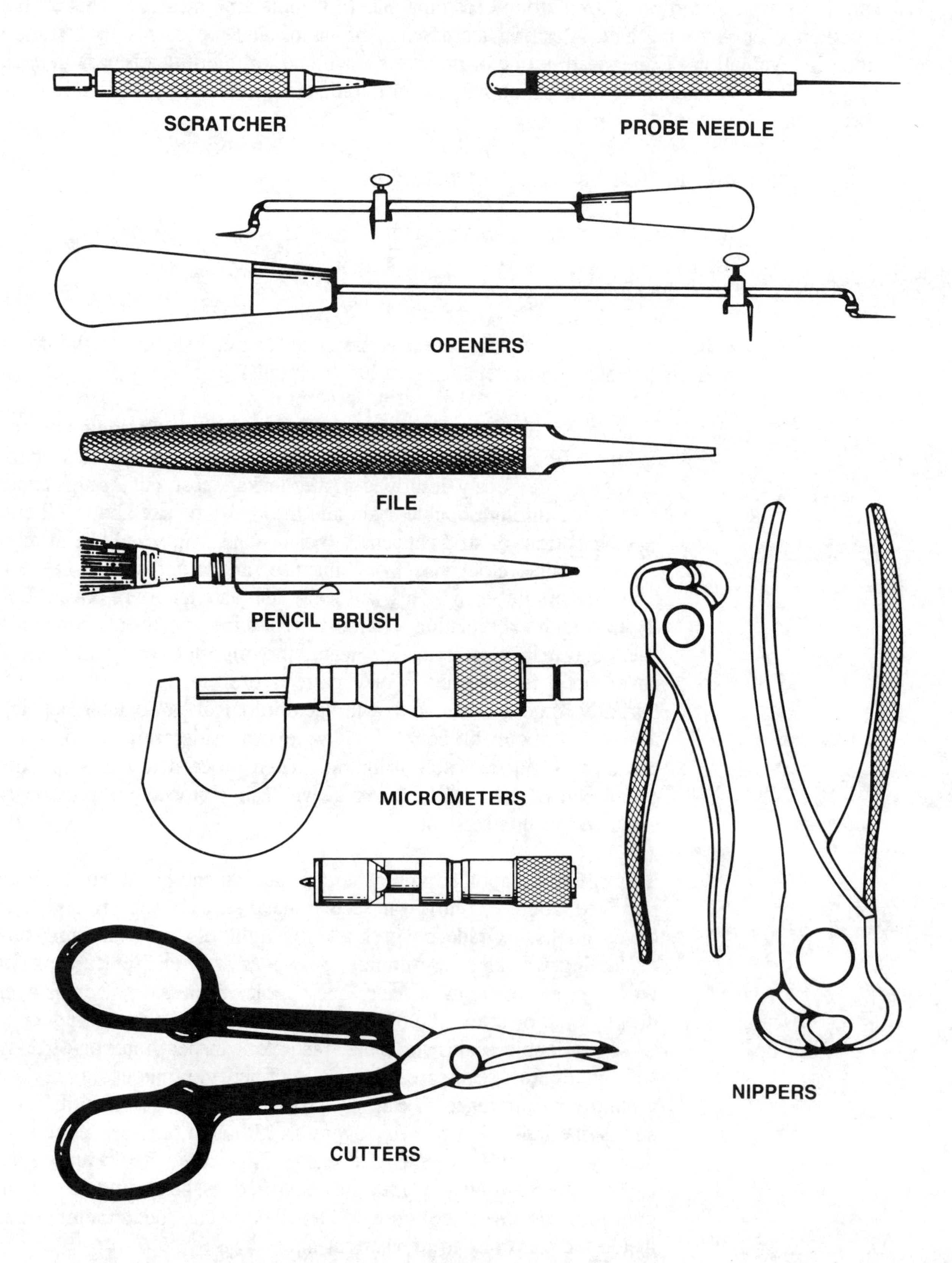

Figure 31

c) Break open and observe both laps. Note solder voids and channels, and evaluate appearance of solder for discoloration and staining.

d) Identify hot breaks and cold breaks in solder. Observe flange area of lap closely for defects. (Also see 3) f) for details.)

e) Examine side seam construction, pulling apart solder bond with side seam breaker as follows: Place can over body horn of appropriate can diameter and pull down on breaker arm to expand can and open solder bond. Observe side seam for solder voids, channels, and discolored areas.

f) Examine laps for heavy solder, as evidenced by excessive solder adjacent to lap inside and outside can, or on body hook, causing severely crawled lap or cutover. Break open lap and see if excessive solder is still present in lap.

3) General comments

a) As mentioned previously, take countersink measurements while ends of can are still intact. Such measurements taken on swelled cans are seldom meaningful because of possible distention of end caused by the swelling. However, measurements that indicate deep countersinks are useful, since they represent the condition as is.

b) Severe wrinkling immediately adjacent to cross-over impression indicates jumped seam (jump-over) and constitutes possible leakage point in this critical area.

c) The portion of the cover hook that intersects the side seam is called the juncture area, and is about 1 cm wide. In an ideally made 3-piece can, there is no reduction in the length of the cover hook at the juncture, even though there is a depression in the cover hook at this point. In most commercial 3-piece cans, a reduction in the length of the cover hook at the side seam is present in varying degrees. The shortening of the cover hook at the sieve seam is caused by the 2 additional thicknesses of body flange at the side seam. These additional layers of metal prevent the cover hook from being tucked up under the body hook to the same extent that it is tucked up under the body hook away from the side seam.

It is essential to have a good juncture with long enough cover and body hooks. Without a good juncture, the overlap at the side seam may be critically short, making the can less abuse-resistant. As with the "tightness rating" in which the rating is based on the wrinkle-free metal of the hook length, the "juncture rating" is based on the amount of droop-free metal for the existing hook length away from the juncture.

Because the cover hook is frequently shorter at the juncture in the form of a droop, the juncture is rated on a percentage scale, starting at 100% as ideal and decreasing in 25% increments. If the cover hook in the juncture is even with the rest of the cover hook, it is rated 100%. If the cover hook reduction at the juncture is about 1/4 the length of the cover hook, the juncture is rated 75%. If the cover hook at the juncture is 1/2 the length of the rest of the cover hook, the juncture is rated 50%. Similarly, if the cover hook length at the juncture is only 1/4 the length of the rest of the cover hook, it is rated 25%.

d) A severe condition of crawled laps creates an area of very little overlap at the cross-over, and is a possible leakage point. This condition is usually, but not always, accompanied by an external droop at the cross-over.

e) Cover hooks made with double cold reduced (2 CR) ends may look wrinkled, but these wrinkles may be reverse wrinkles, which do not indicate a loose seam. Compound wrinkles do not give wavy cut edges to the cover hook and are not used to establish tightness ratings.

f) Discolored or stained solder may indicate that leakage had previously taken place through the lap. It is not necessary to have a solderless channel for leakage to occur at the lap. This can happen as a result of either a hot break or a cold break in the solder. Both of these conditions are conducive to container leakage. A hot break is identified by solder that appears relatively smooth and occurs in the soldering operation during manufacture when the lap opens up before the solder solidifies completely. A cold break in the solder gives it a round, mottled appearance; if leakage has occurred at this point, the area shows a dark discoloration. Cold breaks may occur at almost any time after the solder has hardened, and are difficult for the manufacturing plants to detect, since they usually break after the can manufacturing operation. Cold breaks are normally due to weak soldering bonds or poor manufacturing techniques.

In general, if pressure testing shows leakage at the cross-over, if the fluorescein test indicates a path through the flange area, and if no other double seam or lap defects are found at the cross-over, the most likely cause of leakage is a solder break at the flange area of the lap. It is also possible to have no solder at this area. This defect is easier to detect visually than the solder break. The leakage path may be confirmed by a fluorescein path or by a typical stained or darkened appearance in the solder. A condition known as "islands" is often observed in the side seam. An island is an isolated area in the side seam fold that is void of solder, but without a connecting solderless path, or break, leading to the outside of the can. This condition is not necessarily associated with leakage, but does indicate a weak side seam.

b. Micrometer measuring system

1) Materials

Use micrometer especially made for measuring double seams and reading to nearest 0.0025 cm. Be sure to adjust micrometer properly. When micrometer is set at zero position, zero graduation on movable barrel should match exactly with index line on stationary member. If zero adjustment is more than 1/2 space from index line at this setting, adjust it.

2) Procedure

a) Make seam measurements on round cans, at minimum of 3 points about 120° apart, around circumference of can, beginning about 2.5 cm to one side of cross-over (or at least 1.3 cm away from cross-over).

b) Obtain the 5 required measurements:

- Seam thickness
- Seam width (length, height)
- Body hook length
- Cover hook length
- Tightness (observation for wrinkle, see d), below)

c) The two optional measurements are:

- Countersink depth
- Overlap, calculated by the following formula:

$$\text{Overlap} = \mathbf{CH} + \mathbf{BH} + \mathbf{T} - \mathbf{W}$$

where **CH** = cover hook length, **BH** = body hook length, **T** = cover plate thickness, and **W** = seam width (height, length).

d) Grade tightness (wrinkle) of a double seam by examining cover hook wrinkle according to percentages illustrated in Figures 32 and 33. Drawing shows cover hook with 0-100% tightness, with wrinkle number shown below it. Tightness can also be indicated by flatness of cover hook; that is, cover hook should not appear round. Make this observation on cover hook removed from seam that has been sectioned with seam saw. This method is good verification for wrinkle method but should not be a substitute for it. Tightness may be expressed in terms of percentage of cover hook not included in wrinkle or by rating number equivalent to distance up cover hook. Both procedures are listed below. Percentage method is preferred.

Tightness, expressed in terms of wrinkles:

No. 0 - Smooth, no wrinkle

No. 1 - Wrinkle up to 1/3 distance from edge

No. 2 - Wrinkle up to 1/2 distance from edge

No. 3 - Wrinkle up to more than 1/2 distance from edge

Tightness ratings expressed in % of cover hook not included in the wrinkle (preferred method):

100 - Equivalent to a No. 0

90 - Equivalent to between No. 0 and No. 1

70 - Equivalent to a No. 1

50 - Equivalent to a No. 2

Less than 50 - Equivalent to a No. 3

e) Rate the juncture as previously described.

f) Free space (FS) measurement may be used to determine seam condition of 2-piece oblong and oval cans, as follows:

FS = **ST - (2 BPT + 3 CPT)**

where **ST** = seam thickness, **BPT** = body plate thickness, and **CPT** = cover plate thickness. However, specifications have not been well established at this time.

Tightness (Wrinkle) Rating in Percent

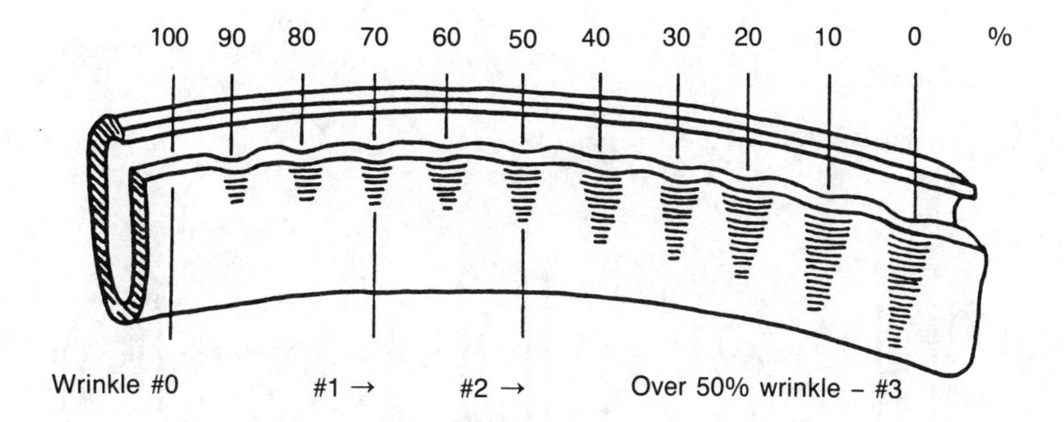

Figure 32

Evaluation of Tightness by Flatness of the Cover Hook

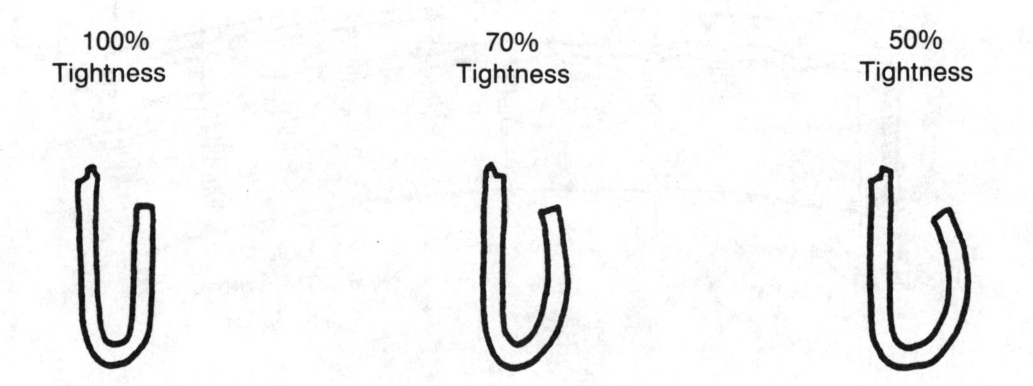

Figure 33

c. Seam projector measuring system

As an alternative to the use of the micrometer, or as a verification, cross-sections of double seams may be examined visually by a seam projector. A section of the double seam is cut in the form of a metal strip that remains attached to the can body and that is then placed in the projector. From the image projected on the screen, the seam width, hook lengths, and overlap dimensions may be measured with a specially calibrated caliper. General seam formation, and, in some instances, seam tightness may be observed. The seam projector method facilitates examination of the critical overlap area at the cross-over; this is especially valuable for examining 3-piece soldered No. 10 cans, which are particularly vulnerable to leakage at this point.

1) Materials

a) Seam projector (Wilkens-Anderson)

b) Waco saw (Wilkens-Anderson)

c) Micrometer

2) Procedure

a) Obtain the 4 required measurements:

- Body hook length
- Overlap
- Seam thickness
- Tightness (observation for wrinkle)

b) The 3 optional measurements are:

- Width (length, height)
- Cover hook length
- Countersink depth

c) Measure each double seam characteristic at 2 different locations on each double seam, excluding the cross-over.

d) Cut cross-sections through double seams with Waco seam saw. Polish cross-section surface with fine Emery cloth to ensure bright surface that will project clear image on screen. Place polished section in clamp on side of projector, look into shadow box, and observe image. Bring calipers in instrument into position. Make note of any looseness, tightness, or other malformations. Using specially calibrated calipers, carefully measure and record width, cover hook, body hook, and overlap on image. Repeat this procedure in all 4 different locations along double seam. To properly evaluate seam for degree of looseness, strip cover hook from can, and visually grade for wrinkle formation. Observe absence or discontinuities of sealing compound after cover hook has been removed from double seam. The sealing compound should form a complete 360° circle around the edge of the lid.

e) Overlap percentage is a measure of both how well end hook and body hook overlap and how well hooks match each other in length. It is also a ratio of the existing distance between the body hook and the cover hook compared to the distance the hooks would lap for the given seam. Overlap percentage is measured directly when seam projector is used with nomograph placed on viewing screen (see Figure 34). Percentage can be calculated from seam length, body hook, cover hook, and body and end plate thicknesses.

$$\text{Overlap \%} = \frac{\mathbf{100 \times (BH + CH + EPT - W)}}{\mathbf{W - (2\,EPT + BPT)}}$$

where **BH** = body hook length, **CH** = cover hook length, **EPT** = end plate thickness, **W** = width (seam height, length), and **BPT** = body plate thickness.

Use minimum value found for each measurement (maximum value for W), to approximate lowest possible overlap percentage. Overlap may also be measured in millimeters or thousandths inches by using specially calibrated calipers as in d).

f) Open calipers as wide as they will go and place nomograph card on screen. Position nomograph card so that image appears on it and reference lines of nomograph are parallel to hook images. Adjust position of nomograph to place zero line on inside of body hook radius of image and then move it forward or backward until the 100 line is on inside of end hook radius. Now, move nomograph, keeping reference lines parallel to hooks and allowing no forward or backward motion, until zero line is at end of end hook; read nomograph at end of body hook. This value is percent overlap.

g) Rate the juncture as previously described.

D. RECORD SHEET

See Annexes 17 and 18.

Nomograph for Use with Seam Projector

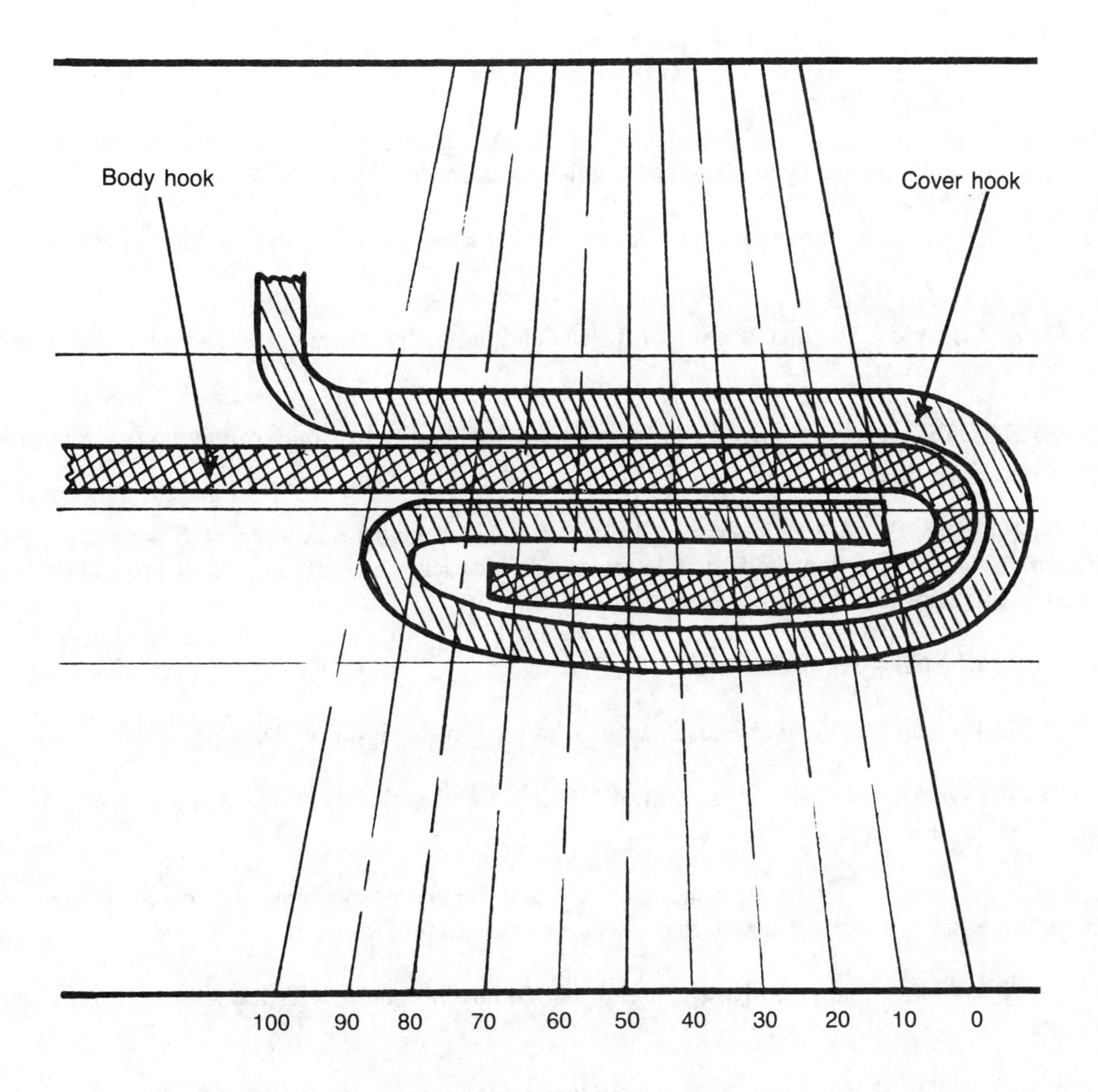

Direct reading of percent overlap
(This example shows 75% overlap.)

Figure 34

GLOSSARY

The terminology of container seal and double seam formation and evaluation is specialized and may be confusing to the inexperienced person. In the foregoing, an attempt was made to define terms as they are introduced. However, for convenient reference and completeness, a listing of seam terms and definitions follows.

BASE PLATE PRESSURE - The force of the base plate that holds the can body and end against the chuck during the double seaming operation. In general, it has the following effect on the seam formation:

Low pressure - short body hook
High pressure - long body hook

BODY - The principal part of a container, usually the largest part in one piece comprising the sides. The body may be cylindrical, rectangular, or another shape.

BODY HOOK - The flange of the can body that is turned down in the formation of the double seam.

BOTTOM SEAM - The double seam of the can end put on by the can manufacturer, also known as factory end seam.

CABLE CUTS - Cuts or grooves worn into can ends and bodies by cables of the runway conveyor system.

CAN, SANITARY - Full, open-top, 2-piece drawn can and 3-piece can with double seamed bottom. The cover or top end is double-seamed on by packer after filling. Ends are compound-lined. Also known as packer's can or open-top can.

CANNER'S END - See packer's end.

CAP TILT - Cap should be essentially level with the transfer bead or shoulder.

CHIPPED GLASS FINISH - A defect in which a piece of glass has broken away (chipped) from the finish surface.

CHUCK - Part of a closing machine that fits inside the end countersink and acts as an anvil to support the cover and body against the pressure of the seaming rolls.

CHUCK WALL - The part of the can end that comes in contact with the seaming chuck (see Figure 30).

COCKED CAP - Cap not level because cap lug is not properly seated under glass lug.

CODE CUT - A fracture in the metal of a can end due to improper code embossing.

COMPOUND - A sealing material consisting of a water or solvent dispersion or solution of rubber and placed in the curl of the can end. The compound aids in effecting a hermetic seal by filling spaces or voids in the double seam.

COUNTERSINK DEPTH - The measurement from the top edge of the double seam to the end panel adjacent to the chuck wall.

COVER - See packer's end.

COVER HOOK - The part of the double seam formed from the curl of the can end. Wrinkling and other visual defects shall be observed by stripping off the cover hook.

CRACKED GLASS FINISH - A defect normally termed "split finish" which is an actual break in the glass over the sealing surface of the finish.

CRAWLED LAPS - Also known as "creep"; occurs when two layers of metal are bent and the outer layer looks shorter because it has a greater radius to traverse than the inner layer, which has a smaller radius, perhaps being bent almost double.

CROSS-OVER - The portion of a double seam at the juncture with the side seam of the body.

CROSS-SECTION - A section cut through the double seam for the purpose of evaluating the seam.

CRUSHED LUG - Lug on cap forced over glass lug, causing the cap lug not to seat under glass lug.

CURL - The extreme edge of the cover that is turned inward after the end is formed. In metal can double seaming,the curl forms the cover hook of the double seam. For the closure for glass containers, the curl is the rolled portion of metal at the bottom of the closure skirt (may be inward or outward).

CUTOVER - A break in the metal at the top of the inside portion of the double seam due to a portion of the cover being forced over the top of the seaming chuck. This condition usually occurs at the cross-over. This is also known as a "cut through" by some manufacturers. These manufacturers refer to a cutover as the same condition without the break.

CUT THROUGH - Gasket damage caused by excessive vertical pressure.

DEADHEAD - An incomplete double seam resulting from the seaming chuck spinning in the end's countersink during the double seaming operation. Also known as a spinner, skidder, or slip.

DELAMINATION - Any separation of plies that results in questionable pouch integrity; delamination of more than 0.16 cm (1/16 in) (widest point) anywhere outside the food product is unacceptable.

DOUBLE SEAM - The closure formed by interlocking and compressing the curl of the end and the flange of the can body. It is commonly produced in 2 operations. The first operation roll pre-forms the metal to produce the 5 thicknesses or folds, and the second operation roll presses and flattens them together to produce double seam tightness.

DROOP - A smooth projection of the double seam outside and below the bottom of the normal seam. Usually occurs at the side seam lap area.

FACTORY END - See manufacturer's end.

FALSE SEAM - A double seam where a portion of the cover hook and body hook are not interlocked, i.e., no hooking of body and cover hooks.

FINISH - That part of the glass container for holding the cap or closures.

FLANGE - The outward flared edge of the can body cylinder that becomes the body hook in the double seaming operation.

FLEXIBLE CONTAINER - A container, the shape or contour of which, when filled and sealed, is affected by the enclosed product.

HEAVY LAP - A lap containing excess solder. Also called a thick lap.

HOOK, BODY - See body hook.

HOOK, COVER - See cover hook.

IMPROPER POUCH SEAL - A defect (e.g., entrapped food, grease, moisture, voids, fold-over wrinkles) in that area of the closure seal which extends 0.32 cm vertically from the edge of the seal on the food product side and along the full length of the seal.

JUMPED SEAM - See jumpover.

JUMPOVER - A double seam that is not rolled tight enough adjacent to the crossover caused by jumping of the seaming rolls at the lap.

JUNCTURE - Also known as the "cross-over"; the junction of the body side seam and the end double seam, or that point where the 2 seams come together.

KNOCKED-DOWN FLANGE - A common term for a false seam where the bottom of the flange is visible below the double seam. A portion of the body flange is bent back against the body without being engaged with the cover hook.

LAP - The section at the end of the side seam consisting of 2 layers of metal bonded together. As the term implies, the 2 portions of the side seam are lapped together to allow for the double seam, rather than hooked as in the center of the side seam.

LID - See packer's end.

LIP - A projection where the cover hook metal protrudes below the double seam in one or more "V" shapes. Also known as a vee.

LUG CAP - A closure with raised internal impressions that intermesh with identical threads on the finish of the glass container. It is a closure with horizontal protrusions that seat under angled threads on the glass container finish.

MANUFACTURER'S END - The end of the can that is attached by the can manufacturer.

NOTCH - The small cut-out section in the lap designed to facilitate the formation of the body hook at cross-over.

OPEN LAP - A lap that is not properly soldered or has failed by separating or opening because of various strains in the solder.

OVERLAP - The distance the cover hook laps over the body hook.

PACKER'S END - The end of the can attached and coded by the food packer, also known as the canner's end.

PLATE - A general term for tinplate, aluminum, and the steel sheets from which cans are made. It is usually tin plate, which is black plate with tin applied to it.

PRESSURE RIDGE - The impression (chuck impression) around the inside of the can body directly opposite the double seam.

PULL-UP - Term applied to distance measured from the leading edge of the closure lug to the vertical neck ring seam.

SEAM THICKNESS - The maximum dimension of the double seam measured across or perpendicular to the layers of the seam.

SEAM WIDTH (LENGTH OR HEIGHT) - The maximum dimension of the double seam measured parallel to the folds of the seam.

SECURITY - Residual clamping force remaining in the closure application when the gasket has properly seated after processing and cooling.

SEMIRIGID CONTAINER - A container, the shape or contour of which, when filled and sealed, is not affected by the enclosed product under normal atmospheric temperature and pressure, but which may be deformed by external mechanical pressure of less than 0.7 kg/cm^2 (10 psi), i.e., normal firm finger pressure.

SIDE SEAM - The seam joining the 2 edges of the body blank to form a can body.

SKIDDER - A can having an incompletely finished double seam due to the can slipping in the seaming chuck. In this defect, part of the seam will be incompletely rolled out. The term has the same meaning as deadhead when referring to seamers that revolve the can; also known as a spinner.

SOFT CRAB - Colloquial term used to describe a breakdown in the packer's can resulting in a hole between the end and the body.

SPINNER - <u>See</u> deadhead and skidder.

STRIPPED CAP - Lug closure applied with too much torque, which causes lugs to pass over glass lugs. May have vacuum but has no security value.

TIGHTNESS - The degree to which the double seam is compressed by the second operation roll. Tightness is determined primarily by the degree of freedom from wrinkles in the cover hook. The tightness rating is a percentage, and it ranges from 100 down to 0 depending on the depth of the wrinkle: 100% indicates no wrinkle and 0% indicates a wrinkle extending completely down the face of the cover hook. A well defined continuous impression around the circumference of the can in the double seam area indicates a tight seam. This impression is known as a pressure ridge.

TOP SEAM - Top of packer's end seam.

UNEVEN HOOK - A body or cover hook that is not uniform in length.

WRINKLE (COVER HOOK) - A waviness occurring in the cover hook from which the degree of double seam tightness is determined.

CHAPTER 16

BACILLUS CEREUS

Bacillus cereus is widely distributed in nature and can be isolated from a wide variety of foods. However, unless it is able to grow, it is not a significant health hazard. Consumption of food containing more than 10^5 viable B. cereus cells per gram has resulted in outbreaks of food poisoning. Foods incriminated in outbreaks of B. cereus poisoning include: boiled and fried rice, cooked pasta, cooked meats, cooked vegetables, soups, salads, puddings, and vegetable sprouts.

Two types of illness have been attributed to the consumption of food contaminated with B. cereus. The "diarrheal syndrome" is characterized by abdominal pain and diarrhea. It has an incubation period of 8 to 16 hours and symptoms last 12 to 24 hours. The "emetic syndrome" is characterized by an accute attack of nausea and vomiting which occurs one to 5 hours after a meal. Diarrhea is not a predominant feature in this type of illness, but it does occur in some cases.

A. TRANSPORTING AND STORAGE OF SAMPLES

Transport and examine samples promptly without freezing, if possible. If samples must be shipped to the laboratory, pack them in insulated shipping containers with enough gel-type refrigerant to maintain them at 6°C or below. Upon receipt in the laboratory, store the samples at 4°C and begin the analysis as soon as possible. If the analysis cannot be started with 4 days after the samples are collected, freeze them rapidly and store them at -20°C until examined. Thaw samples at room temperature and proceed with the analysis as usual. Dehydrated foods may be stored at room temperature and shipped without refrigeration.

B. EQUIPMENT AND MATERIALS

1. See Chapter 1 for preparation of food homogenate

2. Butterfield's phosphate-buffered dilution water (R3) sterilized in bottles to yield final volumes of 450 ± 5 ml and 90 ± 2 ml

3. Pipets, 1, 5, and 10 ml, graduated in 0.1 ml units

4. Spreading rods, hockey stick or hoe-shaped glass rods, 3-4 mm diameter with 45-55 mm spreading area

5. Incubators, 30 ± 2°C and 35 ± 2°C

6. Colony counter, Quebec or equivalent, dark-field model

7. Tally register

8. Marking pen, black felt type

9. Large and small Bunsen burners

10. Wire loops, No. 24 nichrome or platinum wire, 2 mm and 3 mm id

11. Vortex mixer

12. Microscope, microscope slides, and cover slips

13. Culture tubes, 13 x 100 mm, sterile

14. Test tubes, 16 x 125 mm, or spot plate

15. Bottles, 200 ml, sterile

16. Anaerobic jar, BBL GasPak, equipped with hydrogen + CO_2 generator envelopes and catalyst

17. Water bath, 48-50°C

18. Culture tube racks

19. Staining rack

20. Petri dishes, sterile, glass or plastic

C. MEDIA AND REAGENTS

Materials needed for preparing culture media and reagents, including size of culture tubes, etc., to be used for each, are specified in the directions given for their preparation.

1. Mannitol-egg yolk-polymyxin (MYP) agar plates (M56)

2. Egg yolk emulsion, 50% (M27)

3. Polymyxin B solutions for MYP agar and trypticase soy-polymyxin broth

4. Trypticase soy-polymyxin broth (M108)

5. Phenol red glucose broth (M75)

6. Tyrosine agar (M115)

7. Lysozyme broth (M51)

8. Voges-Proskauer medium (M119)

9. Nitrate broth (M63)

10. Nutrient agar for B. cereus (M67)

11. Motility medium (B. cereus) (M57)

12. Trypticase soy-sheep blood agar (M109)

13. Nitrite detection reagents (R23)

14. Voges-Proskauer test reagents (R40)

15. Creatine crystals

16. Gram stain reagents (S6)

17. Basic fuchsin staining solution (S1)

18. Methanol

D. SAMPLE PREPARATION

Using aseptic technique, weigh 50 g of sample into sterile blender jar. Add 450 ml Butterfield's phosphate buffer and blend for 2 min at 18,000-21,000 rpm. Using 1:10 dilution prepared above, make serial dilutions of sample for enumeration of B. cereus as described in E or F, below.

E. MOST PROBABLE NUMBER (MPN) OF B. CEREUS

Inoculate 3-tube MPN series in trypticase soy-polymyxin broth, using 1 ml inoculum of 10^{-1}, 10^{-2}, and 10^{-3} dilutions of sample with 3 tubes at each dilution. (Additional dilutions should also be tested if B. cereus population is expected to exceed 10^{3}/g.) Incubate tubes 48 ± 2 hr at 30°C and check for dense growth, which is typical of B. cereus. Streak tubes showing typical growth onto separate MYP agar plates and incubate plates 24-48 hr at 30°C. Pick one or more eosin pink, lecithinase-positive colonies from each MYP agar plate and transfer to nutrient agar slants for confirmation as B. cereus. Confirm isolates as B. cereus as described in F, below, and calculate MPN of B. cereus/g of sample (see appropriate table in Chapter 21) on the basis of the number of tubes at each dilution in which the presence of B. cereus was confirmed.

F. CONFIRMATION OF B. CEREUS

Pick 5 or more eosin pink, lecithinase-positive colonies from MYP agar plates and transfer to nutrient agar slants. Incubate slants 24 hr at 30°C. Make Gram-stained smears from slants and examine microscopically. B. cereus will appear as large Gram-positive bacilli in short-to-long chains; spores are ellipsoidal and central to subterminal; they do not swell the sporangium. Transfer 3 mm loopful of culture from each slant to 13 x 100 mm tube containing 0.5 ml of sterile phosphate-buffered dilution water and suspend culture in diluent with Vortex mixer. Use suspended cultures to inoculate the following confirmatory media:

1. **Phenol red glucose broth**. Inoculate 3 ml broth with 2 mm loopful of culture. Incubate tubes anaerobically 24 hr at 35°C in GasPak anaerobic jar. Shake tubes vigorously and check for growth as indicated by turbidity and color change from red to yellow, which indicates that acid has been produced anaerobically from glucose.

2. **Nitrate broth**. Inoculate 5 ml broth with 3 mm loopful of culture. Incubate tubes 24 hr at 35°C. Test for presence of nitrite by adding 0.25 ml each of nitrite test reagents A and C to each culture. An orange color, which develops within 10 min, indicates that nitrite is present.

3. **Modified VP medium**. Inoculate 5 ml medium with 3 mm loopful of culture and incubate tubes 48 $\pm$ 2 hr at 35°C. Test for production of acetylmethylcarbinol by pipetting 1 ml culture into 16 x 125 mm test tube and adding 0.6 ml alpha-naphthol solution (R40) and 0.2 ml 40% potassium hydroxide (R40). Shake, and add a few crystals of creatine. Read results after holding for 1 hr at room temperature. Test is positive if pink or violet color develops.

4. **Tyrosine agar**. Inoculate entire surface of tyrosine agar slant with 3 mm loopful of culture. Incubate slants 48 hr at 35°C. Check for clearing of medium near growth, which indicates that tryrosine has been decomposed. Examine negative slants for obvious signs of growth, and incubate for a total of 7 days before discarding.

5. **Lysozyme broth**. Inoculate 2.5 ml of nutrient broth containing 0.001% lysozyme with 2 mm loopful of culture. Also inoculate 2.5 ml plain nutrient broth as positive control. Incubate tubes 24 hr at 35°C. Record growth as positive or negative in lysozyme broth and in nutrient broth control. Incubate negative tubes for additional 24 hr before discarding.

6. **MYP agar**. This test may be omitted if test results were clear-cut with original MYP agar plates and there was no interference from other microorganisms which were present. Mark bottom of a plate into 6-8 equal sections with felt marking pen, and label each section. Inoculate premarked 4 cm sq area of MYP agar plate by gently touching surface of agar with 2 mm loopful of culture. (Six or more cultures can be tested in this manner on one plate.) Let inoculum be absorbed and incubate the plate 24 hr at 35°C. Check plates for lecithinase production as indicated by zone of precipitation surrounding growth. Mannitol is not fermented by isolate if growth and surrounding medium are eosin pink. (Yellow color indicates that acid is produced from mannitol.) B. cereus colonies are usually lecithinase-positive and mannitol-negative on MYP agar.

7. **Record results obtained with the different confirmatory tests**. Tentatively identify as B. cereus those isolates which 1) produce large Gram-positive rods with spores that do not swell the sporangium; 2) produce lecithinase and do not ferment mannitol on MYP agar; 3) grow and produce acid from glucose anaerobically; 4) reduce nitrate to nitrite; 5) produce acetylmethylcarbinol (VP-positive); 6) decompose L-tyrosine; and 7) grow in the presence of 0.001% lysozyme.

These basic characteristics are shared with other members of the B. cereus group, including the rhizoid strains B. cereus var. mycoides, the crystalliferous insect pathogen B. thuringiensis, and the mammalian pathogen B. anthracis. However, these species or varieties can usually be differentiated from B. cereus by determining specific characteristics which are typical of each species or variety. The tests described in G, below, are useful for this purpose and can easily be performed in most laboratories. Strains which give atypical results in these tests will require further testing before they can be classified as B. cereus.

G. TESTS FOR DIFFERENTIATING MEMBERS OF THE B. CEREUS GROUP (Table 27)

The following tests are useful for differentiating typical strains of B. cereus from other members of the B. cereus group, including B. cereus var. mycoides, B. thuringiensis, and B. anthracis.

1. **Motility test.** Inoculate BC motility medium by stabbing down the center with 3 mm loopful of 24 hr culture suspension. Incubate tubes 18-24 hr at 30°C and examine by transmitted light for type of growth along stab line. Motile organisms produce diffuse growth out into medium away from stab. Nonmotile organisms produce growth only in and along stab. Alternatively, add 0.2 ml sterile distilled water to surface of nutrient agar slant and inoculate slant with 3 mm loopful of culture suspension. Incubate slant 6-8 hr at 30°C and suspend 3 mm loopful of liquid culture from base of slant in drop of sterile water on microscope slide. Apply cover glass and examine immediately with microscope for evidence of motility. Report whether or not isolates tested were motile. Most strains of B. cereus and B. thuringiensis are motile by means of peritrichous flagella. B. anthracis, and all except a few strains of B. cereus var. mycoides, are nonmotile. A few B. cereus strains are also nonmotile.

2. **Rhizoid growth.** Pour 18-20 ml nutrient agar into sterile 15 x 100 mm Petri dishes and let dry at room temperature for 1-2 days. Inoculate by gently touching surface of medium near center of each plate with 2 mm loopful of 24 hr culture suspension. Let inoculum be absorbed and incubate plates 48-72 hr at 30°c. Check for development of rhizoid growth, which is characterized by production of colonies with hair or rootlike structures that may extend several centimeters from site of inoculation. Rough galaxie-shaped colonies are often produced by B. cereus strains and should not be confused with typical rhizoid growth, which is the definitive characteristic of B. cereus var. mycoides. Most strains of this variety are also nonmotile.

3. **Test for hemolytic activity.** Mark bottom of a plate into 6-8 equal sections with felt marking pen, and label each section. Inoculate a premarked 4 cm^2 area of trypticase soy-sheep blood agar plate by gently touching medium surface with 2 mm loopful of 24 hr culture suspension. (Six or more cultures can be tested simultaneously on each plate.) Incubate plates 24 hr at 35°C. Examine plates by transmitted light for hemolytic activity. B. cereus cultures usually are strongly hemolytic and produce 2-4 mm zone of complete (beta) hemolysis surrounding growth. Most B. thuringiensis and B. cereus var. mycoides strains are also beta-hemolytic. B. anthracis strains are usually nonhemolytic after 24 hr incubation.

4. **Test for protein toxin crystals.** Inoculate nutrient agar slants with 3 mm loopfuls of 24 hr culture suspensions. Incubate slants 24 hr at 30°C and hold at room temperature for 2-3 days. Prepare smears with sterile distilled water on microscope slides. Air-dry and lightly heat-fix by passing slide through flame of Bunsen burner. Place slide on staining rack and flood with methanol. Let stand 30 s, pour off methanol, and dry slide thoroughly by passing it though flame of Bunsen burner. Return slide to staining rack and flood completely with 0.5% basic fuchsin or TB carbolfuchsin ZN stain (Difco). Heat slide gently from below with small Bunsen burner until steam is seen. Wait 1-2 min and repeat this step. Let stand 30 s, pour off stain, and rinse slide thoroughly in liter of clean tap water. Dry slide without blotting and examine under oil immersion for presence of free spores and darkly stained tetragonal (diamond-shaped) toxin crystals. Crystals are usually

somewhat smaller than spores. Toxin crystals are usually abundant in a 3- to 4-day-old culture of B. thuringiensis but cannot be detected by the staining technique until lysis of the sporangium has occurred. Therefore, unless free spores can be seen, cultures should be held at room temperature for a few more days and re-examined for toxin crystals. B. thuringiensis usually produces protein toxin crystals that can be detected by the staining technique either as free crystals or parasporal inclusion bodies within the exosporium. B. cereus and other members of the B. cereus group do not produce protein toxin crystals.

5. **Interpreting test results**. On the basis of the test results, identify as B. cereus those isolates which are actively motile and strongly hemolytic and do not produce rhizoid colonies or protein toxin crystals. Nonmotile B. cereus strains are also fairly common and a few strains are weekly hemolytic. These nonpathogenic strains of B. cereus can be differentiated from B. anthracis by their resistance to penicillin and gamma bacteriophage. CAUTION: Nonmotile, nonhemolytic isolates which are suspected to be B. anthracis should be submitted to a reference laboratory for identification or destroyed by autoclaving. Acrystalliferous variants of B. thuringiensis and nonrhizoid strains derived from B. cereus var. mycoides cannot be distinguished from B. cereus by the cultural tests.

H. FLOW SHEET

See Figure 35.

I. RECORD SHEET

See Annex 19.

J. PRECAUTIONS AND LIMITATIONS OF METHOD

In determining the presence of B. cereus in foods, the following points should be observed:

1. The confirmatory tests recommended above may, in some instances, be inadequate for distinguishing B. cereus from culturally similar organisms that could occasionally be encountered in foods.

2. The tests described for differentiating the typical strains of B. cereus from other members of the B. cereus group are usually adequate. However, results with atypical strains of B. cereus are quite variable and further testing may be necessary to identify the isolates.

3. At present, no practical tests are available for detecting either the diarrheal or emetic enterotoxins produced by B. cereus. Until such tests are available, cultural tests such as those described in this method must be used for confirming suspect B. cereus isolates.

4. Rough, galaxy-shaped colonies are often produced by B. cereus strains and should not be confused with typical rhizoid growth, which is the definitive characteristic of B. cereus variant mycoides.

5. Nonmotile, nonhemolytic isolates which are suspected to be B. anthracis should be submitted to a reference laboratory for identification or destroyed by autoclaving.

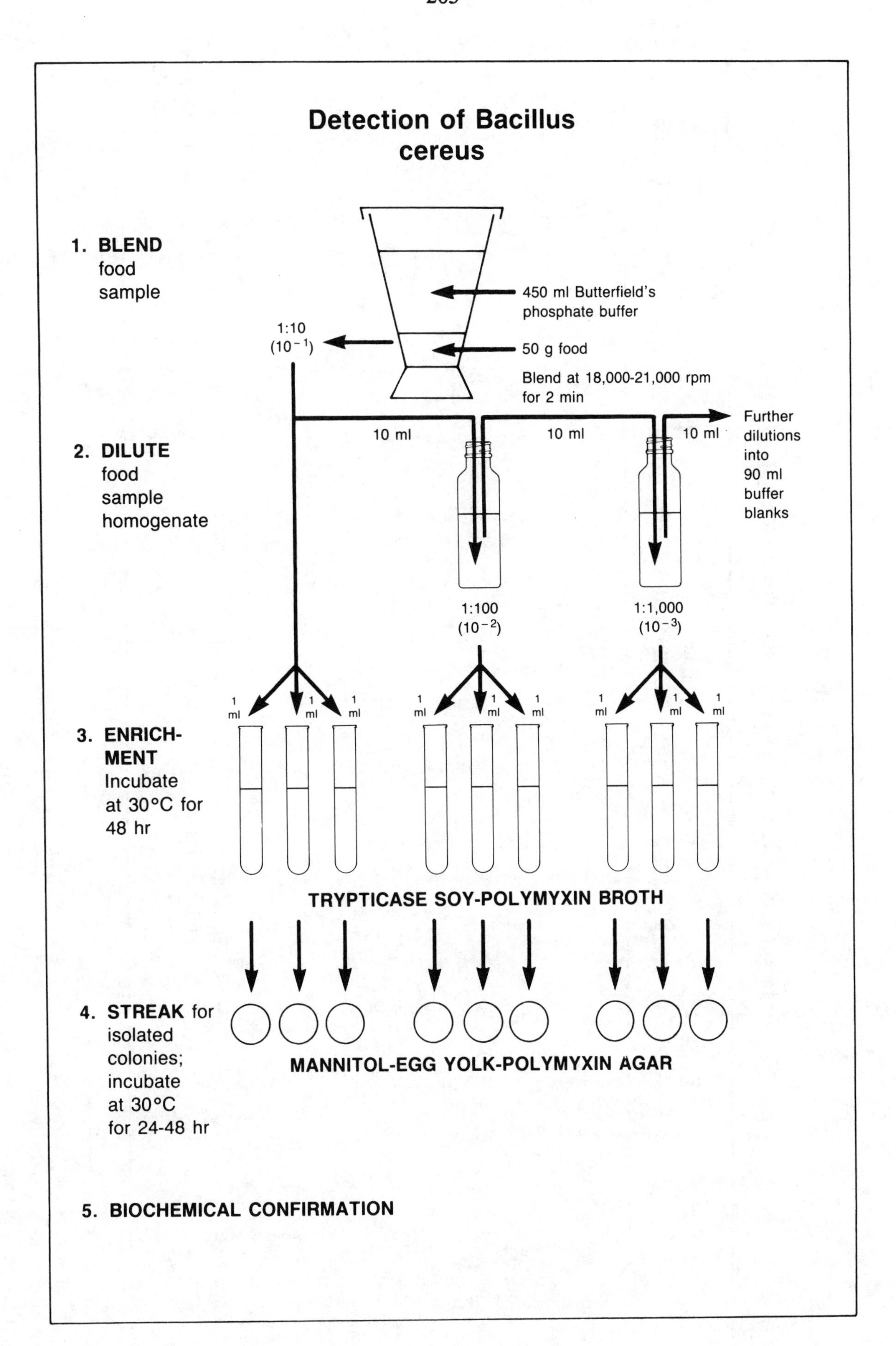

Detection of Bacillus cereus
1. BLEND food sample
450 ml Butterfield's phosphate buffer
1:10 (10⁻¹)
50 g food
Blend at 18,000-21,000 rpm for 2 min
2. DILUTE food sample homogenate
10 ml
10 ml
10 ml
Further dilutions into 90 ml buffer blanks
1:100 (10⁻²)
1:1,000 (10⁻³)
1 ml
3. ENRICH-MENT Incubate at 30°C for 48 hr
TRYPTICASE SOY-POLYMYXIN BROTH
4. STREAK for isolated colonies; incubate at 30°C for 24-48 hr
MANNITOL-EGG YOLK-POLYMYXIN AGAR
5. BIOCHEMICAL CONFIRMATION

Figure 35

Table 27

Differential characteristics of large-celled Group I Bacillus species

Feature	B. megaterium	B. cereus	B. thuringiensis	B. cereus var. mycoides	B. anthracis
Gram reaction	+	+[a]	+	+	+
Catalase	+	+	+	+	+
Motility	+/-	+/-[b]	+/-	-[c]	-
Reduction of nitrate	-[d]	+	+/-	+	+
Tyrosine decomposed	+/-	+	+	+/-	-[d]
Lysozyme-resistant	-	+	+	+	+
Egg yolk reaction	-	+	+	+	+
Anaerobic utilization of glucose	-	+	+	+	+
VP reaction	-	+	+	+	+
Acid produced from mannitol	+	-	-	-	-
Hemolysis (sheep red blood cells)	-	+	+	+	-[d]
Known pathogenicity characteristic		produces enterotoxins	endotoxin crystals pathogenic to insects	rhizoidal growth	pathogenic to animals and humans

[a]+, 90-100% of strains are positive.
[b]+/-, 50-55% of strains are positive.
[c]-, 90-100% of strains are negative.
[d]-, Most strains are negative.

CHAPTER 17

CLOSTRIDIUM PERFRINGENS

Clostridium perfringens causes one of the most common types of human foodborne illness. Symptoms of abdominal cramps, gas, and diarrhea occur about 8-15 hr after ingestion of the contaminated food.

Small numbers of C. perfringens may be commonly found in raw poultry and other meats, raw vegetables, and dehydrated soups. C. perfringens may also be found in cooked poultry and meats since the oxygen level is reduced to sufficiently low levels during the cooking process. Although the vegetative cells would normally be killed during the cooking process, heat-resistant spores could survive, germinate, and multiply to disease-producing levels in foods that had been inadequately refrigerated.

The enumeration of several hundreds of thousands of C. perfringens cells per g of food supports a diagnosis of perfringens food poisoning when verified by clinical and epidemiological investigations. Care must be taken, however, for the proper handling of the sample since refrigeration or freezing may result in a significant reduction of the population of C. perfringens cells.

A. TRANSPORTING OF SAMPLES TO THE LABORATORY

1. Transport and examine samples promptly without freezing, if possible, and hold at about 10°C until examined. If analysis cannot be started within 8 hr or if sample must be shipped to the laboratory, treat it with buffered glycerin-salt solution as described below.

2. Use aseptic technique to prepare sample for storage or shipment. Transfer 25 g portion of sample to sterile plastic bag. Add 25 ml buffered glycerin-salt solution, exclude air from bag, and mix well with glycerin solution. Liquid samples such as gravy should be mixed well with equal volume of double strength buffered glycerin-salt solution.

3. Store glycerin-treated samples immediately at -20° to -32°C in low temperature freezer or with dry ice so that freezing occurs as quickly as possible. Maintain this temperature until analysis is started. Thaw and transfer sample and glycerin-salt solution to sterile blender jar. Add 200 ml peptone dilution fluid to blender jar and proceed with examination.

B. EQUIPMENT AND MATERIALS

1. High speed blender, Waring or equivalent, and 1 liter glass or metal jars with covers; 1 jar required for each sample

2. Anaerobic jars, BBL GasPak, or Oxoid anaerobic jars (Oxoid USA, Columbia, MD 21045 USA) equipped with GasPak H + CO_2 generator envelopes.

3. Incubator, 35 ± 1°C

4. Water bath, 46 ± 0.5°C

5. Colony counter, Quebec or equivalent, dark-field model

6. Pipets, 1.0 ml serological with 0.1 ml graduations, and 10.0 ml with 1.0 ml graduations

7. Petri dishes, sterile 15 x 100 mm

8. Platinum loop, 3 mm id

C. **MEDIA AND REAGENTS**

1. Tryptose-sulfite-cycloserine (TSC) agar (M114)

2. Chopped liver broth (M22) or cooked meat medium (modified) (M24) (chopped liver broth is preferred)

3. Thioglycollate medium (fluid) (M97)

4. Iron milk medium (modified) (M36)

5. Lactose-gelatin medium (for C. perfringens) (M40)

6. Sporulation broth (for C. perfringens) (M92)

7. Motility-nitrate medium, buffered (for C. perfringens) (M58)

8. Spray's fermentation medium (for C. perfringens) (M93)

9. Peptone dilution fluid (R25)

10. Nitrite detection reagents (R23)

11. Glycerin-salt solution (buffered) (R12)

12. Gram stain reagents (S6)

13. Fermentation test papers. Saturate 15 cm Whatman No. 31 filter paper disks with 0.2% bromthymol blue solution adjusted to pH 8-8.5 with ammonium hydroxide. Air-dry the disks and store them for later use.

14. Bromthymol blue solution, 0.04% (R2)

D. **CULTURAL AND ISOLATION PROCEDURES**

1. From the food sample itself, prepare a film, stain by Gram method, and examine smear for large Gram-positive rods.

2. **Plate count of viable C. perfringens**. Aseptically weigh 25 g food sample into sterile blender jar. Add 225 ml peptone dilution fluid and blend 1-2 min at 10,000-12,000 rpm. Using this 1:10 prepared dilution, make serial dilutions from 10^{-1} to 10^{-6} by transferring 10 ml of previous dilution to 90 ml peptone dilution fluid blanks. Pour 6-7 ml tryptose-sulfite-cycloserine (TSC) agar without egg yolk into each of twelve 15 X 100 mm Petri dishes and spread evenly on bottom of rapidly rotating dish. When agar has solidified, aseptically

transfer 1 ml of each dilution of homogenate in duplicate to agar surface in center of each dish. Pour additional 15 ml TSC agar without egg yolk into dish and mix with inoculum by gently rotating dish. When agar has solidified, place plates upright in anaerobic jar. Establish anaerobic conditions and place jar in 35°C incubator for 20-24 hr. After incubation, examine plates for black colonies. Select plates showing 20-200 black colonies. Using Quebec colony counter, count black colonies and calculate number of clostridia cells/per g of food. Save plate for identification tests described in E, below.

3. Prepare chopped liver broth or cooked meat medium by heating 10 min in boiling water or flowing steam and cooling rapidly without agitation. Inoculate 3 or 4 tubes (chopped liver broth or cooked meat medium) with 2 ml of 1:10 homogenate as back-up for preceding plating procedure. Incubate these tubes 24-48 hr at 35°C in standard incubator. They may be disregarded if plate count dilutions are positive.

E. PRESUMPTIVE CONFIRMATION TEST

1. Select 10 typical colonies from TSC agar plates showing 20-200 black colonies. (Select a total of 10 colonies from the duplicate TSC plates of appropriate dilution). Inoculate each colony into tube of freshly deaerated and cooled fluid thioglycollate broth. Incubate in standard incubator 18-24 hr at 35°C. Perform Gram stain on each culture. C. perfringens is a short, thick, Gram-positive bacillus.

2. **Iron-milk presumptive test**. Inoculate modified iron-milk medium with 1 ml of vigorously growing fluid thioglycollate culture and incubate medium at 46°C in a water bath. After 2 hr, check hourly for "stormy fermentation". This reaction is characterized by rapid coagulation of milk followed by fracturing of curd into spongy mass which usually rises above medium surface. Remove positive tubes to prevent spilling over into water bath. For this reason, do not use short tubes for the test. Cultures that fail to exhibit "stormy fermentation" within 5 hr are unlikely to be C. perfringens. An occasional strain may require 6 hr or more, but this should be considered a questionable result to be confirmed by further testing.

F. COMPLETED CONFIRMATION TEST

1. Stab inoculate motility-nitrate, buffered, and lactose-gelatin media with 2 mm loopfuls of pure fluid thioglycollate medium culture or portion of isolated colony from TSC agar plate. Stab lactose-gelatin repeatedly to ensure adequate inoculation and then rinse loop in beaker of warm water before flaming to avoid splattering. Incubate inoculated media 24 hr at 35°C. Examine lactose-gelatin medium culture for gas production and color change from red to yellow, which indicates acid production. Chill tubes 1 hr at 5°C and check for gelatin liquefaction. If medium gels, incubate additional 24 hr at 35°C and recheck for gelatin liquefaction.

2. Inoculate sporulation broth with 1 ml fluid thioglycollate medium culture and incubate 24 hr at 35°C. Make Gram-stained smear from sporulation broth and examine microscopically for spores. Store sporulated cultures at 4°C if further testing of isolates is desired.

3. C. perfringens is nonmotile. Examine tubes of motility-nitrate medium by transmitted light for type of growth along stab line. Nonmotile organisms grow only in and along stab. Motile organisms usually produce diffuse growth out into the medium, away from the stab.

4. C. perfringens reduces nitrates to nitrites. To test for nitrate reduction, add 0.5 ml reagent A and 0.2 ml reagent B (R23) to culture in buffered motility-nitrate medium. Violet color which develops within 5 min indicates presence of nitrites. If no color develops, add a few grains of powdered zinc metal and let stand a few minutes. A negative test (no violet color) after zinc dust is added indicates that nitrates were completely reduced. A positive test with the addition of zinc dust indicates that the organism is incapable of reducing nitrates.

5. Tabulate results. C. perfringens is provisionally identified as a nonmotile, Gram-positive bacillus which produces black colonies in TSC agar, reduces nitrates to nitrites, produces acid and gas from lactose, and liquefies gelatin with 48 hr. Some strains of C. perfringens can be expected to exhibit poor sporulation in sporulation medium. Organisms suspected to be C. perfringens which do not meet the stated criteria should be tested further.

6. Subculture isolates which do not meet all criteria for C. perfringens (i.e., nonmotile, reduces nitrate, ferments lactose with acid and gas, and liquefies gelatin within 48 hr) into fluid thioglycollate medium. Incubate 24 hr at 35°C, make Gram-stained smear, and check for purity and typical cell morphology.

7. Inoculate 1 tube of freshly deaerated Spray's medium, containing 1% salicin, 1 tube containing 1% raffinose, and 1 tube of medium without carbohydrate, with 0.1 ml pure fluid thioglycollate culture. Incubate media 24 hr at 35°C and check medium containing salicin for acid and gas. Test for acid by transferring 2 mm loopful of culture to bromthymol blue test paper (B-13, above). Use only a platinum loop. No change or slight green color indicates that acid was produced. Alternatively, transfer 1.0 ml of culture to test tube or spot plate and add 1 or 2 drops of 0.04% bromthymol blue. A light green or yellow color indicates that acid was produced. Incubate media for another 48 hr and test for production of acid. Salicin is rapidly fermented with production of acid and gas by culturally similar species, but usually is not fermented by C. perfringens. Acid is usually produced from raffinose within 3 days by C. perfringens but is not produced by culturally similar species. A slight change in pH can occur in the medium without fermentation of carbohydrates.

8. Calculate number of C. perfringens cells in sample on basis of percent of colonies tested which are confirmed as C. perfringens. Example: If average plate count of 10^{-5} dilution was 72, and 9 of 10 colonies tested were confirmed as C. perfringens, then the number of C. perfringens cells/g of food is 72 x (9/10) x 100,000 = 6,500,000.

G. FLOW SHEET

See Figure 36.

H. RECORD SHEET

See Annex 20.

Enumeration of Clostridium perfringens

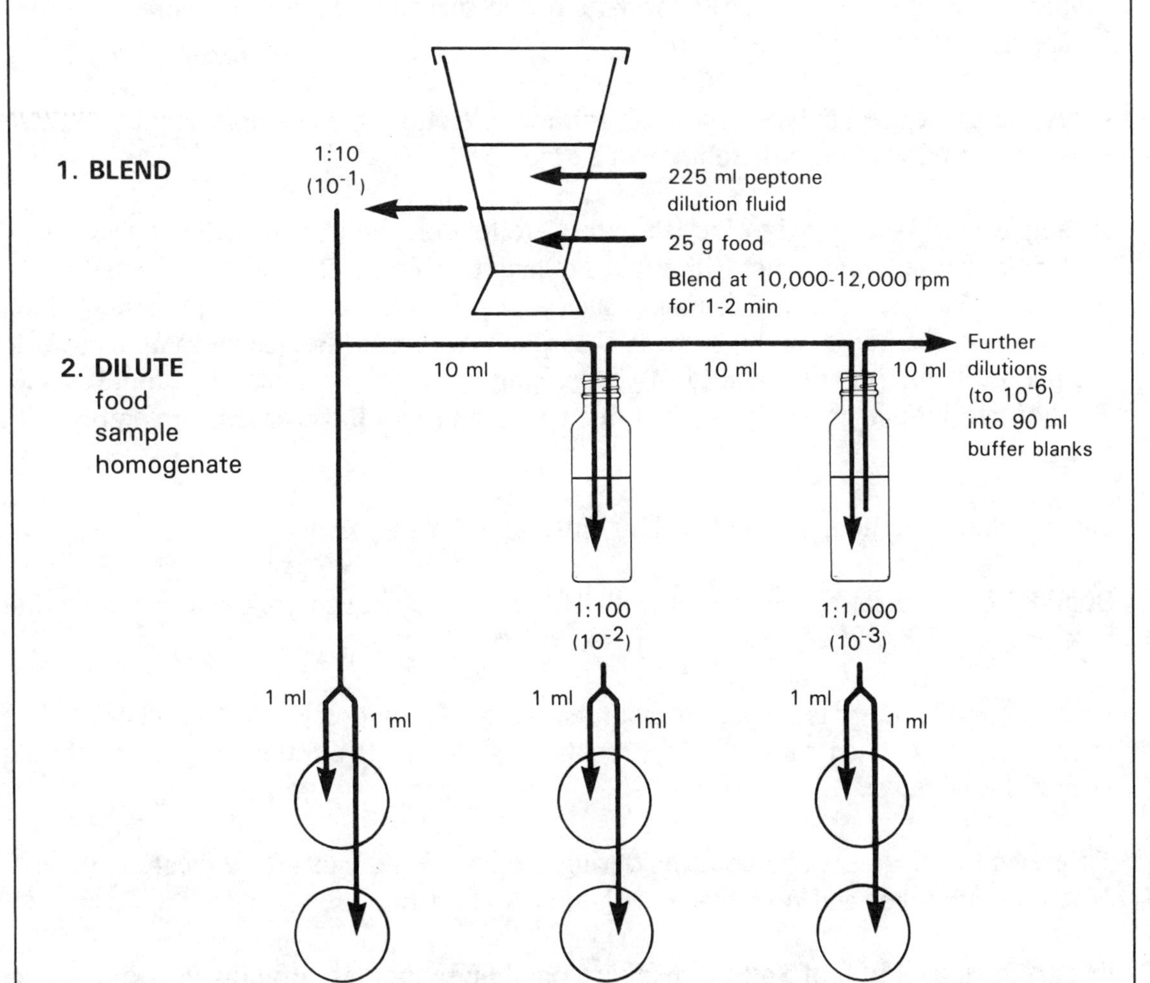

3. **POUR** 6-7 ml TSC agar without egg yolk into each of 12 Petri dishes; after solidification, transfer 1 ml of each dilution in duplicate to center of agar surface; pour additional 15 ml TSC agar without egg yolk and mix with inoculum; as "back-up," inoculate 3 or 4 tubes of chopped liver broth or cooked meat medium with 2 ml of 1:10 homogenate; tubes may be discarded if plate count dilutions are positive.

4. **INCUBATE** upright in anaerobic jar at 35°C for 20-24 hr.

5. **SELECT** plates containing 20-200 black colonies.

6. **PICK** 10 colonies from TSC plates of appropriate dilution and inoculate into fluid thioglycollate broth.

7. **BIOCHEMICAL CONFIRMATION**

Figure 36

I. PRECAUTIONS AND LIMITATIONS OF METHOD

In determining the presence of C. perfringens in foods, the following points should be observed:

1. Sample the entire portion of food (whole roast, chicken, gravy, etc.) or take representative samples of 25 g each from different parts of the suspect food because contamination may be unevenly distributed.

2. If sample cannot be analyzed within 8 hr after collection, then sample should be frozen in sterile buffered glycerin-salt solution at -20 to -32°C.

3. If sample must be shipped to the laboratory, follow procedures as outlined previously and pack frozen sample in contact with dry ice to maintain temperature as low as possible during shipment. Pack sample in a container such as a paint can or in Nalgene bottles which are impervious to carbon dioxide, because absorption of carbon dioxide by sample could lower the pH and diminish the viability of C. perfringens. Sample should be stored at -20 to -32°C on receipt and should be kept at this temperature until examined, preferably within a few days.

4. Sample should be homogenized with as little aeration as possible.

5. Bottom of Petri dish must be coated with TSC agar without egg yolk before adding dilution inocula.

6. In the event that there is no growth with the direct plating procedure, the tubes of chopped liver broth or cooked meat medium containing the 1:10 dilution of sample should be analyzed qualitatively.

7. To prevent spillage of tube contents during the iron milk "stormy fermentation" test, use 16 x 150 mm tubes and read reactions hourly up to 5 hr.

8. In transferring loopful of Spray's medium containing salicin to bromthymol blue test paper, use only a platinum loop.

CHAPTER 18

CLOSTRIDIUM BOTULINUM

Clostridium botulinum is a strictly anaerobic, sporeforming, Gram positive rod-shaped bacterium that produces proteinaceous toxin with a characteristic toxicity. Botulism is a type of food poisoning that results when food, containing this toxin, is ingested.

There are 7 recognized antigenic types: A, B, C, D, E, F, and G. Cultures of 5 of these types apparently produce only one type of toxin. Types C and D cross-react with antitoxins to each other because they produce more than one toxin and have at least one common toxin component. Type C produces predominantly C_1 toxin with lesser amounts of D and C_2, or only C_2. Type D produces predominantly type D toxin along with smaller amounts of C_1 and C_2 toxins.

Cultures producing types C and D are not proteolytic on coagulated egg white or meat and have a common metabolic pattern which sets them apart from the others. All cultures that produce type A toxin and some that produce toxins of types B and F are proteolytic. All type E strains and the remaining B and F strains are nonproteolytic, with carbohydrate metabolic patterns differing from the C and D non-proteolytic groups. Strains that produce type G have not been studied in sufficient detail for satisfactory characterization.

C. botulinum is widely distributed in soils and in sediments of oceans and lakes. The finding of type E in aquatic environments correlates with cases of type E botulinism that were traced to contaminated fish or other seafoods. Types A and B are most common in foods subject to contamination with soil.

A. EQUIPMENT AND MATERIALS

1. Refrigerator
2. Clean dry towels
3. Bunsen burner
4. Sterile can opener (bacteriological or puncture type)
5. Sterile mortar and pestle
6. Sterile forceps
7. Sterile cotton-plugged pipets
8. Mechanical pipeting device
9. Sterile culture tubes (at least a few should be screw-cap tubes)
10. Anaerobic jars (GasPak or Case-nitrogen replacement)
11. Transfer loops

12. Incubator, 35 ± 1°C and 26 ± 1 °C

13. Reserve sample jars, sterile

14. Culture tube racks

15. Microscope slides

16. Microscope, phase contrast or bright field

17. Petri dishes, 15 x 100 mm, sterile

18. Centrifuge tubes

19. Centrifuge, refrigerated, high-speed

20. Syringes, 1.0 or 3.0 ml, sterile, with 25 gauge, 1.6 cm, needles for injecting mice

21. Mice (about 15-20 g)

22. Mouse cages, feed, water bottles, etc.

B. **MEDIA AND REAGENTS**

1. Chopped liver broth (M22) or cooked meat medium (M24)

2. Trypticase-peptone-glucose-yeast extract broth (TPGY) (M101) or with trypsin (TPGYT)

3. Liver-veal-egg yolk agar (M46) or anaerobic egg yolk agar (M4)

4. Alcoholic solution of iodine (4% iodine in 70% ethanol) (R5)

5. Gel-phosphate buffer, pH 6.2, sterile (R11)

6. Pysiological saline solution, sterile (R28)

7. 1 N sodium hydroxide solution (R37)

8. 1 N hydrochloric acid (R14)

9. Absolute ethanol

10. Trypsin solution (prepared from Difco 1:250)

11. Gram stain reagents (S6), crystal violet (S5), or methylene blue (S7) solutions

C. SAMPLE PREPARATION

1. **Preliminary handling**. Refrigerate samples until testing, except for unopened canned foods, which need not be refrigerated unless badly swollen or in danger of bursting. Clean and mark container with laboratory identification codes.

2. **Solid and liquid foods**. Aseptically transfer foods with little or no free liquid to sterile mortar. Add equal amount of gel-phosphate buffer solution and grind with sterile pestle in preparation for inoculation. Alternatively, inoculate small pieces of produce directly into enrichment broth with sterile forceps. Inoculate liquid foods directly into culture media with sterile pipets. Prepare reserve sample; after culturing, aseptically remove reserve portion to sterile sample jar for tests which may be needed later.

D. DETECTION OF VIABLE C. BOTULINUM

1. **Enrichment**. Remove dissolved oxygen from enrichment medium by steaming 10-15 min and cooling quickly without agitation before inoculation.

   a. Inoculate 2 tubes of cooked meat medium with 1-2 g solid or 1-2 ml liquid food per 15 ml enrichment broth. Incubate at 35°C.

   b. Inoculate 2 tubes of TPGY broth as above. Incubate at 26°C. Use TPGYT as an alternative only when the organism involved is strongly suspected of being a nonproteolytic strain of Types B, E, or F.

   c. Introduce inoculum slowly beneath surface of broth. After 5 days of incubation, examine enrichment cultures. Check for turbidity, gas production, and digestion of meat particles. Note the odor.

   Examine cultures microscopically by wet mount under high-power phase contrast, or a smear stained by Gram stain, crystal violet, or methylene blue under bright field illumination. Observe morphology of organisms and note existence of typical clostridial cells, occurrence, and relative extent of sporulation, and location of spores within cells. At this time, test each culture for toxin as described in F, below. Usually, 7-day incubation is period of active growth and produces highest concentration of botulinum toxin. If enrichment culture shows no growth at 7 days, incubate an additional 10 days to detect possible delayed germination of C. botulinum injured spores before discarding sample as sterile. For pure culture isolation, save enrichment culture at peak sporulation and keep under refrigeration.

2. **Isolation of pure cultures**. C. botulinum is more readily isolated from mixed flora of enrichment culture or original specimen if sporulation has been good.

   a. Pre-treatment of specimens for streaking. Add equal volume of filter-sterilized absolute alcohol to 1 or 2 ml of culture in sterile screw-cap tube. Mix well and incubate 1 hr at room temperature. To isolate from sample, take 1 or 2 ml of retained portion; add equal volume of filter-sterilized absolute alcohol in sterile screw-cap tube. Mix well and incubate 1 hr at room temperature. Alternatively, heat 1 or 2 ml of enrichment culture or sample to destroy vegetative cells (80°C for 10-15 min). DO NOT use heat treatment for nonproteolytic types of C. botulinum.

b. Plating of treated cultures. With inoculating loop, streak 1 or 2 loopfuls of ethanol or heat-treated cultures to either or both liver-veal-egg yolk agar and anaerobic egg yolk agar to obtain isolated colonies. If necessary, dilute culture to obtain well separated colonies, and dry the agar plates well before use to prevent spreading of colonies. Incubate streaked plates at 35°C for about 48 hr under anaerobic conditions. A Case anaerobic jar or the GasPak system is adequate to obtain anaerobiosis; however, other systems may be used.

E. SELECTION OF TYPICAL C. BOTULINUM COLONIES

1. **Selection.** Select about 10 well separated typical colonies. These colonies may be raised or flat, smooth or rough. They commonly show some spreading and have an irregular edge. On egg yolk medium, they usually exhibit surface iridescence when examined by oblique light. This luster zone is often referred to as a pearly layer; it usually extends beyond and follows the irregular contour of the colony. Besides the pearly zone, colonies of C. botulinum types C, D and E are ordinarily surrounded by a wide zone (2-4 mm) of yellow precipitate. Colonies of types A and B generally show a smaller zone of precipitation. Considerable difficulty may be experienced in picking toxic colonies since certain other members of the genus Clostridium produce colonies with similar morphological characteristics but do not produce toxins.

2. **Inoculation.** Use sterile transfer loop to inoculate each selected colony into tube of sterile broth. Inoculate C. botulinum type E into TPGY broth. Inoculate other toxin types of C. botulinum into chopped liver broth or cooked meat medium. Incubate as described in D-1, a and b, above, for 5 days. Test for toxin production as described in F, below.

3. **Isolation of pure culture.** Restreak toxic culture in duplicate on egg yolk agar medium. Incubate one plate anaerobically at 35°C. Incubate second plate aerobically at 35°C. If colonies typical of C. botulinum are found only on anaerobic plate (no growth on aerobic plate), the culture may be pure. Failure to isolate C. botulinum from at least one of the selected colonies means that its population in relation to the mixed flora is probably low. Repeated serial transfer through additional enrichment steps may increase the numbers sufficiently to permit isolation. Store pure culture in sporulated state either under refrigeration, on glass beads, frozen, or lyophilized.

F. DETECTION OF BOTULINUM TOXIN

1. **Preparation of food sample.** Remove one portion of sample for detection of viable C. botulinum; remove another portion for toxicity testing. Store remainder of sample in refrigerator. Centrifuge samples containing suspended solids under refrigeration and use supernatant fluid for toxin assay. Extract solid foods with equal volume of gel-phosphate buffer, pH 6.2. Macerate food and buffer with pre-chilled mortar and pestle. Centrifuge macerated sample under refrigeration and use supernatant fluid for toxin assay. Rinse empty containers suspected of having held toxic foods with a few milliliters of gel-phosphate buffer. Use as little buffer as possible to avoid diluting toxin beyond detection.

2. **Determination of toxicity in food samples or cultures**

a. Trypsin treatment. Toxins of nonproteolytic types, if present in sample, may need trypsin activation to be detected. Therefore, trypsin-treat a portion of food supernatant fluid, liquid food, or cooked meat medium cultures before testing for toxin. Do not trypsin-treat TPGY cultures, since this medium already contains trypsin and further treatment may degrade any fully activated toxin present in the culture. Adjust portion of supernatant fluid, if necessary, to pH 6.2 with 1 N NaOH or HCl. Add 0.2 ml saturated aqueous trypsin solution to 1.8 ml of each supernatant fluid to be tested for toxicity. (To prepare trypsin solution, place 1 g of Difco 1:250 trypsin in clean culture tube and add 10 ml distilled water, shake, and warm to dissolve.) Incubate trypsin-treated preparation at 35-37°C for 1 hr with occasional gentle agitation.

b. Toxicity testing. Conduct parallel tests with trypsin-treated materials and untreated duplicates. Dilute a portion of untreated sample fluid or culture to 1:2, 1:10, and 1:100, in gel-phosphate buffer. Make same dilutions of each trypsinized sample fluid or culture. Inject each of separate pairs of mice intraperitoneally (i.p.) with 0.5 ml untreated undiluted fluid and 0.5 ml of each dilution of untreated test sample, using 1.0 or 3.0 ml syringe with 1.6 cm (5/8 in), 25 gauge, needle. Repeat this procedure with trypsin-treated duplicate samples. Heat 1.5 ml of untreated supernatant fluid or culture for 10 min at 100°C. Cool heated sample and inject each of a pair of mice with 0.5 ml undiluted fluid. The mice should not die, because botulinum toxin, if present, will be inactivated by heating.

Observe all mice periodically for 48 hr for symptoms of botulism. Record symptoms and deaths. Typical botulism signs in mice begin usually in first 24 hr with ruffling of fur, followed in sequence by labored breathing, weakness of limbs, and, finally, total paralysis with gasping for breath, followed by death due to respiratory failure. Death of mice without clinical symptoms of botulism is not sufficient evidence that injected material contained botulinum toxin. On occasion, death occurs from other chemicals present in injected fluid or from trauma.

G. FLOW SHEET

See Figure 37.

H. RECORD SHEET

See Annex 21.

Testing for Clostridium botulinum Toxin

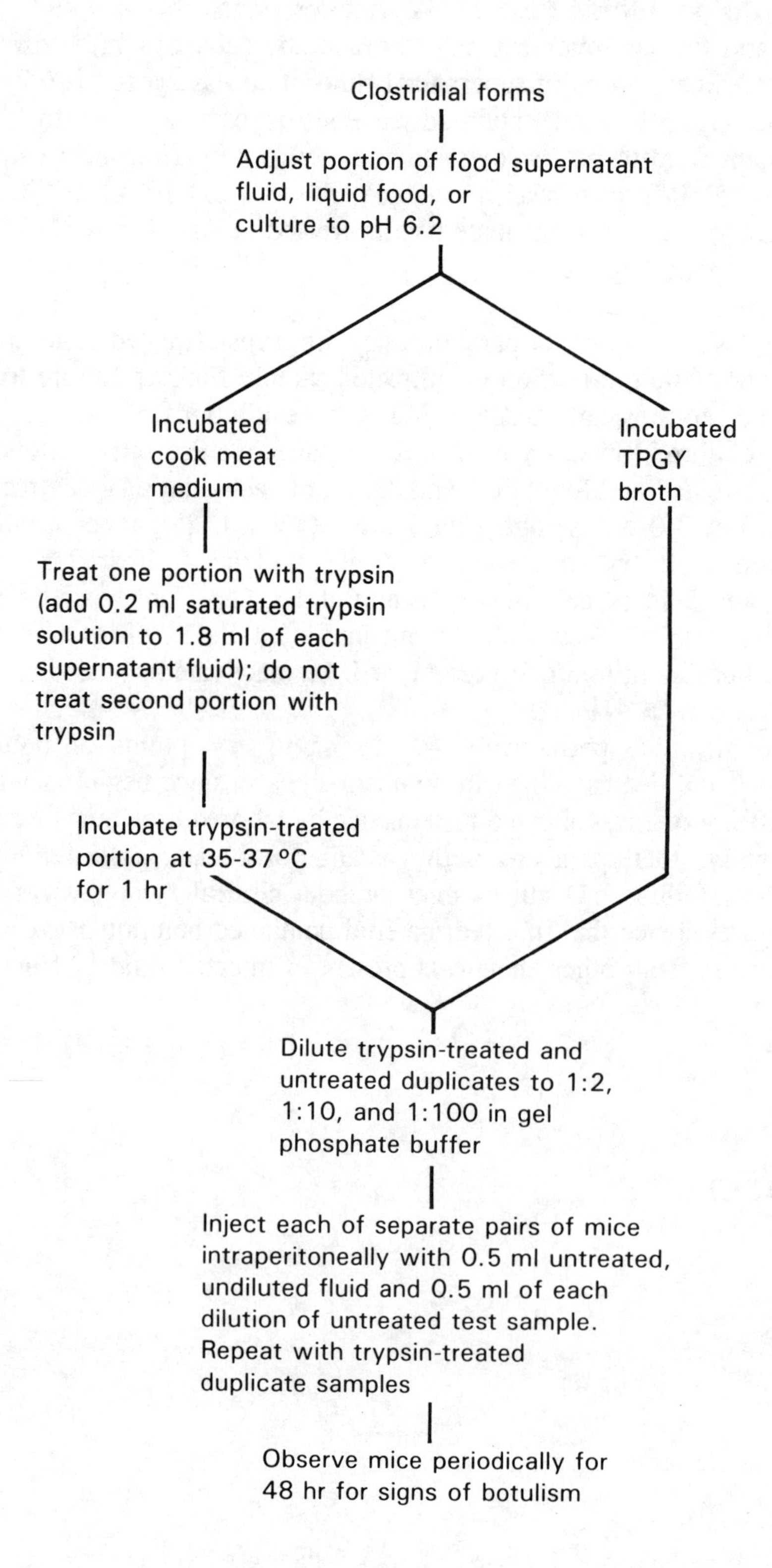

Figure 37

I. PRECAUTIONS AND LIMITATIONS OF METHOD

In determining the presence of C. botulinum in foods, the following points should be observed:

1. Except for unopened canned foods in unswollen cans, all samples should be refrigerated until initiation of analyses.

2. If can is badly swollen, analyst should take precautionary measures against spraying of can contents. For instance, a sterile funnel can be inverted over the can, or the can may be covered with sterile towels.

3. Preferably, the uncoded end of the can should be opened.

4. Do not taste the product under any circumstances.

5. The cooked meat medium and the TPGY broth should be inoculated with the food sample as slowly as possible to minimize the introduction of air into the sample.

6. TPGY culture should not be trypsin treated since this medium already contains trypsin and further treatment may degrade any fully activated toxin in the culture.

7. The first 24 hours are the most important time regarding symptoms and death of mice: 98-99% of animals die within 24 hr.

8. Mice injected with botulinal toxin may become hyperactive before symptoms occur.

9. With cooked meat medium, vortex tubes completely since toxin may adhere to meat particles.

10. Viable C. botulinum, but no toxin in foods, is not proof that the food in question caused botulism. Ingested organisms may be found in the alimentary tract, but they are considered to be unable to multiply and produce toxin in vivo, except in infants.

11. All workers in the laboratory should wear laboratory coats and safety glasses.

12. Use 1% hypochlorite solution to wipe laboratory table tops before and after work.

13. Never pipet by mouth.

14. Centrifuge toxic materials in a hermetically closed centrifuge with safety cups.

15. There should be no eating or drinking in the laboratory when anyone is working.

CHAPTER 19

YEASTS AND MOULDS

IDENTIFICATION OF YEASTS AND MOULDS

A. TYPES OF FUNGI

The yeasts and moulds constitute a very diverse group of organisms. Todate, over 400,000 species of yeasts and moulds are recognized.

Many other types of fungi exist, e.g., the mushrooms, puff balls, mildews, rusts, smuts and primitive water molds known as chytrids. This chapter is concerned only with the yeasts and moulds.

B. CHARACTERISTICS OF YEASTS AND MOULDS

These organisms are eukaryotic. That is, they all have defined chitin and/or cellulose cell walls, nuclei, and other cellular organelles. All yeasts and molds are heterotrophic. Heterotrophs lack chlorophyll and, thus, are dependent upon an outside source to meet their energy needs. This is the reason why they are commonly found as contaminants of foods and feeds - in the field, in storage during and after processing, and in the home.

The basic units of structure of all moulds and some yeasts are multicellular, filamentous, threadlike strands called hyphae. These hyphae collectively are called the mycelium. The sexual and the asexual stages of the molds and of some yeasts are produced in and on the mycelium. It should be mentioned that most foodborne yeasts and moulds do not have a known sexual stage.

C. ENVIRONMENTAL REQUIREMENTS FOR YEAST AND MOULD GROWTH

Growth of these organisms is dependent on several important environmental factors. One of these is temperature. While most foodborne yeasts and moulds are mesophilic (i.e.,optimal growth at 20-28°C), some species are true psychrophiles and can grow at temperatures as low as -2°C. Other species are true thermophiles and will grow at temperatures as high as 60°C. There also exist a number of thermotolerant and psychrotolerant yeasts and moulds which are able to survive, but not proliferate, at excessively high and low temperatures, respectively.

Another important growth factor is moisture. Most foodborne yeasts and moulds can grow in and on substrates that can generate a relative humidity (RH) of 85% or above. However, some moulds, especially storage moulds, can grow at RHs as low as 60-70%. Similarly, certain yeast species can grow on low moisture substrates containing 50-60% glucose.

Still another factor in the growth of these organisms is suitable pH. The pH range for the initiation of yeast and mould growth is quite broad, with values ranging from 2 to above 9. For some species, this range may have lower or higher limits. In the absence of a strong buffer, most of these organisms can alter the pH to one that is more favorable for their growth, usually at pH values of 4 to 6.5.

Finally, another significant factor in the growth of these organisms is oxygen. All mould species are obligate aerobes, although some species can grow at very low oxygen tension. Some species can grow in the head space of canned products, and others can grow in an oxygen-free atmosphere, receiving their oxygen from the substrate itself. But there are no absolutely anaerobic yeasts or moulds. At least some free oxygen must be available.

D. PRINCIPAL YEASTS AND MOULDS OCCURRING IN FOOD AND FEEDS

Most of these organisms, especially the moulds, have no known sexual reproductive stage, although they do reproduce asexually. Thus, they are classified and identified mainly according to their asexual reproductive structures. Fungi, which have no known sexual stage, are classified as "Fungi Imperfecti", or, Deuteromycetes. There are, however, a few mould species occurring in foods and feeds which do have a known sexual stage. This is not true for most yeasts.

1. **Yeasts**. Although there are at least 30 recognized genera of yeasts, only 3 genera are regularly encountered in foods and feeds: Saccharomyces, Candida, and Rhodotorula.

   Saccharomyces is unicellular and produces a yeast-like odor. Colonies are white on malt extract agar and potato dextrose agar. Several species have been used for centuries to make leavened bread, beers, and wines. Saccharomyces has a sexual stage which is rarely detected as it grows. It is not infectious (infectious being defined as the ability of an organism to establish itself within a susceptible host).

   Candida is unicellular, although some species produce threadlike structure, called pseudohyphae, due to the fusing of several cells in a longitudinal pattern. Some species also produce thick-walled, often ornamental, cells called chlamydospores which serve as a means of combating dessication. Colonies are creamy white and odorless on malt extract agar and potato dextrose agar. Candida has no known sexual stage. Some species can be facultative pathogens causing a disease mainly of the gastrointestinal tract and the vagina. The disease is called candidiasis.

   Rhodotorula is totally unicellular. This genus produces pink colonies on malt extract agar and potato dextrose agar. It has no known sexual stage, nor is it pathogenic.

2. **Moulds**. Numerous mould genera and species are encountered in foods and feeds. However, there are 5 genera which appear to be prominent: Aspergillus, Penicillium, Fusarium, Alternaria and Cladosporium. Below are discussed these genera and some of the most frequently occurring species.

   a. The genus Aspergillus. This genus is unique because of the type of conidiophore, or asexual fruiting body, produced (Figure 38). The conidiophore arises from a very thick walled cell in the mycellium called a foot cell. It produces a stalk which terminates to form a bulbous structure called a vesicle. Surrounding the vesicle are either one or 2 sets of pegs called sterigmata. From the outermost set of sterigmata are produced the spores, i.e., the conidia. No other mould genus has a similar conidiophore. This genus is especially prominent in tropical and sub-tropical areas.

Characteristics of the Conidiophore of the Genus *Aspergillus*

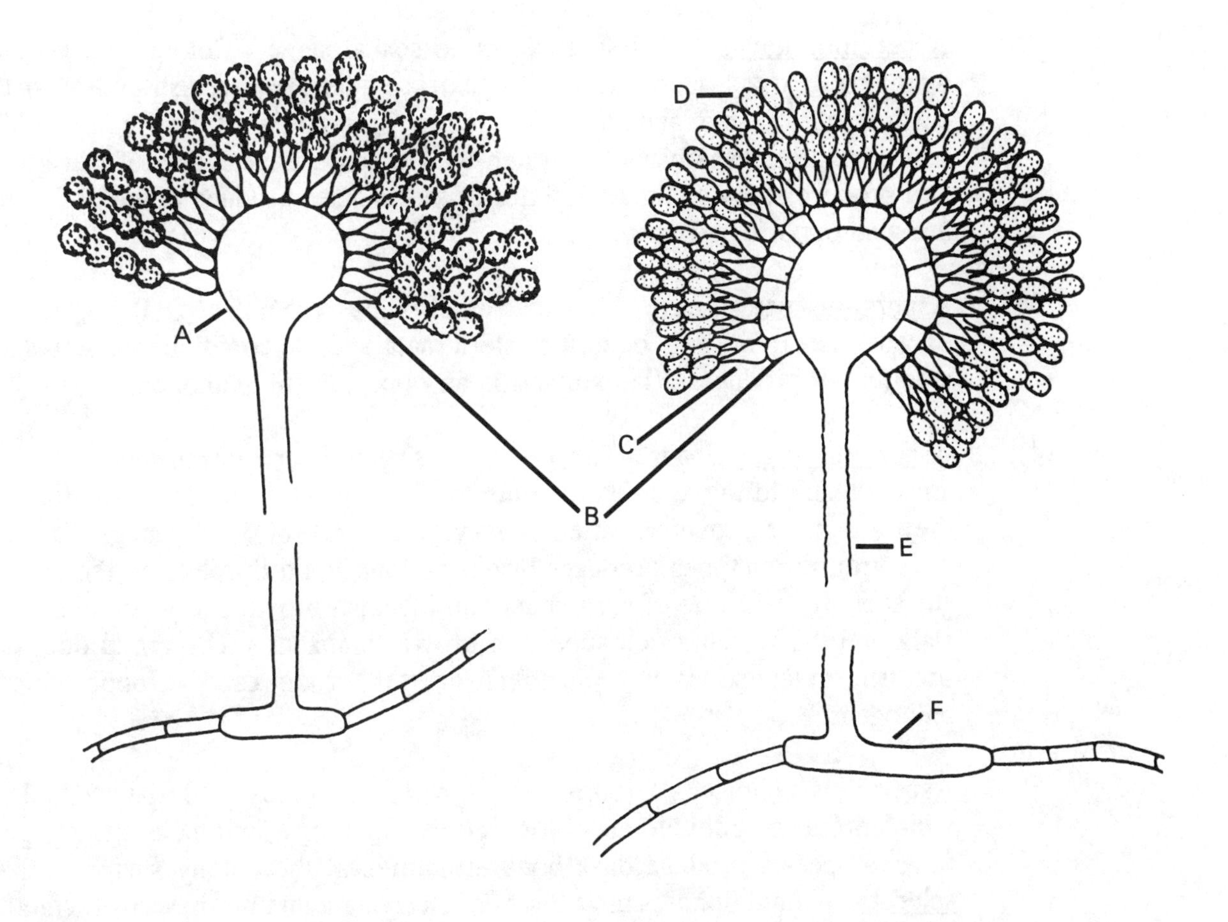

A. Vesicle; B. Primary sterigmata; C. Secondary sterigmata; D. Conidia; E. Stalk; F. Foot cell.

Figure 38

1) Aspergillus glaucus Group. This group of moulds, consisting of 18 species, is perhaps the most frequently occurring fungal contaminant of foods and feeds. Sixteen of the species have a known sexual stage exhibiting bright yellow or orange balls, called cleistothecia, which contain several colorless sacks. Each sack is called an ascus. Within each ascus are 8 colorless sexually produced spores called ascospores. The Aspergillus glaucus Group has an asexual stage that has conidiophores with only one set of sterigmata on a swollen vesicle, and produces blue-gray conidia which are usually roughened. Species of this group are not known to produce mycotoxins nor are they infectious.

2) Aspergillus flavus. This species has no sexual stage. Colonies on Czapek agar, malt extract agar, and potato dextrose agar are greenish yellow at the surface due to the abundant production of conidia of this color. The conidiophores, usually with a rough-walled stalk, have 2 sets of sterigmata and produce spiny conidia. Isolates of this species produce aflatoxin. They are also opportunistic pathogens.

3) Aspergillus parasiticus. This species strongly resembles A. flavus with 2 exceptions. It has only one set of sterigmata and the conidiophore stalks are smooth. It produces aflatoxin and is an opportunistic pathogen.

4) Aspergillus niger Group. This group consists of over a dozen species. They appear black to the unaided eye due to the production of dark conidia, but they are more brownish when observed under low power magnification. This group sometimes produces bright yellow liquid droplets on the colony surface. It has 2 sets of sterigmata and blackish-brown spiny conidia. The stalk and the vesicle often appear in brownish shades. The group does not produce mycotoxins, but members of this group can be opportunistic pathogens.

5) Aspergillus ochraceus Group. This group has at least 10 species, all of which produce yellow to tan colonies on Czapek agar and malt extract agar. Several species produce dark bony structures at the colony surface called sclerotia. Conidiophores have 2 sets of sterigmata and produce oval, smooth conidia. The stalk wall is yellow. Several species produce the mycotoxin, ochratoxin, and can be opportunistic pathogens.

6) Aspergillus fumigatus. This species produces conidiophores with only one set of sterigmata. The sterigmata, the vesicle, and the stalk are bluish-green in color. The spiny conidia are produced in long cigarette-shaped columns. Colonies are bluish-black. This species is a human and animal pathogen, usually invading through the respiratory system. It thrives in tropical and sub-tropical climates. It does not produce toxins.

b. The genus Penicillium. This genus is much more complex than Aspergillus. Very few species have a sexual stage. Thus, species identification is based primarily on conidiophore morphology.

There are 3 types of conidiophores in the genus Penicillium (Figure 39). The first and simplest type of conidiophore consists of a stalk, a small vesicle, and one set of flask-shaped sterigmata which produce the conidia.

The second type of conidiophore consists of a stalk but no vesicle. At the stalk tip are produced a series of closely packed cells called metulae. At the tips of the metulae are produced very elongate, tapered, lance-shaped sterigmata which give rise to the conidia.

The third type of conidiophore is very asymmetrical. It consists of a stalk which branches freely near its tip end. Attached to the tip ends of each branch are metulae which produce flask-shaped sterigmata and conidia. Very few species of Penicillium have a foot cell.

Species of Penicillium are usually highly colored in shades of blue, green, yellow, and orange. The colony surface often contains numerous liquid droplets. The genus Penicillium is not very common in tropical and sub-tropical areas. Below are descriptions of some of the more commonly occurring species.

1) Penicillium frequentans. This is perhaps the most frequently occurring Penicillium species. Its conidiophore consists only of a stalk and sterigmata. A small vesicle is produced. Colonies are dark green on Czapek agar and malt extract agar, and the underside of the colony is orange to brown. Conidia are slightly roughened and are produced in crusts. P. frequentans does not produce toxins, nor is it pathogenic.

2) Penicillium islandicum. This species is sometimes detected in the tropics. Its conidiophores consists of a stalk, metulae and sterigmata. There are no branches. Colonies on Czapek agar or malt extract agar are brightly colored in orange-green shades, the orange color due to pigmented mycelium and droplets and the green color due to conidia production. The underside of the colony is orange-red. Conidia are oval and smooth walled. P. islandicum produces several mycotoxins, including luteoskyrin and islanditoxin. It is not infectious.

3) Penicillium oxalicum. This species is frequently found in the field. It has an asymmetric conidiophore, having branches as well as metulae and sterigmata. Conidia are smooth, oval, and unusually large (over 6 μm in diameter). Colonies are dark greenish-black with a blackish underside. Conidia are formed in crusts. It does not produce toxins, nor is it infectious.

Characteristics of Three Basic Types of Conidiophores Produces by the Genus *Penicillium*

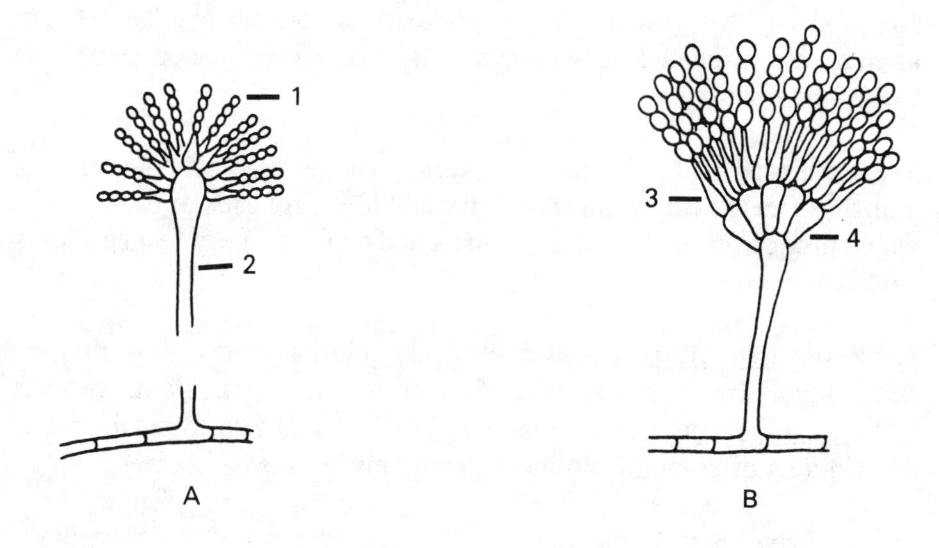

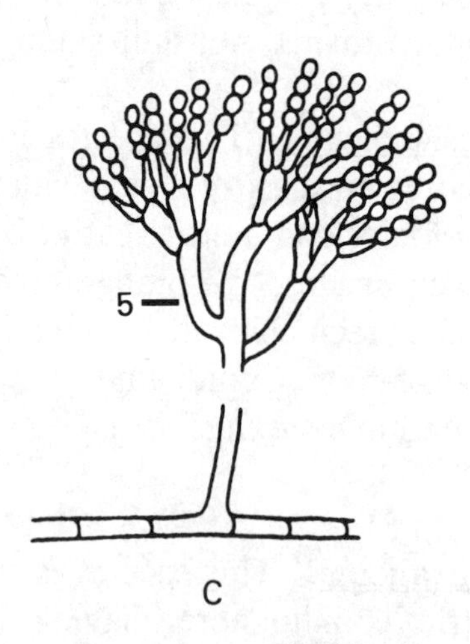

A. Stalk and flask-shaped sterigmata only; B. Stalk, metulae, and lanceolate sterigmata; C. Stalk, branches, metulae, and flask-shaped sterigmata. (1) Conidia; (2) Stalk; (3) Sterigmata; (4) Metulae; (5) Branches.

Figure 39

4) Penicillium cyclopium. This is a very important storage organism. It has an asymmetric, slightly roughened conidiophore and produces smooth to slightly roughened globase conidia. Colonies on Czapek agar or malt extract agar are usually bright blue with colorless to yellow, or even red, droplets. Underside of colony can also be similarly colored. Colony surface texture is usually granular rather than velvety. This species has a history of mycotoxin production, e.g., penicillic acid and the tremorgens, but it is not infectious.

5) Penicillium viridicatum. This is another important storage mould. It has morphological characteristics almost identical with P. cyclopium. However, the colonies are bright green on Czapek agar or malt extract agar. This species produces several mycotoxins including patulin, citrinin, xanthomegnin. It can be an opportunistic pathogen.

c. The genus Alternaria. This genus is very common in moderate climates but is rare in tropical and sub-tropical areas. There are more than 30 species of Alternaria, none of which have a sexual stage. Individual species will not be described here. The principal characteristics of Alternaria are as follows. Colonies on potato dextrose agar are gray on the surface and almost black on the underside. The colonies are wooly. The genus produces very short conidiophores upon which conidia are produced singly or in chains. Conidia are multicellular and have both horizontal and vertical cross walls (Figure 40). Under magnification, conidia are black to deep brown and are either roughened or smooth. The genus does produce some toxic metabolites (tenuazonic acid and alternariol) but is not infectious. The conidia can be an important source of allergy problems.

d. The genus Fusarium. This genus prefers moderate to cool temperatures for growth. There are at least 100 species of Fusarium. Individual species will not be discussed here.

On potato dextrose agar, the genus produces colonies which are usually cottony and white. Some species produce colonies that have red, blue, or yellow tints. A few species do produce a sexual stage.

The conidiophores are very tiny and sometimes non-detectable. Two types of conidia are produced, microconidia and macroconidia (Figure 41). The microconidia are small 1-2-celled spores that are rectangular or pear-shaped. Most, but not all, species of Fusarium produce microconidia. The macroconidia are larger, canoe-or comma-shaped, spores with several horizontal cross walls. All species of Fusarium produce microconidia which are their chief basis for identification. It is often difficult to get microconidium production on laboratory media. Some Fusarium species produce thick-walled chlamydospores (resting spores) in the mycelium singly, in chains, or in clumps. Several Fusarium species produce mycotoxins (zearalenone and trichothecenes). None are infectious.

Characteristics of Conidiophores and Conidia of the Genus *Alternaria*

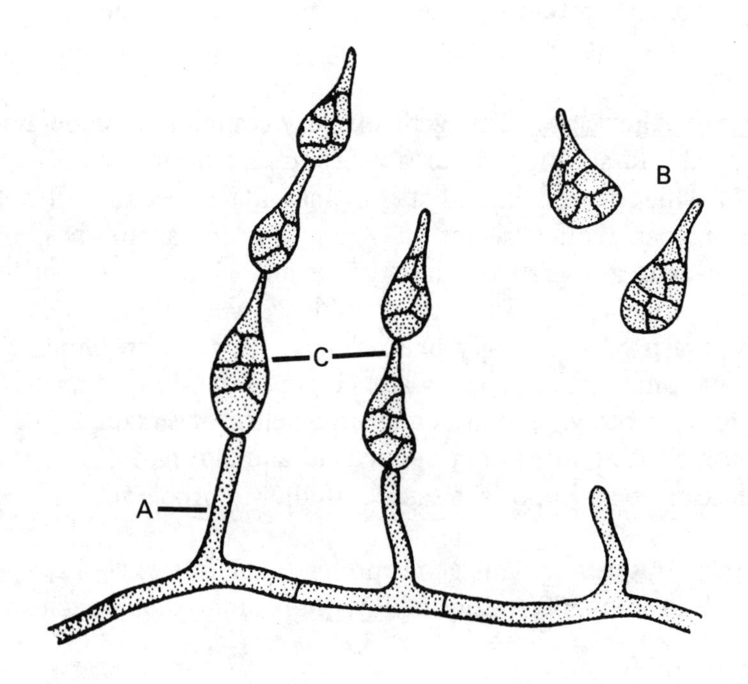

A. Short, simple, dark-pigmented conidiophores arising from dark pigmented septate hyphae. B. Typical dark-pigmented, apically beaked, many-celled conidia; C. Typical conidial chains.

Figure 40

Typical Reproductive Spores of the Genus *Fusarium*

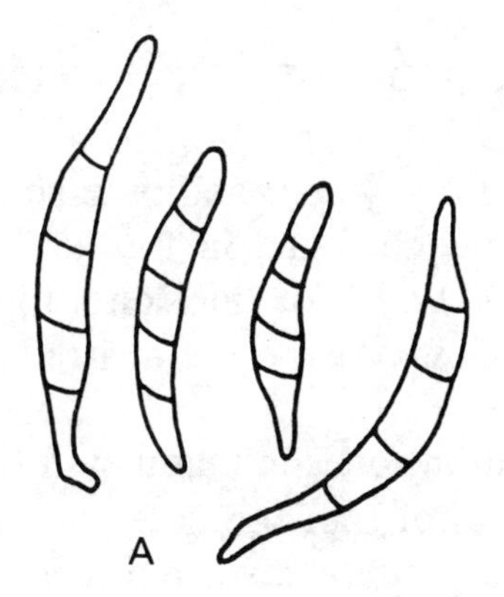

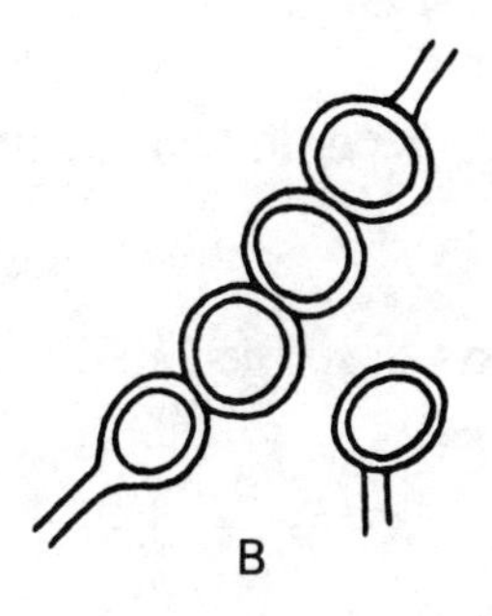

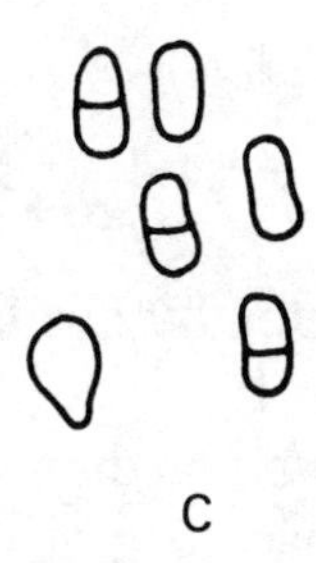

A. Elongate, many-celled, lightly to nonpigmented macrocondidia; B. Thick-walled chlamydospores produced within the mycelium singly or in chains; C. 1- to 2-celled colorless micronidia.

Figure 41

e. The genus Cladosporium. There are few species in this genus, but the genus itself is very common in foods and feeds. A rather short, slightly branched, brownish colored conidiophore is produced (Figure 42). It gives rise to irregularly shaped 1-2-celled brownish conidia. Cladosporium colonies have a shiny green-black color on the surface and on the underside of the colony. These organisms do not produce toxins nor are they pathogenic.

ENUMERATION OF YEASTS AND MOULDS - DILUTION PLATING TECHNIQUE

Because of their slow growth and relatively poor ability to compete with bacteria successfully, the yeasts and moulds are most likely to be found in the foods in which the environment is less favorable for bacteria growth, e.g., low pH, low moisture, high salt or sugar content, low storage temperature, the presence of antibiotics, or exposure to irradiation.

Yeasts and moulds present a problem in foods in that they discolor food surfaces, cause off-odors and off-flavors, cause various degrees of spoilage, alter substrates allowing for the outgrowth of pathogenic bacteria, and are able to produce mycotoxins in certain instances. Methods for the enumeration of yeasts and moulds are presented below.

A. EQUIPMENT AND MATERIALS

1. Basic equipment (and appropriate techniques) for preparation of a food sample homogenate as described in Chapters 1 and 2

2. Equipment for plating samples as described in Chapter 2

3. Incubator set at 22-25°C

4. Arnold steam chest

5. pH meter

B. MEDIA AND REAGENTS

1. Potato dextrose agar (M80), commercially available in dehydrated form

2. Potato dextrose-salt agar (M80). Same medium as above, amended with 75 g NaCl. This medium requires 20 g agar rather than 15 g agar per liter.

3. Malt extract agar (M54), commercially available in dehydrated form

4. Plate count agar (standard methods) (M77)

5. Antibiotic solution(s), see C-2a, below

6. Tartaric acid solution, 10%, sterile, see C-2b, below

Characteristics of Conidiophores and Conidia of the Genus *Cladosporium*

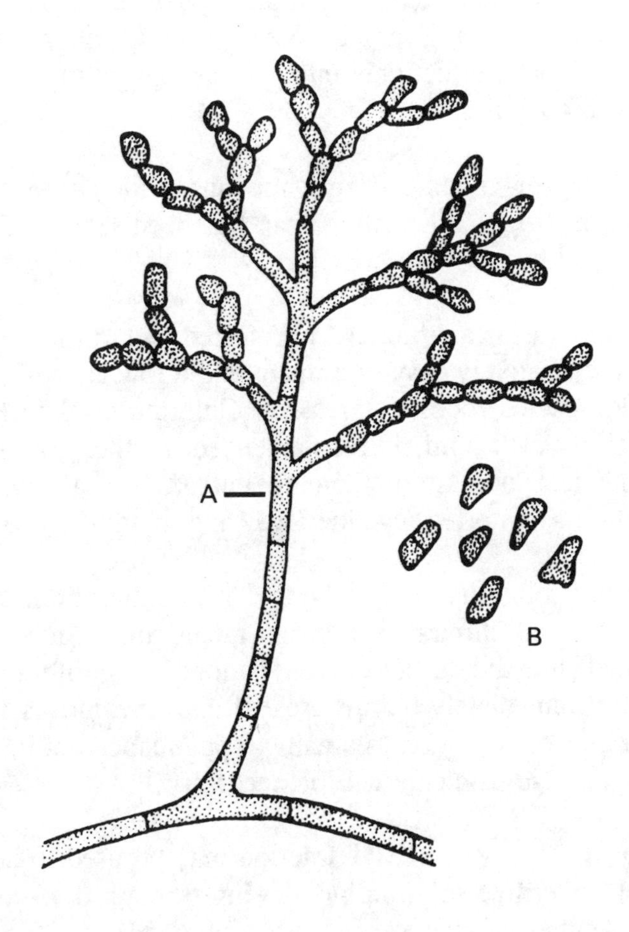

A. Typical dark-pigmented conidiophore of the genus *Cladosporium* arising from septate, dark-pigmented hyphae, and producing chains of dark-pigmented conidia; B. Typical irregularly shaped 1- to 2-celled conidia.

Figure 42

C. ANALYSIS OF SAMPLES

1. Prepare sterile agar medium (250 ml portions in prescription bottles or flasks, autoclaved 15 min at 121°C and 15 psi). Temper to 45 ± 1°C in water bath. Prepare medium well in advance and let solidify before re-melting and tempering. Do not re-melt solidified medium more than once or under pressure. An Arnold steam chest is recommended. Once medium has been tempered, it can be held for 2-3 hr before use, provided water level of water bath is 2-3 cm above surface of agar in aliquot container. Medium of choice is potato dextrose agar, although other media listed above may be used. Potato dextrose-salt agar is especially useful for analyzing samples containing "spreader" moulds (Mucor, Rhizopus, etc.) since the added NaCl effectively inhibits their growth but readily allows detection of other yeast-mould propagules.

2. To inhibit bacterial growth, amend agar medium with either antibiotics or sterile 10% tartaric acid solution (to be done after agar has been tempered and immediately before pouring plates) as follows:

   a. Antibiotics. Use of antibiotics is preferred to tartaric acid solution because stock solutions are relatively easy to prepare and a low agar pH, inhibitory to some yeast and mould species, does not result. Chlortetracycline.HCl, at an agar medium concentration of 40 ppm, is recommended. Other antibiotics may be used (e.g., chloramphenicol and streptomycin) but should always be used at the same concentration as chlortetracycline.HCl, and in addition to it.

      Prepare stock solutions by dissolving 1 g of antibiotic in 100 ml of sterile distilled water and filtering through a 0.45 um membrane. Store stock solutions in dark at 4-8°C. Shelf life should not exceed 1 month. Equilibrate stock solutions to room temperature immediately before use. If agar medium is in 250 ml aliquots, add 1 ml of 100 ml stock solution to obtain 40 ppm concentration. If medium aliquots are greater or less, adjustments will be necessary.

   b. Tartaric acid solution. A 10% solution may be used to adjust agar medium to pH 3.5 ± 0.1. Sterilize solution by filtering through 0.45 μm membrane. Titrate to determine amount of solution needed to adjust pH to 3.5. Type and aliquot volume of medium will affect amount of solution needed. After adding solution to medium, verify pH by letting a portion of medium solidify and checking with pH meter. Do this for every new lot of medium prepared.

3. Prepare food homogenate (Chapter 1) and make appropriate dilutions (Chapter 2). Dilutions to 10^{-6} dilution should suffice.

4. Use sterile cotton-plugged pipet to place 1 ml portions of sample dilutions into prelabeled 15 x 100 mm Petri plates (plastic or glass), and immediately add 20-25 ml tempered agar medium containing either antibiotic(s) or tartaric acid solution. Mix contents by gently swirling plates clockwise, then counterclockwise, taking care to avoid spillage on dish lid. Add agar within 1-2 min after adding dilution. Otherwise, dilution may begin to adhere to dish bottom (especially if sample is high in starch content and dishes are plastic) and may not mix uniformly. Plate each dilution in triplicate, using wide bore pipets. From preparation of first sample dilution to pouring of final plate, no more than 20 min, preferably 10 min, should elapse.

5. Incubate plates in dark at 22-25°C. Do not stack plates higher than 3 and do not invert. Let plates remain undisturbed until time for counting.

6. Count plates after 5 days of incubation. Do not count plates after 3 days since handling of plates could result in secondary growth from dislodged spores, making 5 day counts invalid. Count plates containing 10-150 colonies. If mainly yeasts are present, plates with 150 colonies are usually countable. However, if substantial amounts of mould are present, depending on the type of mould, the upper countable limit may have to be lowered at the discretion of the analyst. Report results in colonies/g or colonies/ml based on an average count of the triplicate set. Round off counts to 2 significant figures. If third digit is 6 or above, round off to digit above (e.g., 456= 460); if 4 or below, round off to digit below (e.g., 454 = 450). If third digit is 5, round off to digit below if first 2 digits are an even number (e.g., 445 = 440); round off to digit above if first 2 digits are an odd number (e.g., 455 = 460)

D. FLOW SHEET

See Figure 43.

E. RECORD SHEET

See Annex 22

F. PRECAUTIONS AND LIMITATIONS OF METHOD

If determining the yeast and mould counts of a food sample, the following points should be observed:

1. The solidified potato dextrose agar should not be re-melted more than once or under pressure.

2. Melted agar should be used within 2-3 hr.

3. In addition to chlortetracycline.HCl, other antibiotics such as chloramphenicol or streptomycin may be used. These antibiotics should be used at the same concentration as chlortetracycline.HCl. Moreover, these antibiotics are used in addition to, rather than in place of, the chlortetracycline.HCl.

4. Antibiotic stock solutions must be equilibrated to room temperature before being added to melted potato dextrose agar.

5. Melted agar should be added to Petri plate within 1-2 min after adding food sample dilution. Otherwise, dilution may adhere to bottom of plate, especially if plate is plastic.

6. No more than 10 min should elapse between preparation of first dilution and pouring of final plate.

7. Plates should not be counted at 3 days since handling could result in displaced spores initiating secondary growth, thus invalidating the counts at 5 days.

Enumeration of Yeasts and Molds

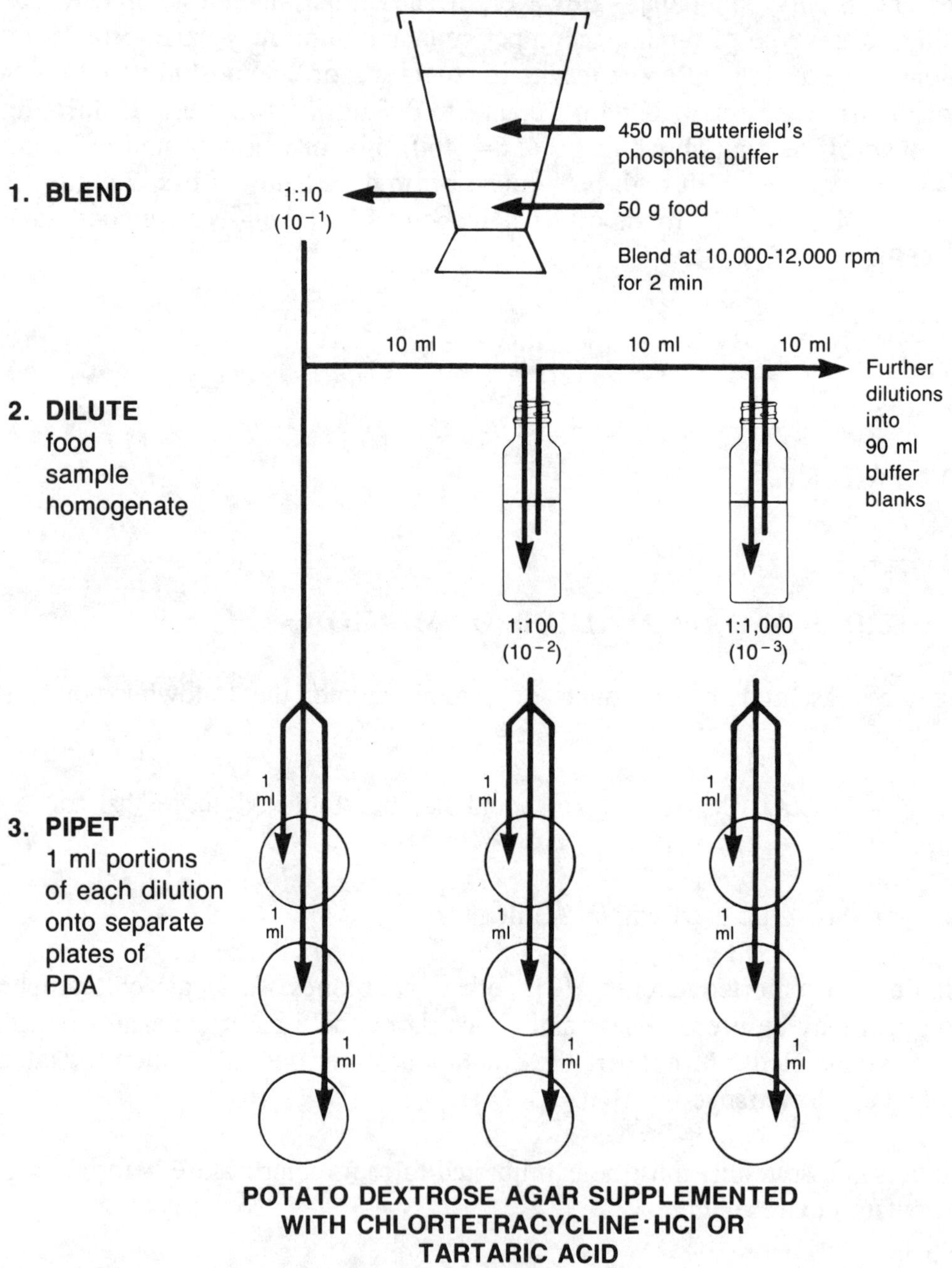

1. **BLEND**

450 ml Butterfield's phosphate buffer

50 g food

Blend at 10,000-12,000 rpm for 2 min

1:10 (10^{-1})

2. **DILUTE** food sample homogenate

10 ml 10 ml 10 ml

Further dilutions into 90 ml buffer blanks

1:100 (10^{-2}) 1:1,000 (10^{-3})

3. **PIPET** 1 ml portions of each dilution onto separate plates of PDA

1 ml 1 ml 1 ml

POTATO DEXTROSE AGAR SUPPLEMENTED WITH CHLORTETRACYCLINE·HCl OR TARTARIC ACID

4. **INCUBATE** in the dark at 22-25°C for 5 days

5. **COUNT** plates containing 10-150 colonies

Figure 43

ENUMERATION OF YEASTS AND MOULDS IN FOODS - DIRECT PLATING

(for Foods Such as Dried Beans, Nuts, Whole Spices, Coffee and Cocoa Beans Which Can Be Handled with a Forceps)

A. EQUIPMENT AND MATERIALS

1. Freezer, -20°C
2. Beakers, sterile, 150 ml
3. Forceps, sterile
4. Arnold steam chest
5. Water bath, 45 $\pm$ 1°C
6. Incubator, 22-25°C

B. MEDIA AND REAGENTS

1. Potato dextrose-salt agar (M80)
2. Antibiotic solution, see C-2a, above
3. NaOCl solution, 5% (commercial bleach solution is adequate)
4. Sterile distilled water

C. ANALYSIS OF NON-SURFACE-DISINFECTED (NSD) FOODS

1. **Before plating.** Hold samples at -20°C for 72 hr to kill mites and other insects that might interfere with analysis.

2. **Preparation of agar plates.** Use potato dextrose-salt agar (containing 75 g NaCl/liter). NaCl inhibits growth of mould "spreaders" and prevents germination of viable seeds which otherwise could cause Petri dish lid and stack disorientation. To tempered agar, add 40 ppm chlortetracycline.HCl (C-2a, above). Into 15 x 100 mm Petri plates (plastic or glass), pour about 30 ml medium and let solidify. Because of prolonged incubation time, more medium is needed for direct plating than for dilution plating in each Petri dish. Prepare 10 plates for each sample to be analyzed. Plates may be prepared in advance, but period between preparation and use should not exceed 24 hr.

3. **Plating of sample.** From each sample, transfer about 50 g into sterile 150 ml beaker. Using 95% ethanol-flamed forceps, place intact food items on surface of solidified agar, 5 items per plate (one in plate center and one in each quadrant), 50 items total per sample. Flame forceps between plating of each item. Use several forceps alternately to avoid overheating. Do not plate visibly mouldy or otherwise blemished items.

4. **Incubation of plates.** Align plates in stacks of 10; identify top and bottom plate of each stack with sample number plus date of plating. Incubate stacks, undisturbed, in dark at 22-25°C for 14-21 days.

5. **Reading of plates.** Determine occurrence of mould in percentages (e.g., if mould emerged from all 50 food items, mouldiness is 100%, if from 32 items, mouldiness is 64%). Determine percent occurrence of individual mould genera and species in like manner. Several Aspergillus species (or species complexes) plus most other foodborne mould genera may be identified directly on above medium by experienced analysts with low power (10-30X) magnification.

D. ANALYSIS OF SURFACE-DISINFECTED (SD) FOODS

Perform disinfection in clean laboratory sink, not stainless steel, free from any acid residues, with tap water running (to inhibit chlorine gas generation). Using rubber gloves, transfer about 50 g of sample into sterile 150 ml beaker. Cover with 5% NaOCl solution for 1 min, swirling beaker contents gently but constantly in clockwise-counterclockwise motion. Decant 5% NaOCl solution and give beaker contents three 1 min sterile distilled water rinses. Prepare plates, plate sample, incubate, and read plates as in C, 2-5, above. Comparison of NSD and SD results from same sample will indicate whether mouldiness was due mainly to surface contamination or to internal invasion and growth.

CHAPTER 20

MEDIA, REAGENTS, DILUENTS AND STAINS

Although this chapter provides the formulation of all the media used by the various methods described in this manual, the use of commercially available dehydrated media is preferred for method uniformity and analyst convenience. However, if a commercial medium is not available, the medium must be made from its individual ingredients.

Methods are usually written in generic (noncommercial) terms so that a particular commercial brand of medium is seldom recommended. Occasionally, however, a particular name is recommended when its superiority to competing brands has been demonstrated. In those instances, the specified brand must be used.

No more than a year's supply of media should be ordered at one time. Climates with high humidities may rapidly cause caking of opened dehydrated media which may affect method performance. In addition, alkaline media may absorb carbon dioxide and change the pH. Whenever possible, media should be ordered with an expiration date on the bottle. If an expiration date is not specified, then the media should not be used beyond one year after its receipt in the laboratory. Media that have become visibly altered by discoloration, clumping, or moisture accumulation, should be discarded.

Chemical compounds are widely used as media components, selective agents, indicators, and stains in various microbiological procedures. Because chemical impurities can either inhibit or stimulate microbial growth, or can otherwise produce an undesirable reaction, only chemicals meeting the specifications of a certifying organization such as the American Chemical Society, or equivalent organization, should be used. Dyes from commercial sources should receive special attention because of the variability from lot to lot in percentage of dye, dye complex, insolubles, and inert substances present. Only dyes certified by the Biological Stain Commission, or equivalent organization, should be used. Unlike most dehydrated media, chemicals are seldom provided with an expiration date. Thus, it is recommended that no more than a 2-year supply be ordered at one time.

The guidelines for determining the shelf life of rehydrated media are not as definite as those for dehydrated media. Several factors determine the length of time that rehydrated media may be stored: volume of dispensed medium, storage temperature, type of container and container closure, and the specific nature of the medium itself. Agar or broth that has been dispensed in tightly closed screw-cap tubes may be stored up to 3 months at 4°C. Agar or broth that has been dispensed in tubes with snap cap closures may be stored up to one week at this same temperature. Refrigerated tubes containing inverted fermentation vials must be examined at room temperature for false-positive gas bubbles before use. Moisture loss in tubes of broth is determined by marking the original fluid level in several tubes of each batch of freshly prepared media. If the estimated loss exceeds 10% of the original volume, these tubes should be discarded. Poured plates with loose-fitting covers in sealed plastic bags may be stored up to 2 weeks at 4°C, whereas large volumes of agar in tightly closed screw-cap flasks or bottles may be stored up to 3 months at this temperature. Without exception, all rehydrated media should be protected from exposure to light.

The holding guidelines discussed above will not apply in all situations. For instance, some selective media may lose their selectivity upon aging and must be made relatively fresh, or on the day of use, whereas other media are required to be made a certain number of days before use. In these instances the general guidelines recommended for holding prepared media are superseded.

There are no uniform guidelines for the storage and shelf life of chemical reagent solutions. Some reagents must be prepared only on the day of use while others may be prepared and stored a few days, or even several weeks, before use; some reagent solutions must be stored under refrigeration, while others may be stored at room temperature (24-26°C); some solutions must be stored in glass-stoppered, air-tight bottles, whereas this precaution would not be critical for other reagent solutions. The one precautionary measure, however, which all of these reagent solutions have in common is that they should be protected from exposure to excessive heat and light. Thus, one should follow the storage conditions recommended by the method being used. If these conditions are not specified, adequate controls must be included to ensure proper performance of the chemical reagent solutions.

The media and reagents described herein have been evaluated and found to be effective for use in a particular method. These media and reagents should be used exactly as specified. Introduction of any modification, however seemingly minor, may affect the results given by the respective method.

MEDIA

M1. Acid Broth

Proteose peptone	5 g
Yeast extract	5 g
Dextrose	5 g
K_2HPO_4	4 g
Distilled water	1 litre

Dissolve ingredients and dispense 12-15 ml portions into 20 x 150 mm tubes. Autoclave 15 min at 121°C. Final pH, 5.0.

M2. Aeromonas hydrophila Medium (AHM)

Proteose peptone	5 g
Yeast extract	3 g
Tryptone	10 g
L-Ornithine-HC1	5 g
Mannitol	1 g
Inositol	10 g
Sodium thiosulfate	0.4 g
Ferric ammonium citrate	0.5 g
Bromcresol purple	0.02 g
Agar	3 g
Distilled water	1 litre

Mix to dissolve; adjust to pH 6.7; heat to boil, dispense 5 ml quantities into 13 x 100 mm tubes, and autoclave at 121°C for 12 min.

M3. Alkaline Peptone Water

Peptone	10 g
NaCl	10 g
Distilled water	1 litre

Adjust pH so that value after sterilization is 8.5 $\pm$ 0.2. Dispense into screw cap tubes. Autoclave 10 min at 121°C.

M4. Anaerobic Egg Yolk Agar

Agar base

Fresh eggs	2
Yeast extract	5 g
Tryptone	5 g
Proteose peptone	20 g
NaCl	5 g
Agar	20 g
Distilled water	1 litre

Autoclave 15 min at 121°C. Final pH, 7.0 ± 0.2.

Treatment of eggs. Wash 2 fresh eggs with stiff brush and drain. Soak in 0.1% mercuric chloride solution for 1 hr. Pour off mercuric chloride solution and replace with 70% ethanol. Soak 30 min. Crack eggs aseptically. Retain yolks. Drain contents of yolk sacs into sterile stoppered graduate and discard sacs. Add yolk to equal volume of sterile 0.85% saline. Invert graduate several times to mix.

Preparation of medium. To 1 litre melted medium (48-50°C) add 80 ml yolk-saline mixture, and mix. Pour plates immediately. After solidification, dry 2-3 days at ambient temperature or at 35°C for 24 hr. Check plates for contamination before use. After drying, plates may be stored for a short period in refrigerator.

M5. Andrade's Carbohydrate Broth and Indicator

Base

Beef extract	3 g
Peptone or gelysate	10 g
NaCl	10 g
Distilled water	1 litre

Adjust pH to 7.2 ± 0.2. Autoclave 15 min at 121°C.

Andrade's indicator

Acid fuchsin	0.2 g
Distilled water	100 ml
1 N NaOH	16 ml

Allow to decolorize before use. Add 1-2 ml NaOH if necessary. Add 10 ml indicator to 1 litre base.

Carbohydrate stock solution. Prepare dextrose, lactose, sucrose, and mannitol in 10% solutions. Prepare dulcitol, salicin, and other carbohydrates in 5% solutions. Sterilize by filtration through 0.20 μm membrane. Dilute sugar solutions 1:10 in base with Andrade's indicator to give recommended concentration. Mix gently.

M6. **Arginine Dihydrolase Medium (Sutter)**

Peptone	2 g
NaCl	5 g
K_2HPO_4	0.3 g
Bromthymol blue	0.03 g
L-Arginine	10 g
Distilled water	1 litre

Heat gently to dissolve. Dispense 5 ml portions into 13 x 100 mm screw-cap tubes. Autoclave 10 min at 121°C. Final pH, 6.0 ± 0.2.

M7. **Baird-Parker Medium**

Basal medium

Tryptone	10 g
Beef extract	5 g
Yeast extract	1 g
Sodium pyruvate	10 g
Glycine	12 g
Lithium chloride.$6H_2O$	5 g
Agar	20 g

Autoclave 15 min at 121°C. Final pH, 7.0 ± 0.2 If desired for immediate use, maintain melted medium at 48-50°C before adding enrichment. Otherwide, store solidified medium at 4 ± 1°C for 48 hr. Melt medium before use.

Enrichment. Bacto EY tellurite enrichment.

Complete medium. Aseptically add 5 ml prewarmed (45-50°C) Bacto EY tellurite enrichment to 95 melted base. Mix well (avoiding bubbles) and pour 15-18 ml portions into sterile 15 x 100 mm Petri dishes. The medium must be densely opaque. Dry plates before use.

M8. **Bile Esculin Agar**

Beef extract	3 g
Peptone	5 g
Esculin	1 g
Oxgall	40 g
Ferric citrate	0.5 g
Agar	15 g
Distilled water	1 litre

Heat with agitation to dissolve. Dispense into tubes, autoclave 15 min at 121°C, and slant until solidified. Final pH, 6.6 ± 0.2.

M9. **Bismuth Sulfite Agar (Wilson and Blair)**

Polypeptone (or peptone)	10 g
Beef extract	5 g
Dextrose	5 g
N_2HPO_4 (anhydrous)	4 g
$FeSO_4$ (anhydrous)	0.3 g
Bismuth sulfite (indicator)	8 g
Brilliant green	0.025 g
Agar	20 g
Distilled water	1 litre

Mix thoroughly and heat with agitation. Boil about 1 min to obtain uniform suspension. (Precipitate will not dissolve.) Cool to 45-50°C. Suspend precipitate by gentle agitation, and pour 20 ml portions into sterile 15 x 100 mm Petri dishes. Let plates dry about 2 hr with lids partially removed; then close plates. Final pH, 7.6 ± 0.2. DO NOT AUTOCLAVE. Prepare plates on day before streaking and store in dark. Selectivity decreases in 48 hr.

M10. **Blood Agar Base (Infusion Agar)**

Heart muscle, infusion from	375 g
Thiotone	10 g
NaCl	5 g
Agar	15 g
Distilled water	1 litre

Heat gently to dissolve. Autoclave 20 min at 121°C. Final pH, 7.3 ± 0.2.

M11. **Brain Heart Infusion (BHI) Agar (0.7%)**
(for staphylococcal enterotoxin)

Prepare a suitable quantity of brain heart infusion broth (M12). Adjust pH to 5.3 with 1 N HC1. Add agar to give 0.7% concentration. Dissolve by minimal boiling. Dispense 25 ml portions into 25 x 200 mm tet tubes. Autoclave 10 min at 121°C.

M12. **Brain Heart Infusion (BHI) Broth and Agar**

Calf brain infusion	200 g
Beef heart infusion	250 g
Proteose peptone or gelysate	10 g
NaCl	5 g
$Na_2HPO_4.12H_2O$	2.5 g
Dextrose	2 g
Distilled water	1 litre

Dissolve ingredients in distilled water with gentle heat. Dispense broth into bottles or tubes for storage. Autoclave 15 min at 121°C. Final pH, 7.4 ± 0.2.

To prepare brain heart infusion agar, add 15 g agar to 1 litre BHI broth. Heat to dissolve agar before dispensing into bottles or flasks. Autoclave 15 min at 121°C.

M13. **Brilliant Green Bile (BGB) 2% Broth**

Peptone	10 g
Lactose	10 g
Oxgall	20 g
Brilliant green	0.0133 g
Distilled water	1 litre

Dissolve peptone and lactose in 500 ml distilled water. Add 20 g dehydrated oxgall dissolved in 200 ml distilled water. The pH of this solution should be 7.0 - 7.5. Mix and add water to make 975 ml. Adjust pH to 7.4. Add 13.3 ml 0.1% aqueous brilliant green in distilled water. Add distilled water to make 1 litre. Dispense into 20 x 150 mm tubes containing inverted 10 x 75 mm fermentation vials, making certain that the fluid level covers the vials completely. Autoclave 15 min at 121°C. Final pH, 7.2 ± 0.1.

M14. **Bromcresol Purple Broth**

Base

Peptone	10 g
Beef extract	3 g
NaCl	5 g
Bromcresol purple	0.04 g
Distilled water	1 litre

Add 5 g carbohydrate per litre of base. Autoclave 10 min at 121°C. Final pH, 7.0 ± 0.2. For use with *V. parahaemolyticus*, add 25 g NaCl per litre.

M15. **Bromcresol Purple Dextrose Broth (BCP)**

Dextrose	10 g
Beef extract	3 g
Peptone	5 g
Bromcresol purple (1.6% in ethanol)	2 ml
Distilled water	1 litre

Dissolve ingredients in distilled water. Dispense 12-15 ml into tubes. Autoclave 15 min at 121°C. Final pH, 7.0 ± 0.2.

M16. **Brucella Broth and Semi-Solid Agar**

Tryptone or trypticase	10 g
Thiotone or peptamine	10 g
Dextrose	1 g
Yeast autolysate	2 g
NaCl	5 g
$NaHSO_3$	0.1 g
Distilled water	1 litre

Suspend ingredients in 1 litre distilled water and mix thoroughly. Autoclave 15 min at 121°C. Final pH, 7.0 $\pm$ 0.2.

To prepare Brucella semi-solid agar, add 1.8 g agar and 10 ml neutral red solution (see M17) to above ingredients. Heat with agitation to dissolve agar. Boil 1 min. Autoclave 15 min at 121°C. Final pH, 7.0 $\pm$ 0.2.

M17. **Campylobacter Biochemical Screening Agar**

Base

Tryptone	10 g
Peptamine	10 g
Glucose	1 g
Yeast extract	2 g
NaCl	5 g
Sodium bisulfite	0.1 g
Agar	1.8 g
Distilled water	1 litre

Dissolve ingredients in distilled water by heating to boiling. Cool slightly and divide into four 250 ml portions.

Neutral red solution

Dissolve 0.2 g neutral red in ethanol in a 100 ml volumetric flask. Add distilled water to bring volume up to etched mark. Add 2.5 ml neutral red solution to 3 of the 250 ml volumes of base, above.

Biochemcials

Glycine, 1%
NaCl, 3.5%
Cysteine, 0.02%
Nitrite, 1%

Add 2.5 g glycine to one of the 250 ml portions of base with neutral red solution, 0.75 g NaCl to a second portion with neutral red solution, and 0.05 g cysteine to a third portion with neutral red portion. Add 2.5 g KNO_3 to the fourth portion without neutral red solution.

Adjust pH values to 7.0 $\pm$ 0.2 and dispense 7-10 ml per 16 x 125 mm tube. Autoclave 15 min at 121°C.

M18. Campylobacter Confirmation Broth

Formulation No. 1

Campylobacter enrichment broth (base)	5 ml
FBP concentrate-low concentration (see M19)	0.1 ml

Aseptically combine components in a sterile 50 ml Erlenmeyer flask.

Formulation No. 2

Campylobacter enrichment broth (base)	5 ml
FBP concentrate-low concentration (see M19)	0.1 ml
Fetal bovine serum	0.5 ml

Filter serum through a 0.22 μm membrane. Aseptically combine components in a sterile 50 ml Erlenmeyer flask. Filtered serum may be stored in polypropylene tubes at -20°C.

M19. Campylobacter Enrichment Broth

Base

Lab-Lemco powder (Oxoid L29)	10 g
Peptone	10 g
NaCl	5 g
Yeast extract	6 g
Distilled water	950 ml

Dispense into appropriate-sized flasks or bottles. Autoclave 15 min at 121°C. Final pH, 7.5 ± 0.2.

FBP concentrate-high concentration

Ferrous sulfate	6.25 g
Sodium metabisulfite	6.25 g
Sodium pyruvate	6.25 g

Combine ingredients in a 100 ml volumetric flask and bring level to etched mark with distilled water. Sterilize by filtration through 0.45 μm membrane and freeze (-20°C) in 15-25 ml portions in sterile polypropylene containers for up to 2 months. Thawed concentrate may be refrozen one time. Protect from light.

FBP concentrate-low concentration

Ferrous sulfate	0.125 g
Sodium metabisulfite	0.125 g
Sodium pyruvate	0.125 g

Combine ingredients in a 5 ml volumetric flask and bring level to etched mark with distilled water. Sterilize by filtration through 0.45 μm membrane and freeze (-20°C) 0.5 ml portions in sterile polypropylene containers for up to 2 months.

Lysed horse or sheep blood

Only use fresh blood and lyse upon receipt. To lyse blood, gently mix and pour approximately 100 ml per sterile polypropylene bottle. Place in freezer. After blood has frozen, thaw, and refreeze. Frozen storage inhibits hemoglobin's absorption of oxygen. Use blood up to 6 months. It may be thawed and refrozen several times.

Antibiotic Supplement No. 1

Sodium cefoperazone	30 mg
Trimethoprim lactate	12.5 mg
Vancomycin	10 mg
Cycloheximide	100 mg

Dissolve 0.75 g sodium cefoperazone in 100 ml distilled water and filter through 0.45 μm membrane. This solution can be stored up to 5 days at 4°C, 3 weeks at -20°C (in polypropylene containers) and 5 months at -70°C. Store the powder at -20°C.

Dissolve 0.3125 g trimethoprim lactate in 100 ml distilled water and filter through 0.45 μm membrane. This solution may be stored up to one year at 4°C.

Dissolve 0.25 g vancomycin in 100 ml distilled water and filter through 0.45 μm membrane. This solution may be stored up to 2 months at 4°C.

Dissolve 2.5 g cycloheximide in 20-30 ml 70% ethanol in 100 ml volumetric flask. Bring volume up to graduated mark with distilled water and sterilize by filtration through 0.45 μm membrane. This solution may be stored up to 1 year at 4°C. Do not freeze.

Antibiotic Supplement No.2

Sodium cefoperazone	15 mg
Trimethoprim lactate	12.5 mg
Vancomycin	10 mg
Cycloheximide	100 mg

Dissolve 0.375 g sodium cefoperazone in 100 ml distilled water and filter through 0.45 μm membrane.

Trimethoprim lactate, vancomycin, and and cycloheximide solutions are prepared as in Antibiotic Supplement No. 1.

Antibiotic Supplement No. 3

Rifampicin	10 mg
Sodium cefoperazone	15 mg
Trimethoprim lactate	12.5 mg
Cycloheximide	100 mg

Dissolve 0.25 g rifampicin in 50 ml 70% ethanol in 100 ml volumetric flask. Bring volume up to graduated mark with distilled water and sterilize by filtration through 0.45 μm membrane. The rifampicin solution and powder may be stored at -20°C. The solution is stable up to one year. Sodium cefoperazone, trimethoprim lactate, and cycloheximide solutions are prepared as in Antibiotic Supplement No. 2.

Complete medium

Base	950 ml
FBP concentrate-high concentration	4 ml
Lysed horse or sheep blood	50 ml
Individual antibiotic solutions comprising appropriate antibiotic supplement (except as noted below)	4 ml

NOTE:

Sodium cefoperazone at an initial concentration of 30 mg/litre enrichment broth may be inhibitory to Campylobacter organisms. Thus, add the total volume of 4.0 ml in 2 portions. Add a volume of 2.0 ml of the sodium cefoperazone solution to the enrichment medium initially. Add a second volume of 2.0 ml of this antibiotic solution after the 4 hr preenrichment period (for samples refrigerated for 10 days or less) or after the 3 hr preenrichment period (for samples that have been stored for more than 10 days or have been heated or frozen).

Aseptically combine components and dispense into appropriate containers.

M20. Campylobacter Isolation Agar

Base

Nutrient Broth No. 2	25 g
Charcoal, bacteriological grade	4 g
Casein hydrolysate	3 g
Sodium deoxycholate	1 g
Casein hydrolysate	3 g
Ferrous sulfate	0.25 g
Sodium pyruvate	0.25 g
Agar	12 g
Yeast extract	2 g
Distilled water	1 litre

Autoclave 15 min at 121°C. Final pH, 7.4 ± 0.2. After autoclaving and prior to dispensing agar into plates, add an additional 100 ml warmed sterile distilled water and 4.4 ml of the sodium cefoperazone solution and 4.4 ml of the cycloheximide solution, below. Scrape the foam off the last plate poured with a sterile tongue depressor. Flaming the plate will precipitate the charcoal. Keep agar plates protected from light during storage.

Antibiotic supplement

Sodium cefoperazone'	30 mg
Cycloheximide	100 mg

Dissolve 0.75 g sodium cefoperazone in 100 ml distilled water. Sterilize by filtration through 0.45 µm membrane.

Dissolve 2.5 g cycloheximide in 20-30 ml 70% ethanol in 100 ml volumetric flask. Bring volume up to graduated mark with distilled water and sterilize by filtration through 0.45 µm membrane.

M21. Cefsulodin-Irgasan Novobiocin (CIN) Agar or Yersinia Selective Agar (YSA)

A. Basal medium

Special peptone	20 g
Yeast extract	2 g
Mannitol	20 g
Pyruvic acid (Na salt)	2 g
NaCl	1 g
$MgSO_4.7H_2O$ (10 mg/ml)	1 ml
Agar	12 g
Distilled water	756 ml

B. Irgasan (Ciba-Geigy) solution

Irgasan 0.40% in 95% ethanol	1 ml

May be stored at -20°C up to weeks.

C. Desoxycholate solution

Sodium desoxycholate	0.5 g
Distilled water	200 ml

Bring to boil with stirring; cool to 50-55°C.

D. Sodium hydroxide, 5 N — 1 ml

E. Neutral red, 3 mg/ml — 10 ml

F. Crystal violet, 0.1 mg/ml — 10 ml

G. Cefsulodin (Abbott Labs) 1.5 mg/ml — 10 ml
May be stored at -70°C. Thaw to room temperature just before use.

H. Novobiocin, 0.25 mg/ml — 10 ml

I. Strontium chloride, 10%; filter-sterilized — 10 ml

Preparation

Add ingredients for solution A (basal medium) to water and bring to boil with stirring. Cool to about 80°C (10 min in 50°C water bath). Add solution B (Irgasan) and mix well. Cool to 50-55°C. Add solution C (desoxycholate); solution should remain clear. Add solutions D through H. Slowly add solution I with stirring. Adjust pH to 7.4 with 5 N NaOH. Dispense 15-20 ml into each Petri dish.

Commercially prepared dehydrated Yersinia selective agar (Difco) with supplements may be substituted. Follow manufacturer's instructions for preparation.

M22. **Chopped Liver Broth**

Fresh beef liver	500 g
Peptone	10 g
K_2HPO_4	1 g
Soluble starch	1 g
Distilled water	1 litre

Grind liver into the water. Heat to boiling and simmer 1 hr. Cool, adjust pH to 7.0, and boil 10 min. Filter through cheesecloth and press out excess liquid. Add other ingredients and adjust pH to 7.0. Add water to make 1 litre. Filter through coarse filter paper. Store broth and meat separately in freezer. To 18 x 150 or 20 x 150 mm test tubes, add chopped liver to depth of 1.2-2.5 cm and 10-12 ml of broth. Autoclave 15 min at 121°C.

M23. **Christensen's Urea Agar**

Base

Peptone	1 g
NaCl	5 g
Dextrose	1 g
KH_2PO_4	2 g
Phenol red (6 ml of 1:500 solution)	0.012 g
Agar	15 g
Distilled water	900 ml

Urea concentrate

Urea	20 g
Distilled water	100 ml

Adjust pH to 6.8 $\pm$ 0.1. Sterilize by filtration. Dissolve agar in 900 ml distilled water, add other ingredients, and autoclave 15 min at 121°C. Cool to 50 $\pm$ 2°C; then add urea concentrate. Mix and distribute to sterile 16 x 125 mm tubes. Slant to obtain deep butt.

M24. Cooked Meat Medium

Beef heart	454 g
Proteose peptone	20 g
Dextrose	2 g
NaCl	5 g
Distilled water	1 litre

Follow directions as in M22, or suspend 12.5 g commercial dehydrated cooked medium in 100 ml cold distilled water. Mix and let stand 15 min to wet particles thoroughly. Or distribute 1.25 g into 20 x 150 mm test tubes, add 10 ml cold distilled water, and mix thoroughly to wet all particles. Autoclave 20 min at 121°C. Final pH, 7.2.

M25. Decarboxylase Basal Medium (Agrinine, Lysine, Ornithine)

Base

Peptone or gelysate	5 g
Yeast extract	3 g
Dextrose	1 g
Bromcresol purple	0.02 g
Distilled water	1 litre

For arginine broth, add 5 g L-arginine to 1 litre base; for lysine (Falkow) broth, add 5 g L-lysine to 1 litre base; for ornithine, add 5 g L-ornithine to 1 litre base. Adjust pH so that value after sterilization is 6.5 $\pm$ 0.2. Autoclave 10 min at 121°C. For control, use unsupplemented base.

M26. EC Medium

Tryptose or trypticase	20 g
Bile salts No. 3	1.5 g
Lactose	5 g
K_2HPO_4	4 g
KH_2PO_4	1.5 g
NaCl	5 g
Distilled water	1 litre

Distribute 8 ml portions into 16 x 150 mm test tubes containing inverted 10 x 75 mm fermentation vials. Autoclave 15 min at 121°C. Final pH, 6.9 $\pm$ 0.2.

M27. Egg Yolk Emulsion, 50%

Wash fresh eggs with a stiff brush and drain. Soak eggs 1 hr in 0.1% $HgCl_2$. Drain solution, replace with 70% ethanol, and soak 30 min. Drain ethanol. Crack eggs aseptically and discard whites. Remove egg yolks with sterile syringe or wide-mouth pipet. Place yolks in sterile container and mix aseptically with equal volume of sterile 0.85% saline. Store at 4°C until use.

M28. **Esculin Agar, Modified**

Infusion agar	40 g
Esculin	1 g
Ferric citrate	0.5 g
Distilled water	1 litre

Heat with agitation to dissolve agar. Cool to 55°C. Adjust pH to 7.0 ± 0.2. Dispense 4 ml portions to 13 x 100 mm tubes. Autoclave 15 min at 121°C. Slant tubes.

M29. **Gelatin Agar Medium**

Gelatin	30 g
Trypticase	10 g
Yeast extract	1 g
Sodium taurocholate	5 g
NaCl	10 g
Agar	15 g
Distilled water	1 litre

Dissolve by boiling and adjust pH to 8.5. Autoclave 15 min at 121°C.

M30. **Gelatinase Solution, 5%**

Gelatinase	5 g
Distilled water	100 g

Suspend gelatinase in distilled water. Centrifuge 10 min at 9,500 rpm and sterilize by filtration through 0.45 μm membrane. Dispense 100 ml portions into bottles.

M31. **Glucose Salt Teepol Broth (GSTB)**

	Single strength	Double strength
Beef extract	3 g	6 g
Peptone	10 g	20 g
NaCl	30 g	60 g
Glucose	5 g	10 g
Methyl violet	0.002 g	0.004 g
Teepol™	4 ml	8 ml
Distilled water	1 litre	1 litre

Dispense 10 ml single strength portions to 20 x 150 mm tubes. Dispense 10 ml double strength portions to tubes large enough to receive 10 ml of sample. If larger quantities of sample are to be examined (25 g), use screw-cap jars containing 225 ml single strength broth. Autoclave 15 min at 121°C. Final pH, 7.4 ± 0.2.

M32. GPS Medium

Gelatin	10 g
NaCl	10 g
K_2HPO_4	5 g
Distilled water	1 litre
Agar (if solid medium preferred)	10 g

Heat with agitation to dissolve components. Autoclave 15 min at 121°C.

M33. Heart Infusion Agar

Heart infusion	500 g
Tryptose	10 g
Sodium chloride	5 g
Agar	15 g
Distilled water	1 litre

Heat gently to dissolve. Fill 16 x 150 mm tubes about 1/3 full. Autoclave 15 min at 121°C. Before medium solidifies, incline tubes to obtain 4-5 cm slants and 2-3 cm butts. Final pH, 7.4 ± 0.2.

M34. Hektoen Enteric (HE) Agar

Peptone	12 g	Sodium thiosulfate	5 g
Yeast extract	3 g	Ferric ammonium citrate	1.5 g
Bile salts	9 g	Bromthymol blue	0.064 g
Lactose	12 g	Acid fuchsin	0.1 g
Sucrose	12 g	Agar	13.5 g
Salicin	2 g	Distilled water	1 litre
NaCl	5 g		

Heat to boiling with frequent agitation to dissolve. Boil no longer than 1 min. Do not overheat. Cool in water bath. Pour 20 ml portions into sterile 15 x 100 mm Petri dishes. Let dry 2 hr with lids partially removed. Final pH, 7.6 ± 0.2. Do not store more than 1 day.

M35. Hugh-Leifson Glucose Broth (HLGB)

Peptone	2 g
Yeast extract	0.5 g
NaCl	30 g
Dextrose	10 g
Bromcresol purple	0.015 g
Agar	3 g
Distilled water	1 litre

Heat with agitation to dissolve agar. Adjust pH to 7.4 ± 0.2. Autoclave 15 min at 121°C.

M36. **Iron Milk Medium (Modified)**

Fresh whole milk	1 litre
Ferrous sulfate.$7H_2O$	1 g
Distilled water	50 ml

Dissolve ferrous sulfate in 50 ml distilled water. Add slowly to 1 litre milk and mix with magnetic stirrer. Dispense 11 ml medium into 16 x 150 mm culture tubes. Autoclave 15 min at 118°C. Prepare fresh medium before use.

M37. **Kligler Iron Agar**

Polypeptone peptone	20 g
Lactose	20 g
Dextrose	1 g
NaCl	5 g
Ferric ammonium citrate	0.5 g
Sodium thiosulfate	0.5 g
Agar	15 g
Phenol red	0.025 g
Distilled water	1 litre

Heat with agitation to dissolve. Dispense into 13 x 100 mm screw-cap tubes and autoclave 15 min at 121°C. Cool and slant to form deep butts. Final pH, 7.4 $\pm$ 0.2.

M38. **Koser's Citrate Medium**

$NaNH_4HPO_4.4H_2O$	1.5 g
K_2HPO_4	1 g
$MgSO_4.7H_2O$	0.2 g
Sodium citrate.$2H_2O$	3 g
Distilled water	1 litre

Dispense into test tubes as desired. Autoclave 15 min at 121°C. Final pH, 6.7 $\pm$ 0.2. This formulation is listed in Official Methods of Analysis of the AOAC. It differs from the composition of commercially available dehydrated media. The latter have been found to be satisfactory.

M39. **Lactose Broth**

Beef extract	3 g
Peptone	5 g
Lactose	5 g
Distilled water	1 litre

For E. coli: Dissolve ingredients and dispense 10 ml portions into 20 x 150 mm tubes containing inverted 10 x 75 mm fermentation vials. Autoclave 15 min at 121°C. Final pH, 6.9 $\pm$ 0.2.

For Salmonella: Dispense 225 ml portions into 500 ml Erlenmeyer flasks. After autoclaving 15 min at 121°C and just before use, aseptically adjust volume to 225 ml. Final pH, 6.9 $\pm$ 0.2.

M40. Lactose-Gelatin Medium (for C. perfringens)

Tryptose	15 g
Yeast extract	10 g
Lactose	10 g
Phenol red (as solution)	0.05 g
Gelatin	120 g
Distilled water	1 litre

Heat to dissolve tryptose, yeast extract, and lactose in 400 ml water. Suspend gelatin in 600 ml water and heat at 50-60°C with agitation to dissolve. Mix 2 solutions. Adjust pH to 7.5 ± 0.2. Add phenol red and mix. Dispense 10 ml portions into 16 x 150 mm screw-cap tubes. Autoclave 10 min at 121°C. If not used within 8 hr, deaerate by heating at 50-70°C for 2-3 hr before use.

M41. Lauryl Tryptose (LT) Broth

Tryptose or trypticase	20 g
Lactose	5 g
K_2HPO_4	2.75 g
KH_2PO_4	2.75 g
NaCl	5 g
Sodium lauryl sulfate	0.1 g
Distilled water	1 litre

Dispense 10 ml portions into 20 x 150 mm tubes containing inverted 10 x 75 mm fermentation tubes. Autoclave 15 min at 121°C. Final pH, 6.8 ± 0.2.

M42. Lauryl Tryptose - MUG (LT-MUG) Broth

Tryptose or trypticase	20 g
Lactose	5 g
K_2HPO_4	2.75 g
KH_2PO_4	2.75 g
NaCl	5 g
Sodium lauryl sulfate	0.1 g
4-Methylumbelliferyl beta-D-glucuronide (MUG)	50 mg
Distilled water	1 litre

Prepare LT broth and add MUG. Dissolve with gentle heat, if necessary. Dispense 10 ml portions to 20 x 150 mm tubes containing inverted 10 x 75 mm fermentation vials. Autoclave 15 min at 121°C. Final pH, 6.8 ± 0.2.

M43. Levine's Eosin-Methylene Blue (L-EMB) Agar

Peptone	10 g
Lactose	10 g
K_2HPO_4	2 g
Agar	15 g
Eosin Y	0.4 g
Methylene blue	0.065 g
Distilled water	1 litre

Boil to dissolve peptone, phosphate, and agar in 1 litre of water. Add water to make original volume. Dispense in 100 ml or 200 ml portions and autoclave 15 min at not over 121°C. Final pH, 7.1 ± 0.2. Before use, melt, and to each 100 ml portion add (a) 5 ml sterile 20% lactose solution, (b) 2 ml aqueous 2% eosin Y solution, and (c) 4.3 ml 0.15% aqueous methylene blue solution. When using complete dehydrated product, boil to dissolve all ingredients in 1 litre water. Dispense in 100 or 200 ml portions and autoclave 15 min at 121°C. Final pH, 7.1 ± 0.2.

M44. Listeria Enrichment Broth (LEB)

TSB-YE (see M107) supplemented with:

Acriflavin HC1 (Sigma)	15 mg/L
Nalidixic acid (sodium salt) (Sigma)	40 mg/L
Cycloheximide (Sigma)	50 mg/L

Add the three supplementary ingredients aseptically to TSB-YE after autoclaving and just before use. Make acriflavin and nalidixic supplements as 0.5% stock solutions in distilled water. Make cycloheximide supplement as 1.0% stock solution in a 40% solution of ethanol in water. Filter-sterilize all 3 supplementary ingredients. Add 0.68 ml acriflavin stock solution, 1.8 ml nalidixic stock solution, and 1.15 ml cycloheximide stock solution to 225 ml TSB-YE to achieve the stated amounts in mg/L. LEB is commercially available. (Note: Cycloheximide is very toxic. Care should be exercised in its disposal, especially if large volumes are used. Refer to a current edition of the Merck Index (Merck Co. Inc., Rahway, NJ, USA) for information on its stability properties in relation to disposal methods.)

M45. Lithium Chloride-Phenylethanol-Moxalactam (LPM) Medium

Phenylethanol agar (Difco)	35.5 g
Glycine anhydride (Note: not glycine)	10 g
Lithium chloride	5 g
Moxalactam stock solution, 1% in pH 6.0 phosphate buffer	2 ml
Distilled water	1 litre

Sterilize medium (without moxalactam) at 121°C for 15 min. Cool to 48-50°C and add filter-sterilized moxalactam solution.

Moxalactam stock solution consists of 1 g moxalactam salt (ammonium or sodium) in 100 ml 0.1 M potassium phsophate buffer, pH 6.0. Store filter-sterilized stock solution frozen in 2-ml aliquots.

Moxalactam (E. Lilly Co.) is retailed by Sigma Chemical Co. The LPM medium is most effective in the Henry illumination system when poured thin, i.e., 12-15 ml per standard Petri dish. The greater tendency of thin agar to dry is avoided by refrigeration and/or rapid use of the plates. LPM basal medium is commercially available as a powder.

M46. **Liver-Veal Agar**

Liver, infusion from	50 g	Gasein, isoelectric	2 g
Veal, infusion from	500 g	NaCl	5 g
Proteose peptone	20 g	Sodium nitrate	2 g
Neopeptone	1.3 g	Gelatin	20 g
Tryptone	1.3 g	Agar	15 g
Dextrose	5 g	Distilled water	1 litre
Starch, soluble	10 g		

Heat with agitation to dissolve. Autoclave 15 min at 121°C. Fial pH 7.3 $\pm$ 0.2.

For **Liver-Veal-Egg Yolk Agar**: To each 500 ml of melted liver-veal agar, add 40 ml egg yolk-saline suspension (see M27). Mix thoroughly and pour 20 ml portions into sterile 15 x 100 mm Petri dishes. Dry plates at room temperature for 2 days or at 35°C for 24 hr. Check plates for sterility and store sterile plates in refrigerator.

M47. **Long-term Preservation Medium**

Yeast extract, 0.3%	3 g
Peptone	10 g
NaCl	30 g
Agar	3 g
Distilled water	1 litre

Heat to dissolve ingredients. Dispense 4 ml portions to 30 x 100 mm screw-cap tubes. Autoclave 15 min at 121°C. Cool and tighten caps for storage. No pH adjustment is necessary.

M48. **Lysine Arginine Iron Agar**

Peptone	5 g
Yeast extract	3 g
Glucose	1 g
L-Lysine	10 g
L-Arginine	10 g
Ferric ammonium citrate	0.5 g
Sodium thiosulfate	0.04 g
Bromcresol purple	0.02 g
Agar	15 g

Adjust pH to 6.8. Heat to boiling and dispense 5 ml into each 13 x 100 mm screw-cap culture tube. Autoclave at 121°C for 12 min. Cool tubes in slanted position. (This medium may also be prepared by supplementing Difco lysine iron agar with 10 g L-arginine per litre.)

M49. **Lysine Decarboxylase Broth (Falkow)**
(for Salmonella)

Gelysate or peptone	5 g
Yeast extract	3 g
Glucose	1 g
L-Lysine	5 g
Bromcresol purple	0.02 g
Distilled water	1 litre

Heat until dissolved. Dispense 5 ml portions into 16 x 125 mm screw-cap tubes. Autoclave loosely capped tubes 15 min at 121°C. Screw the caps on tightly for storage and after inoculation. Final pH, 6.5-6.8.

M50. **Lysine Iron Agar (Edwards and Fife)**

Gelysate or peptone	5 g
Yeast extract	3 g
Dextrose	1 g
L-Lysine	10 g
Ferric ammonium citrate	0.5 g
Sodium thiosulfate (anhydrous)	0.04 g
Bromcresol purple	0.02 g
Agar	15 g
Distilled water	1 litre

Heat to dissolve ingredients. Dispense 4 ml portions into 13 x 100 mm screw-cap tubs. Autoclave 12 min at 121°C. Let solidify in slanted position to form 4 cm butts and 2.5 cm slants. Final pH, 6.7 $\pm$ 0.2.

M51. **Lysozyme Broth**

Base. Prepare nutrient broth as recommended. Dispense 99 ml portions to 170 ml bottles. Autoclave 15 min at 121°C. Cool to room temperature before use.

Lysozyme solution. Dissolve 0.1 g lysozyme in 65 ml sterile 0.01 N HC1. Heat to boiling for 20 min. Dilute to 100 ml with sterile 0.01 N HC1. Alternatively, dissolve 0.1 g lysozyme in 100 ml distilled water. Sterilize by filtration through 0.45 μm membrane. Test for sterility before use. Add 1 ml lysozyme solution to 99 ml nutrient broth. Mix and dispense 2.5 ml portions to sterile 13 x 100 mm tubes.

M52. MacConkey Agar

Proteose peptone or polypeptone	3 g
Peptone or gelysate	17 g
Lactose	10 g
Bile salts No. 3 (or bile salts mixture)	1.5 g
NaCl	5 g
Neutral red	0.03 g
Crystal violet	0.001 g
Agar	13.5 g
Distilled water	1 litre

Suspend ingredients and heat with agitation to dissolve. Boil 1-2 min. Autoclave 15 min at 121°C, cool to 45-50°C, and pour 20 ml portions into sterile 15 x 100 mm Petri dishes. Dry at room temperature with lids closed. DO NOT USE WET PLATES. Final pH, 7.1 ± 0.2.

M53. Malonate Broth

Yeast extract	1 g
$(NH_4)_2SO_4$	2 g
K_2HPO_4	0.6 g
KH_2PO_4	0.4 g
NaCl	2 g
Sodium malonate	3 g
Dextrose	0.25 g
Bromthymol blue	0.025 g
Distilled water	1 litre

Dissolve by heating, if necessary. Dispense 3 ml portions into 13 x 100 mm test tubes. Autoclave 15 min at 121°C. Final pH, 6.7 ± 0.2.

M54. Malt Extract Agar

Malt extract	30 g
Agar	20 g
Distilled water	1 litre

Boil to dissolve ingredients. Autoclave 15 min at 121°C. Dispense 20-25 ml into sterile 15 x 100 mm Petri dishes. Final pH, 5.5 ± 0.2.

M55. Malt Extract Broth

Malt extract	15 g
Distilled water	1 litre

Adjust pH to 4.7 ± 0.2. Dispense into tubes. Autoclave 15 min at 121°C. Do not overheat.

M56. Mannitol-Egg Yolk-Polymyxin Agar

Base

Beef extract	1 g
Peptone	10 g
Mannitol	10 g
NaCl	10 g
Phenol red	0.025 g
Agar	15 g
Distilled water	900 ml

Heat with agitation to dissolve agar. Adjust pH so that value after sterilization is 7.2 ± 0.2. Dispense 225 ml portions to 500 ml Erlenmeyer flasks. Autoclave 15 min at 121°C. Cool to 50°C.

Polymyxin B solution, 0.1%. Dissolve 500,000 units sterile polymyxin B sulfate powder in 50 ml sterile distilled water. Store in 85 ml bottle at 4°C.

Egg yolk emulsion, 50% (see M27)

Final medium. To 225 ml melted base add 2.5 ml polymyxin B solution and 12.5 ml egg yolk emulsion. Mix and dispense 18 ml portions to sterile 15 x 100 mm Petri dishes. Dry plates at room temperature for 24 hr before use.

M57. Motility Medium (for B. cereus)

Trypticase	10 g
Yeasts extract	2.5 g
Dextrose	5 g
Na_2HPO_4	2.5 g
Agar	3 g
Distilled water	1 litre

Heat with agitation to dissolve agar. Dispense 100 ml portions to 170 ml bottles. Autoclave 15 min at 121°C. Final pH, 7.4 ± 0.2. Cool to 50°C. Aseptically dispense 2 ml portions to sterile 13 x 100 mm tubes. Store at room temperature 2 days before use.

M58. Motility-Nitrate Medium, Buffered (for C. perfringens)

Beef extract	3 g
Peptone (Difco)	5 g
KNO_3	1 g
Na_2HPO_4	2.5 g
Agar	3 g
Galactose	5 g
Glycerin (reagent grade)	5 ml
Distilled water	1 litre

Dissolve all ingredients except agar. Adjust pH to 7.3 ± 0.1. Add agar and heat to dissolve. Dispense 11 ml portions into 16 x 150 mm tubes. Autoclave 15 min at 121°C. If not used within 4 hr, heat 10 min in boiling water or flowing steam. Chill in cold water.

M59. **Motility Test Medium (Semi-solid)**

Beef extract	3 g
Peptone or gelysate	10 g
NaCl	5 g
Agar	4 g
Distilled water	1 litre

Heat with agitation and boil 1-2 min to dissolve agar. Dispense 8 ml portions into 16 x 150 screw-cap tubes. Autoclave 15 min at 121°C. Final pH, 7.4 ± 0.2.

Fore treatment of Salmonella cultures giving negative flagellar (H) test. Dispense 20 ml portions into 20 x 150 mm screw-cap tubes, replacing caps loosely. Autoclave 15 min at 121°C. Cool to 45°C after autoclaving. Tighten caps, and refrigerate at 5-8°C. To use, remelt in boiling water or flowing steam, and cool to 45°C. Aseptically dispense 20 ml portions into sterile 15 x 100 mm Petri plates. Cover plates and let solidify. USe same day as prepared. Final pH, 7.4 ± 0.2.

M60. **MOX Agar**

Blood agar base (M10)	40 g
0.25 M Magnesium chloride (50.83 g $MgCl_2.6H_2O$/litre); filter-sterilized	80 ml
0.25 M Sodium oxalate (33.5 g/litre); filter-sterilized	80 ml
1 M D-Glucose (180 g/litre); filter-sterilized	10 ml
Distilled water	830 ml

Dissolve blood agar base in distilled water. Autoclave at 121°C for 15 min. Temper to 50-55°C. Add filter-sterilized solutions and stir. Dispense in sterile Petri plates.

M61. **MR-VP Medium**

Buffered peptone	7 g
Glucose	5 g
K_2HPO_4	5 g
Distilled water	1 litre

Dissolve ingredients in 800 ml water with gentle heat. Filter, cool to 20°C, and dilute to 1 litre. Autoclave 12-15 min at 121°C. Final pH, 6.9 ± 0.2.

For use with V. parahaemolyticus, add 30 g NaCl per litre.

For Salmonella: Dispense 10 ml into 16 x 150 mm test tubes, and autoclave 12-15 min at 121°C.

M62. **Mueller-Hinton Agar**

Beef, infusion from	300 g
Acidicase peptone (BBL) or casamino acids (Difco)	17.5 g
Starch	1.5 g
Agar	17 g
Distilled water	1 litre

Heat to boiling for 1 min. Autoclave 15 min at 116°C. Final pH, 7.3 ± 0.2.

M63. **Nitrate Broth**

Beef extract	3 g
Peptone	5 g
KNO_3 (nitrite-free)	1 g
Distilled water	1 litre

Dissolve ingredients. Dispense 5 ml portions into 16 x 125 mm tubes. Autoclave 15 min at 121°C. Final pH, 7.0 0.2.

M64. **Nitrate Reduction Medium**

Prepare Nitrate Reduction Medium as described in M63. For nitrate reduction test, follow directions described in R23.

M65. **Nonfat Dry Milk (Reconstituted)**

Nonfat dry milk	100 g
Distilled water	1 litre

Suspend 100 g dehydrated nonfat dry milk in 1 litre distilled water. Swirl until dissolved. Dispense 225 ml portions into 500 ml Erlenmeyer flasks. Autoclave 15 min at 121°C. After sterilization and just before use, aseptically adjust final volume to 225 ml.

M66. **Nutrient Agar**

Beef extract	3 g
Peptone	5 g
Agar	15 g
Distilled water	1 litre

Heat to boiling to dissolve ingredients. Dispense into tubes or flasks. Autoclave 15 min at 121°C. Final pH, 6.8 ± 0.2. If used as base for blood agar, add 8 g NaCl to prevent hemolysis of blood cells.

M67. **Nutrient Agar (for B. cereus)**

For slants, prepare nutrient agar and dispense 6.5 ml portions into 16 x 125 mm screw-cap tubes. Autoclave 15 min at 121°C. Slant tubes until medium solidifies. For plates, dispense 100-500 ml portions into bottles or flasks and autoclave. Cool to 50°C and dispense 18-20 ml into sterile 15 x 100 mm Petri dishes. Dry plates for 24-48 hr at room temperature before use.

M68. **Nutrient Broth**

Beef extract	3 g
Peptone	5 g
Distilled water	1 litre

Heat to dissolve. Dispense 10 ml portions into tubes or 225 ml portions into 500 ml Erlenmeyer flasks. Autoclave 15 min at 121°C. Final pH, 6.8 ± 0.2.

M69. **Oxford Medium**

Columbia blood agar base	39.0 g
Esculin	1.0 g
Ferric ammonium citrate	0.5 g
Lithium chloride	15.0 g
Cycloheximide	0.4 g
Colistin sulfate	0.02 g
Acriflavin	0.005 g
Cefotetan	0.002 g
Fosfomycin	0.010 g
Distilled water	1 litre

Add the 55.5 g of the first four components (basal medium) to 1 litre of distilled water. Bring gently to boil to dissolve completely. Sterilize by autoclaving at 121°C for 15 min. Cool to 50°C and aseptically add supplement (see below), mix, and pour into sterile Petri dishes. To prepare supplement, dissolve cycloheximide, colistin sulfate, acriflavin, cefotetan, and fosfomycin in 10 ml of 1:1 mixture of ethanol and distilled water. Filter-sterilize supplement before use. Oxford basal medium and supplements are available commercially.

M70. **Oxidative-Fermentative (OF) Test Medium**

Base

Peptone	2 g
NaCl	5 g
K_2HPO_4	0.3 g
Bromthymol blue	0.03 g
Agar	3 g
Distilled water	1 litre

Heat with agitation to dissolve agar. Dispense 3 ml portions to 13 x 100 mm tubes. Autoclave 15 min at 121°C. Cool to 50°C.

Glucose stock solution. Dissolve 10 g glucose in 90 ml distilled water. Sterilize by filtration through 0.22 μm membrane. For half the tubes, add 0.3 ml stock solution to 2.7 ml base in tube. For the other half of the tubes, do not add stock solution. Mix gently the tubes containing glucose and cool at room temperature. For each test culture, inoculate one tube containing glucose and one tube without glucose. Layer one tube with sterile mineral oil. Incubate 48 hr at 35°C.

M71. **Peptone, 0.1%**

Peptone	1 g
Distilled water	1 litre

Heat gently to dissolve. Dispense into appropriate containers and autoclave at 121°C for 15 min. Final pH, 7.0.

M72. **Peptone-Beef Extract-Glycogen (PBG) Agar**

Peptone	10 g
Beef extract	10 g
Glycogen	4 g
NaCl	5 g
Sodium lauryl sulfate	0.1 g
Bromthymol blue	0.1 g
Agar	15 g
Distilled water	1 litre

Heat with agitation to dissolve agar. Dispense into suitable flasks. Autoclave 15 min at 121°C. Final pH, 7.0 ± 0.1.

M73. **Peptone Sorbitol Bile Broth**

Na_2HPO_4	8.23 g
NaH_2PO_4	1.2 g
Bile salts No. 3	1.5 g
NaCl	5 g
Sorbitol	10 g
Peptone	5 g
Distilled water	1 litre

Dispense 100 ml into Wheaton bottles. Autoclave 15 min at 121°C. Final pH, 7.6 ± 0.2.

M74. **Phenol Red Carbohydrate Broth**

Trypticase or proteose peptone No. 3	10 g
NaCl	5 g
Beef extract (optional)	1 g
Phenol red (7.2 ml of .025% phenol red solution)	0.018 g
Distilled water	1 litre
Carbohydrate* (See page 264)	

*Dissolve 10 g carbohydrate (except use 5 g for dulcitol) in this basal broth. Dispense 2.5 ml portions into 13 x 100 mm test tubes containing inverted 6 x 50 mm fermentation tubes. Autoclave 10 min at 118°C. Final pH, 7.4 ± 0.2. Alternatively, dissolve ingredients, omitting carbohydrate, in 800 ml distilled water with heat and occasional agitation. Dispense 2.0 ml portions into 13 x 100 mm test tubes containing inverted fermentation tubes. Autoclave 15 min at 118°C and let cool. Dissolve carbohydrate in 200 ml distilled waer and sterilize by passing solution through bacteria-retaining filter. Aseptically add 0.5 ml sterile filtrate to each tube of sterilized broth after cooling to less than 45°C. Shake gently to mix. Final pH, 7.4 ± 0.2.

M75. **Phenol Red Glucose Broth**

Proteose peptone No. 3	10 g
NaCl	5 g
Beef extract (optional)	1 g
Dextrose	5 g
Phenol red (7.2 ml of 0.25% solution)	0.018 g
Distilled water	1 litre

Dispense 2.5 ml portions into 13 x 100 mm tubes. Autoclave 10 min at 118°C. Final pH, 7.4 ± 0.2.

M76. **Phenylalanine Deaminase Agar**

Yeast extract	3 g
L-Phenylalanine or	1 g
DL-Phenylalanine	2 g
Na_2HPO_4	1 g
NaCl	5 g
Agar	12 g
Distilled water	1 litre

Heat gently to dissolve agar. Tube and autoclave 10 min at 121°C. Incline tubes to obtain long slant. Final pH 7.3 ± 0.2.

M77. **Plate Count Agar**

Tryptone	5 g
Yeast extract	2.5 g
Glucose	1 g
Agar	15 g
Distilled water	1 litre

Heat to dissolve ingredients. Dispense into suitable tubes or flasks. Autoclave 15 min at 121°C. Final pH, 7.0 ± 0.2.

For viable yeasts and molds: Dispense 20-25 ml portions into sterile 15 x 100 mm Petri dishes.

M78. **PMP Broth**

Na_2HPO_4	7.9 g
NaH_2PO_4	1.1 g
Peptone	2.5 g
D-Mannitol	2.5 g
Distilled water	1 litre

Adjust pH to 7.6. Autoclave at 121°C for 15 min.

M79. **Potassium Cyanide (KCN) Broth**

Potassium cyanide	0.5 g
Proteose peptone No. 3 or polypeptone	3 g
NaCl	5 g
KH_2PO_4	0.225 g
Na_2HPO_4	5.64 g
Distilled water	1 litre

Dissolve ingredients and autoclave 15 min at 121°C. Cool and refrigerate at 5-8°C. Final pH, 7.6 ± 0.2. Dissolve 0.5 g KCN stock solution in 100 ml sterile distilled water cooled to 5-8°C. Using bulb pipetter, add 15 ml cold KCN stock solution to 1 litre cold, sterile base. DO NOT PIPET BY MOUTH. Mix and aseptically dispense 1.0-1.5 ml portions to 13 x 100 mm sterile tubes. Using aseptic technique, stopper tubes with No. 2 corks impregnated with paraffin. Prepare corks by boiling in paraffin about 5 min. Place corks in tubes so that paraffin does not flow into broth but forms a seal between rim of tubes and cork. Store tubes at 5-8°C no longer than 2 weeks before use.

M80. **Potato Dextrose Agar**

Potato infusion	200 ml
Dextrose	20 g
Agar	20 g
Distilled water	1 litre

To prepare potato infusion, boil 200 g sliced, unpeeled potatoes in 1 litre distilled water for 30 min. Filter through cheesecloth, saving effluent, which is potato infusion. Mix in other ingredients and boil to dissolve. Autoclave 15 min at 121°C. Dispense 20-25 ml portions into sterile 15 x 100 mm Petri dishes. Final pH, 5.6 ± 0.2. Medium should not be re-melted more than once.
For potato dextrose salt agar, prepare potato dextrose agar, as above, and add 75 g NaCl per litre.

M81. **Purple Carbohydrate Broth**

Proteose peptone No. 3	10 g
Beef extract (optional)	1 g
NaCl	5 g
Bromcresol purple	0.02 g
Distilled water	1 litre

Prepare as for phenol red carbohydrate broth (M74). Final pH, 6.8 ± 0.2.

M82. Purple Carbhohydrate Fermentation Broth Base

Purple broth base (Becton Dickinson)	15 g
Distilled water	900 ml

Add purple broth base to distilled water. Dispense 9 ml to 16 x 125 mm tubes containing a Durham tube. Autoclave at 121°C for 15 min. Prepare all carbohydrates, except esculin, as sterile 5% solutions. Filter-sterilize. Add 1 ml carbohydrate solution to 9 ml broth base to yield 0.5% carbohydrate in broth. Add esculin directly into base broth to make a 0.5% solution and autoclave 15 min at 115°C. A 5% solution of esculin at room temperature is a gel that cannot be pipetted.

M83. Pyrazinamidase Agar

Tryptic soy agar (M102)	30 g
Yeast extract	3 g
Pyrazine-carboxamide	1 g
0.2 M Tris-maleate, pH 6.0	1 litre

Heat to boiling; dispense 5 ml in 16 x 125 mm tubes. Autoclave at 121°C for 15 min. Cool slanted.

M84. Rappaport-Vassiliadis Medium

Broth base

Tryptone	5 g
NaCl	8 g
KH_2PO_4	1.6 g
Distilled water	1 litre

Magnesium chloride solution

$MgCl_2.6H_2O$	400 g
Distilled water	1 litre

Malachite green oxalate solution

Malachite green oxalate	0.4 g
Distilled water	100 ml

To prepare the complete medium, combine 1,000 ml broth base, 100 ml magnesium chloride solution, and 10 ml malachite green oxalate solution (total volume of complete medium is 1,110 ml). Broth base must be prepared on same day that components are combined to make complete medium. Magnesium chloride solution may be stored in dark bottle at room temperature for up to one year. To prepare solution, dissolve entire contents of $MgCl_2.6H_2O$ from newly opened container according to formula, becaue this salt is very hygroscopic. Malachite green oxalate solution may be stored in dark bottle at room temperature for up to 6 months. Merck analytically pure malachite green oxalate (Catalog Number 1398) is recommended because other brands may not be equally effective. Dispense 10 ml volumes of complete medium into 16 x 150 mm test tubes. Autoclave 15 min at 115°C. Final pH, 5.5 $\pm$ 0.2. Store in refrigerator and use within 1 month.

M85. Sabouraud's Dextrose Broth and Agar

Polypeptone or neopeptone	10 g
Dextrose	40 g
Distilled water	1 litre

Dissolve completely and dispense 40 ml portions into screw-cap bottles. Final pH, 5.8. Autoclave 15 min at 118-121°C. Do not exceed 121°C.

For Sabouraud's dextrose agar, prepare broth as above and add 15-20 g agar, depending on gel strength desired. Final pH, 5.6 ± 0.2. Dispense into tubes for slants and bottles or flasks for pouring plates. Autoclave 15 min at 118-121°C.

M86. Salt Trypticase Broth (STB)

Trypticase	10 g
Yeast extract	3 g
Distilled water	1 litre

Add 0, 60, 80, and 100 g NaCl per litre base to prepare 0, 6, 8, and 10% broth for salt tolerance tests. Autoclave 15 min at 121°C. Final pH 7.5 ± 0.2.

M87. Selenite Cystine Broth

Medium 1

Tryptone or polypeptone	5 g
Lactose	4 g
Sodium selenite ($NaHSeO_3$)	4 g
Na_2HPO_4	10 g
L-Cystine	0.01 g
Distilled water	1 litre

Heat to boiling to dissolve. Dispense 10 ml portions into sterile 16 x 150 mm test tubes. Heat 10 min in flowing steam. DO NOT AUTOCLAVE. Final pH, 7.0 ± 0.2. The medium is not sterile. Use same day as prepared.

Medium 2 (North-Bartram modification)

Polypeptone	5 g
Lactose	4 g
Sodium selenite ($NaHSeO_3$)	4 g
Na_2HPO_4	5.5 g
KH_2PO_4	4.5 g
L-Cystine	0.01 g
Distilled water	1 litre

Heat with agitation to dissolve. Dispense 10 ml portions to sterile 16 x 150 mm test tubes. Heat 10 min in flowing steam. DO NOT AUTOCLAVE. Use same day as prepared.

M88. **Sheep Blood Agar**

Blood agar base (Oxoid No. 2)	93 ml
Sheep blood, defibrinated	5 ml

Rehydrate and sterilize as recommended by manufacturer. Agar and blood should both be at 45-46°C before adding blood and pouring plates. Sheep blood agar plates are available commercially.

M89. **Shigella Broth**

Base

Tryptone	20 g
K_2HPO_4	2 g
KH_2PO_4	2 g
NaCl	5 g
Dextrose	1 g
Tween 80	1.5 ml
Distilled water	1 litre

Autoclave 15 min at 121°C. Final pH, 7.0 ± 0.2.

Novobiocin solution. Weigh 50 mg novobiocin into 1 litre distilled water. Sterilize by filtration through 0.45 μm membrane. Add 2.5 ml concentrate to 225 ml base.

M90. **SIM Medium**

Peptone	30 g
Beef extract	3 g
Peptonized iron	0.2 g
Sodium thiosulfate	0.025 g
Agar	3 g
Distilled water	1 litre

Heat gently with occasional agitation. Boil 1-2 min until agar dissolves. Dispense 6 ml medium in 16 x 125 mm screw-cap tubes. Sterilize in the autoclave for 15 min at 121°C. Allow the medium to solidify in a vertical position. Do not slant.

M91. **Simmons Citrate Agar**

Sodium citrate.$2H_2O$	2 g
NaCl	5 g
K_2HPO_4	1 g
$NH_4H_2PO_4$	1 g
$MgSO_4$	0.2 g
Bromthymol blue	0.08 g
Agar	15 g
Distilled water	1 litre

Heat gently with occasional agitation. Boil 1-2 min until agar dissolves. Fill 13 x 100 or 16 x 150 mm screw-cap tubes 1/3 full. Autoclave 15 min at 121°C. Before medium solidifies, incline tubes to obtain 4-5 cm slants and 2-3 cm butts. Final pH, 6.9 ± 0.2.

M92. Sporulation Broth (for C. perfringens)

Polypeptone	15 g
Yeast extract	3 g
Starch, soluble	3 g
$MgSO_4$ (anhydrous)	0.1 g
Sodium thioglycollate	1 g
Na_2HPO_4	11 g
Distilled water	1 litre

Adjust pH 7.8 ± 0.1. Dispense 15 ml portions into 20 x 150 mm screw-cap tubes. Autoclave 15 min at 121°C.

M93. Spray's Fermentation Medium (for C. perfringens)

Tryptone	10 g
Neopeptone	10 g
Agar	2 g
Sodium thioglycollate	0.25 g
Distilled water	1 litre

Dissolve all ingredients except agar and adjust pH to 7.4 ± 0.2. Add agar and heat with agitation to dissolve. Dispense 9 ml portions into 16 x 125 mm tubes. Autoclave 15 min at 121°C. Before use, heat in boiling water or flowing steam for 10 min. Add 1 ml of sterile 10% carbohydrate solution to 9 ml base.

M94. Staphylococcus Agar No. 110

Gelatin	30 g
Trypticase peptone	10 g
Yeast extract	2.5 g
Lactose	2 g
Mannitol	10 g
NaCl	75 g
K_2HPO_4	5 g
Agar	15 g
Distilled water	1 litre

Suspend ingredients in 1 litre distilled water. Mix thoroughly. Heat with agitation and boil 1 min. Dispense into suitable flasks or tubes. Autoclave 15 min at 121°C. Final pH, 7.0 ± 0.2.

M95. T_1N_1 Medium

Trypticase	10 g
NaCl	10 g
Agar (if solid medium is preferred)	20 g
Distilled water	1 litre

Heat, if necessary, to dissolve ingredients. Dispense into 16 x 125 mm screw-cap tubes (if tubed medium is required). Autoclave 15 min at 121°C. Slant tubes until cool or let medium cool to 50°C and pour into 15 x 100 mm Petri dishes. Final pH, 7.2 ± 0.2.

M96. Tetrathionate Broth

Tetrathionate broth base

Polypeptone	5 g
Bile salts	1 g
Calcium carbonate	10 g
Sodium thiosulfate.$5H_2O$	30 g
Distilled water	1 litre

Suspend ingredients in 1 litre distilled water, mix, and heat to boiling. (Precipitate will not dissolve completely.) Cool to less than 45°C. Store at 5-8°C. Final pH, 8.4 ± 0.2.

Iodine-Potassium Iodide (I-KI) solution

Potassium iodide	5 g
Iodine, resumblimed	6 g
Distilled water, sterile	20 ml

Dissolve potassium iodide in 5 ml sterile distilled water. Add iodine and stir to dissolve. Dilute to 20 ml.

Brilliant green solution

Brilliant green dye, sterile	0.1 g
Distilled water, sterile	100 ml

On day of use, add 20 ml I-KI solution and 10 ml brilliant green solution to 1 litre base. Resuspend precipitate by gentle agitation and aseptically dispense 10 ml portions into 20 x 150 or 16 x 150 mm sterile test tubes. Do not heat medium after addition of I-KI and dye solutions.

M97. Thioglycollate Medium (Fluid) (FTG)

L-Cystine	0.5 g
Agar (granulated)	0.75 g
NaCl	2.5 g
Dextrose	5 g
Yeast extract	5 g
Tryptone	15 g
Sodium thioglycollate or thioglycollic acid	0.5 g
Resazurin, sodium solution (1:1,000), fresh	1 ml
Distilled water	1 litre

Mix L-cystine, NaCl, dextrose, yeast extract, and tryptone with 1 litre water. Heat in Arnold steamer or water bath until ingredients are dissolved. Dissolve sodium thioglycollate or thioglycollic acid in solution and adjust pH so that value after sterilization is 7.1 $\pm$ 0.2. Add sodium resazurin solution, mix, and autoclave 20 min at 121°C. If commercial media are used, dispense 10 ml portions to 16 x 150 mm tubes and autoclave 15 min at 121°C.

M98. Thiosulfate-Citrate-Bile Salts-Sucrose (TCBS) Agar

Yeast extract	5 g	NaCl	10 g
Peptone	10 g	Ferric citrate	1 g
Sucrose	20 g	Bromthymol blue	0.04 g
Sodium thiosulfate. $5H_2O$	20 g	Thymol blue	0.04 g
Sodium citrate.$2H_2O$	10 g	Agar	15 g
Sodium cholate	3 g	Distilled water	1 litre
Oxgall	5 g		

Heat to dissolve ingredients. Boil 1-2 min. DO NOT AUTOCLAVE. Dispense 20 ml into 15 x 100 mm Petri dishes. Final pH, 8.6.

M99. Toluidine Blue-DNA Agar

Deoxyribonucleic acid (DNA)	0.3 g
Agar	10 g
$CaCl_2$ (anhydrous)	1.1 mg
NaCl	10 g
Toluidine blue O	0.083 g
Tris (hydroxmymethyl) aminomethane	6.1 g
Distilled water	1 litre

Dissolve tris (hydroxymethly) aminomethane in 1 litre distilled water. Adjust pH to 9.0. Add the remaining ingredients except toluidine blue O and heat to boiling to dissolve. Dissolve toluidine blue O in medium. Dispense to rubber-stoppered flasks. Sterilization is not necessary if used immediately. The sterile medium is stable at room temperature for 4 months and is satisfactory after several melting cycles.

M100. Triple Sugar Iron (TSI) Agar

Medium 1

Polypeptone	20 g
NaCl	5 g
Lactose	10 g
Sucrose	10 g
Glucose	1 g
$Fe(NH_4)_2(SO_4)_2 \cdot 6H_2O$	0.2 g
$Na_2S_2O_3$	0.2 g
Phenol red	0.025 g
Agar	13 g
Distilled water	1 litre

Medium 2

Beef extract	3 g
Yeast extract	3 g
Peptone	15 g
Proteose peptone	5 g
Glucose	1 g
Lactose	10 g
Sucrose	10 g
$FeSO_4$	0.2 g
NaCl	5 g
$Na_2S_2O_3$	0.3 g
Phenol red	0.024 g
Agar	12 g
Distilled water	1 litre

These two media are interchangeable for general use. For use with V. parahaemolyticus, add 25 g NaCl per litre to either formula.

Suspend ingredients of Medium 1 in distilled water, mix thoroughly, and heat with occasional agitation. Boil about 1 min to dissolve ingredients. Fill 16 x 150 mm tubes 1/3 full and cap or plug to maintain aerobic conditions. Autoclave Medium 1 for 15 min at 118°C. Prepare Medium 2 in the same manner as Medium 1, except autoclave 15 min at 121°C. Before the media solidify, incline tubes to obtain 4-5 cm slant and 2-3 cm butt. Final pH, 7.3 ± 0.2 for Medium 1 and 7.4 ± 0.2 for Medium 2.

M101. Trypticase (Tryptic)-Peptone-Glucose-Yeast Extract Broth with Trypsin (TPGYT)

Base

Trypticase	50 g
Peptone	5 g
Yeast extract	20 g
Dextrose	4 g
Sodium thioglycollate	1 g
Distilled water	1 litre

Dissolve solid ingredients of base and dispense: 15 ml in 20 x 150 mm tubes or 100 ml in 170 ml prescription bottles. Autoclave tubes 10 min at 121°C and bottles 15 min at 121°C. Final pH, 7.0 ± 0.2. Refrigerate at 5°C. Add trypsin immediately before use.

Trypsin solution

Trypsin (1:250)	1.5 g
Distilled water	100 ml

Stir trypsin in water to suspend. Let particles settle and filter-sterilize supernatant through 0.45 μm membrane. Before use, steam or boil base for 10 min to expel dissolved oxygen. Add 1 ml trypsin to each 15 ml of broth or 6.7 ml trypsin to 100 ml of broth.

M102. Trypticae (Tryptic) Soy Agar

Trypticase peptone	15 g
Phytone peptone	5 g
NaCl	5 g
Agar	15 g
Distilled water	1 litre

Heat with agitation to dissolve agar. Boil 1 min. Dispense into suitable tubes or flasks. Autoclave 15 min at 121°C. Final pH, 7.3 ± 0.2. For use with V. parahaemolyticus, add 25 g NaCl.

M103. Trypticase (Tryptic) Soy Agar with 0.6% Yeast Extract (TSA-YE)

Trypticase soy agar (Becton Dickinson)	40 g
Yeast extract (Becton Dickinson)	6 g
Distilled water	1 litre

M104. Trypticase (Tryptic) Soy Broth

Trypticase peptone	17 g
Phytone peptone	3 g
NaCl	5 g
K_2HPO_4	2.5 g
Glucose	2.5 g
Distilled water	1 litre

Heat with gentle agitation to dissolve. Dispense 225 ml into 500 ml Erlenmeyer flasks. Autoclave 15 min at 121°C. Final pH, 7.3 ± 0.2. For use with V. parahaemolyticus, add 25 g NaCl.

For trypticase soy broth without glucose, prepare as above, but omit 2.5 g dextrose.

M105. Trypticase (Tryptic) Soy Broth with Ampicillin (TSBA)

Trypticase peptone	17 g
Phytone peptone	3 g
NaCl	5 g
K_2HPO_4	2.5 g
Dextrose	2.5 g
Distilled water	1 litre

Heat with gentle agitation to dissolve, and dispense. Autoclave 15 min at 121°C. Final pH, 7.3 ± 0.2.

After cooling to 40-50°C, and filter-sterilized ampicillin to give final concentration of 30 mg/L.

M106. **Trypticase (Tryptic) Soy Broth with 10% Sodium Chloride and 1% Sodium Pyruvate**

Trypticase or tryptose (pancreatic digest of casein)	17 g
Phytone (papaic diget of soya meal)	3 g
NaCl	100 g
K_2HPO_4	2.5 g
Dextrose	2.5 g
Sodium pyruvate	10 g
Distilled water	1 litre

Dehydrated trypticase or tryptic soy broth is satisfactory with 95 g NaCl and 10 g sodium pyruvate added per litre. Adjust to pH 7.3. Heat gently, if necessary. Dispense 10 ml into 16 x 150 mm tubes. Autoclave 15 min at 121°C. Final pH, 7.3 $\pm$ 0.2. Store up to 1 month at 4 $\pm$ 1°C.

M107. **Trypticase (Tryptic) Soy Broth with 0.6% Yeast Extract (TSB-YE)**

Trypticase soy broth (Becton Dickinson)	30 g
Yeast extract (Becton Dickinson)	6 g
Distilled water	1 litre

Dissolve ingredients in 1 litre water. Heat gently to dissolve. Autoclave 15 min at 121°C. Final pH, 7.3 $\pm$ 0.2. Cool to 50°C and dispense 20 ml portions to 15 x 100 mm Petri dishes.

M108. **Trypticase (Tryptic) Soy-Polymyxin Broth**

Prepare trypticase soy broth (M104) and dispense 15 ml portions into 20 x 150 mm tubes. Autoclave 15 min at 121°C.

Polymyxin B solution, 0.15%. Dissolve 500,000 units sterile polymyxin B sulfate powder in 33.3 ml sterile distilled water. Store solution in 85 ml bottle at 4°C. Bottle use, add 0.1 ml sterile 0.15% polymyxin B sulfate solution to 15 ml medium, and mix.

M109. **Trypticase (Tryptic) Soy-Sheep Blood Agar**

Prepare trypticase soy agar (M102). Sterilize as recommended and cool to 50°C. Add 5 ml defibrinated sheep blood to 100 ml agar. Mix and dispense 20 ml portions to 15 x 100 mm Petri dishes. (Commercial trypticase soy-blood agar plates are satisfactory.)

M110. **Trypticase (Tryptic) Soy-Tryptose Broth**

Trypticase soy broth (commercial, dehydrated)	15 g
Tryptose broth (commercial, dehydrated)	13.5 g
Yeast extract	3 g
Distilled water	1 litre

Dissolve ingredients in 1 litre water. Heat gently to dissolve. Dispense 5 ml portions into 16 x 150 mm test tubes. Autoclave 15 min at 121°C. Final pH, 7.2 $\pm$ 0.2.

M111. **Tryptone (Tryptophane) Broth, 1%**

Tryptone or trypticase	10 g
Distilled water	1 litre

Dissolve and dispense 5 ml portions into 16 x 125 or 16 x 150 mm test tubes. Autoclave 15 min at 121°C. Final pH, 6.9 ± 0.2. For use with V. parahaemolyticus, add 30 g NaCl.

M112. **Tryptone Yeast Extract Agar**

Tryptone	10 g
Yeast extract	1 g
*Carbohydrate	10 g
Bromcresol purple	0.04 g
Agar	2 g
Distilled water	1 litre

Dissolve agar with heat and gentle agitation. Adjust pH to 7.0 ± 0.2. Fill 16 x 125 mm tubes 2/3 full. Autoclave 20 min at 115°C. Before use, steam medium 10-15 min. Solidify by placing tubes in ice water.

*Glucose and mannitol are the carbohydrates used for identification of S. aureus.

M113. **Tryptose Broth and Agar for Serology**

Tryptose (Difco)	20 g
NaCl	5 g
Dextrose	1 g
Agar	15 g
Distilled water	1 litre

For broth, omit agar from formulation.

M114. **Tryptose-Sulfite-Cycloserine (TSC) Agar**

Tryptose	15 g
Yeast extract	5 g
Soytone	5 g
Ferric ammonium citrate (NF Brown Pearls)	1 g
Sodium metabisulfite	1 g
Agar	20 g
Distilled water	900 ml

Heat with agitation to dissolve. Adjust pH to 7.6 ± 0.2. Dispense 250 ml portions to 500 ml flasks. Autoclave 15 min at 121°C. Maintain medium at 50°C before use. Dehydrated SFP agar base (Difco) is satisfactory for base.

D-cycloserine solution. Dissolve 1 g D-cycloserine (white crystalline powder) in 200 ml of distilled water. Sterilize by filtration and store at 4°C until use.

Final medium. For pour plates, add 20 ml of D-cycloserine solution to 250 ml base. To prepare prepoured plates containing egg yolk, also add 20 ml of 50% egg yolk emulsion (M27). Mix well and dispense 18 ml into 15 x 100 mm Petri dishes. Cover plates with a towel and let dry overnight at room temperature before use.

M115. **Tyrosine Agar**

Base. Prepare nutrient agar (M66). Dispense 100 ml portions into 170 ml bottles. Autoclave 15 min at 121°C. Cool to 48°C.

Tyrosine suspension. Suspend 0.5 g L-tyrosine in 10 ml distilled water in 20 x 150 mm culture tube. Mix thoroughly with Vortex mixer. Autoclave 15 min at 121°C.

Final medium. Combine 100 ml base with sterile tyrosine suspension. Mix thoroughly by gently inverting bottle 2 or 3 times. Aseptically dispense 3.5 ml into 13 x 100 mm tubes with frequent mixing. Slant tubes and cool rapidly to prevent separation of tyrosine.

M116. **Urea Broth**

Urea	20 g
Yeast extract	0.1 g
KH_2PO_4	9.1 g
Na_2HPO_4	9.5 g
Phenol red	0.01 g
Distilled water	1 litre

Dissolve ingredients in distilled water. DO NOT HEAT. Sterilize by filtration through 0.45 μm membrane. Aseptically dispense 1.5-3.0 ml portions to 13 x 100 mm sterile test tubes. Final pH, 6.8 $\pm$ 0.2.

M117. **Urea Broth (Rapid)**

Urea	20 g
Yeast extract	0.1 g
KH_2PO_4	0.091 g
Na_2HPO_4	0.095 g
Phenol red	0.01 g
Distilled water	1 litre

Prepare as for urea broth (M116), above.

M118. **Veal Infusion Agar and Broth**

Veal, infusion from	500 g
Proteose peptone No. 3	10 g
NaCl	5 g
Agar	15 g
Distilled water	1 litre

Heat with agitation to dissolve agar. Dispense 7 ml portions into 16 x 150 mm tubes. Autoclave 15 min at 121°C. Incline tubes to obtain 6 cm slant. Final pH, 7.3 ± 0.2.

For veal infusion broth, prepare as above, but omit the 15 g agar. Autoclave 15 min at 121°C. Final pH, 7.4 ± 0.2.

M119. **Voges-Proskauer Medium (Modified)**

Proteose peptone	7 g
NaCl	5 g
Dextrose	5 g
Distilled water	1 litre

Dissolve ingredients in water and adjust pH, if necessary. Dispense 5 ml portions into 20 x 150 mm tubes. Autoclave 10 min at 121°C. Final pH, 6.5 ± 0.2.

M120. **Wagatsuma Agar**

Yeast extract	3 g
Peptone	10 g
NaCl	70 g
K_2HPO_4	5 g
Mannitol	10 g
Crystal violet	0.001 g
Agar	15 g
Distilled water	1 litre

Heat with agitation to dissolve agar. Adjust pH to 8.0 ± 0.2. Steam 30 min. Do not autoclave. Cool to 50°C. Wash human or rabbit red blood cells 3 times with physiological saline. Add 5% by volume of this suspension to the cooled medium. Mix and pour into Petri dishes. (Plates must be dried thoroughly before use.) Use plates promptly.

M121. **Xylose Lysine Desoxycholate (XLD) Agar**

Yeast extract	3 g	Ferric ammonium citrate	0.8 g
L-lysine	5 g	Sodium thiosulfate	6.8 g
Xylose	3.75 g	NaCl	5 g
Lactose	7.5 g	Agar	15 g
Sucrose	7.5 g	Phenol red	0.08 g
Sodium dexoycholate	2.5 g	Distilled water	1 litre

Heat with agitation just until medium boils. Do not overheat. Pour into plates when medium has cooled to 50°C. Let dry about 2 hr with covers partially removed. Then close plates. Final pH, 7.4 ± 0.2. Do not store more than 1 day.

M122. **Yersinia Selective Agar (YSA) Base (with antimicrobic supplement)**

Yeast extract	2 g
Peptone	17 g
Proteose peptone	3 g
Mannitol	20 g
Sodium desoxycholate	0.5 g
Sodium cholate	0.5 g
NaCl	1 g
Sodium pyruvate	2 g
Magnesium sulfate, heptahydrate	10 mg
Agar	13.5 g
Neutral red	30 mg
Crystal violet	1 mg
Irgasan	4 mg
Distilled water	1 litre

Heat to boiling to dissolve completely. Sterilize 15 min at 121°C. AVOID OVERHEATING. Cool to 45-50°C and aseptically add 10 ml rehydrated Yersinia antimicrobic supplement CN. Mix thoroughly and dispense into sterile Petri dishes.

Yersinia antimicrobic supplement (CN) (Formula per 10 ml vial)

Cefsulodin	4 mg
Novobiocin	2.5 mg

REAGENTS AND DILUENTS

R1. 5N Acetic Acid

Acetic acid, glacial (density 1.049 g/ml)	28.63 ml
Distilled water	71.37 ml

R2. Bromthymol Blue Indicator, 0.04%

Bromthymol blue	0.2 g
0.01 N NaOH	32 ml

Dissolve bromthymol blue in NaOH. Dilute to 500 ml with distilled water.

R3. Bufferfield's Phosphate Buffer

Stock solution

KH_2PO_4	34 g
Distilled water	500 ml

Adjust pH to 7.2 with 1 N NaOH. Bring volume to 1 litre with distilled water. Sterilize 15 min at 121°C. Store in refrigerator.

Dissolution blanks

Take 1.25 ml of above stock solution and bring volume to 1 litre with distilled water. Dispense into bottles to 90 $\pm$ 1 ml. Sterilize 15 min at 121°C.

R4. Catalase Test Reagent

Pour 1 ml 3% hydrogen peroxide over growth on slant culture. Gas bubbles indicate positive test. Alternatively, emulsify colony in 1 drop 30% hydrogen peroxide on glass slide. Immediate bubbling is positive catalase test. If colony is taken from blood agar plate, any carry-over of red blood cells can give false positive reaction.

R5. Disinfectants

(for preparation of canned foods for microbiological analysis)

1. Alcoholic solution of iodine

Potassium iodide	10 g
Iodine	10 g
Ethanol (70%)	500 ml

2. Sodium hypochlorite solution*

Sodium hypochlorite	5.0 - 5.25 g
Distilled water	100 ml

* Laundry bleach, which is 5.25% sodium hypochlorite (NaOCl), may be used.

R6. Ethanol Solution, 70%

Ethanol, 95%	700 ml
Distilled water	250 ml

R7. Ferric Chloride, 10%

$FeCl_3$	10 g
Distilled water	90 ml

R8. Ferrous Ammonium Sulfate, 1%

$Fe(NH_4)_2(SO_4)_2$	1 g
Distilled water to make	100 ml

R9. Formalinized Physiological Saline Solution

Formaldehyde solution (36-38%)	6 ml
NaCl	8.5 ml
Distilled water	1 litre

Dissolve 8.5 g NaCl in 1 litre distilled water. Autoclave 15 min at 121°C. Cool to room temperature. Add 6 ml formaldehyde solution. Do not autoclave after addition of formaldehyde.

R10. Gel Diffusion Agar, 1.2%

NaCl	8.5 g
Sodium barbital	8.0 g
Merthiolate (crystalline)	0.1 g
Noble special agar (Difco)	12.0 g
Distilled water	1 litre

Dissolve NaCl, sodium barbital, and merthiolate in 900 ml distilled water. Adjust pH to 7.4 with 1 N HCl and/or 1 N NaOH. Bring volume to 1 litre. Add Noble agar. Melt agar mixture in Arnold steamer. Filter in steamer, while hot, through 2 layers of analytical grade filter paper. Dispense in small (15 - 25 ml) portions into prescription bottles. Do not remelt more than twice.

R11. Gel-Phosphate Buffer

Gelatin	2 g
Na_2HPO_4	4 g
Distilled water	1 litre

Use gentle heat to dissolve ingredients. Sterilize 20 min at 121°C. Final pH, 6.2.

R12. Glycerin-Salt Solution (Buffered)

Glycerin (reagent grade)	100 ml
K_2HPO_4 (anhydrous)	12.4 g
KH_2PO_4 (anhydrous)	4 g
NaCl	4.2 g
Distilled water	900 ml

Dissolve NaCl and bring volume to 900 ml with water. Add glycerin and phosphates. Adjust pH to 7.2. Autoclave 15 min at 121°C. For double strength (20%) glycerin solution, use 200 ml glycerin and 800 ml distilled water.

R13. Hippurate Solution, 1%

Sodium hippurate	0.1 g
Distilled water	10 ml

Dissolve in distilled water and filter sterilize through 0.45 μm membrane. Store refrigerated or in 0.4 ml aliquots at -20°C.

R14. 1 N Hydrochloric Acid

HCl (concentrated)	89 ml
Distilled water to make	1 litre

R15. Hydrogen peroxide solution, 3%

Hydrogen peroxide solution (commercial, 30%)	10 ml
Distilled water	90 ml

Mix ingredients and store solution under refrigeration.

R16. Kovacs' Reagent

p-Dimethylaminobenzaldehyde	5 g
Amyl alcohol (normal only)	75 ml
HCl (concentrated)	25 ml

Dissolve p-dimethylaminobenzaldehyde in normal amyl alcohol. Slowly add HCl. Store at 4°C. To test for indole, add 0.2-0.3 ml reagent to 5 ml of 24 hr bacteria culture in tryptone broth. Dark red color in surface layer is positive test for indole.

R17. McFarland Nephelometer

Make suspensions of barium sulfate as follows:

1. Prepare 1.0% solution of C.P. (chemically pure) sulfuric acid.

2. Prepare 1.0% solution of C.P. barium chloride.

3. Prepare 10 standards as follows:

Add (ml 1% $BaCl_2$ solution)	to	(ml 1% H_2SO_4 solution)
1		99
2		98
3		97
4		96
5		95
6		94
7		93
8		92
9		91
10		90

4. Seal about 3 ml of each standard suspension of barium sulfate precipitate in small test tube. Select test tubes carefully for uniformity of wall thickness, diameter, and color. (These preparations are available commercially from several laboratory supply firms.)

R18. Mercuric Chloride Solution, 0.1%

Mercuric chloride	1.0 g
Distilled water	1 litre

R19. Methyl Red Indicator

Methyl red	0.10 g
Ethanol, 95%	300 ml
Distilled water to make	500 ml

Dissolve methyl red in 300 ml ethanol. Bring volume to 500 ml with distilled water.

R20. Mineral oil

Autoclave 30 min at 121°C. Use screw-cap containers, about 1/2 full with 20-50 ml.

R21. **Neutral Red Solution**

Neutral red	0.2 g
Ethanol	as needed
Distilled water	100 ml

Dissolve neutral red in small volume of ethanol in 100 ml volumetric flask. Add distilled water to bring volume up to etched mark.

R22. **Ninhydrin Reagent**

Ninhydrin	3.5 g
Acetone/butanol (1:1 mixture)	100 ml

Dissolve ninhydrin completely and store solution water under refrigeration.

R23. **Nitrite Detection Reagents**

A. Sulfanilic acid reagent

Sulfanilic acid	1 g
5 N acetic acid	125 ml

B. N-(1-naphthyl)ethylenediamine reagent

N-(1-naphthyl)ethylenediamine dihydrochloride	0.25 g
5N acetic acid	200 ml

C. alpha-Naphthol reagent

alpha-Naphthol	1 g
5 N acetic acid	200 ml

To prepare 5 N acetic acid, add 28.75 ml glacial acetic acid to 71.25 ml distilled water.

Store reagents in glass-stoppered brown bottles. To perform test, add 0.1-0.5 ml each of reagent A and either reagent B or reagent C (as specified in method) to culture grown in liquid or semi-solid medium. Development of red-violet color with reagents A and B or orange color with reagents A and C indicates that nitrate has been reduced to nitrite. Since color produced with reagents A and B may fade or disappear within a few minutes, record reaction as soon as color appears. If no color develops, test for presence of nitrate by adding small amount of zinc dust. If color develops, nitrate has not been reduced.

R24. **Oxidase Test Reagent**

N,N,N',N'-tetramethyl-p-phenylenediamine.2HCl	1 g
Distilled water	100 ml

This is the preferred reagent. Apply freshly prepared solution directly to culture on agar plate or slant. Oxidase-positive colonies develop pink color and progressively turn purple. Use relatively young cultures for test. Prepare reagent fresh, but it may be used up to 7 days if stored in dark glass bottle under refrigeration. Test may also be performed by rubbing culture on strip of impregnated filter paper. Use platinum wire; wire containing iron gives false-positIve reaction. Alternatively, apply 1% solution of N,N-dimethyl-p-phenylenediamine hydrochloride directly to culture plate or slant. To preserve cultures, complete transfer within 3 min. Reagent is toxic to organisms.

R25. **Peptone Diluent, 0.1%**

Peptone	1 g
Distilled water	1 litre

Autoclave 15 min at 121°C. Final pH, 7.0 ± 0.2.

R26. **Peptone Dilution Fluid**

Peptone	1.0 g
Distilled water	1 litre

Autoclave 15 min at 121°C. Final pH, 7.0 ± 0.2.

R27. **0.02 M Phosphate Saline Buffer (pH 7.3-7.4)**

Prepare stock solutions of 0.2 M mono- and disodium phosphate in 8.5% salt solutions and dilute 1:10 for preparation of 0.02 M phosphate saline buffer.

Stock solution 1

Na_2HPO_4 (anhydrous) (reagent grade)	28.4 g
NaCl (reagent grade)	85.0 g
Distilled water to make	1 litre

Stock solution 2

$NaH_2PO_4.H_2O$ (reagent grade)	27.6 g
NaCl (reagent grade)	85.0 g
Distilled water to make	1 litre

To obtain 0.02 M phosphate-buffered saline (0.85%), make 1:10 dilutions of each stock solution. For example:

Stock solution	150 ml	Stock solution	210 ml
Distilled water	450 ml	Distilled water	90 ml
Approximate pH, 8.2		Approximate pH, 5.6	

Using pH meter, titer diluted solution 1 to pH 7.3-7.4 by adding about 65 ml of diluted solution 2. Use resulting 0.02 phosphate saline buffer solution in the lysostaphin susceptibility test on S. aureus.

NOTE: Do not titer 0.2 M phosphate buffer to pH 7.3-7.4 and then dilute to 0.02 M strength. This results in a drop in pH of approximately 0.25. Addition of 0.85% saline after pH adjustment also results in a drop of approximately 0.2 units.

R28. **Physiological Saline Solution 0.85% (Sterile)**

NaCl	8.5 g
Distilled water	1 litre

Dissolve 8.5 g NaCl in water. Autoclave 15 min at 121°C. Cool to room temperature.

R29. **0.5% Potassium Hydroxide in 0.5% Sodium Chloride**

NaCl	5 g
KOH	5 g
Distilled water to make	1 litre

R30. **Potassium Hydroxide Solution, 40%**

KOH	40 g
Distilled water to make	100 ml

R31. **Saline Solution, 0.5% (Sterile)**

NaCl	5 g
Distilled water	1 litre

Dissolve NaCl in water. Autoclave at 121°C for 15 min.

R32. **Slide Preserving Solution**

Prepare 1% acetic acid solution. Add 1 ml glycerin to each 100 ml of solution.

R33. **Sodium Chloride Dilution Water, 3%**

NaCl	30 g
Distilled water	1 litre

Sterilize at 121°C for 15 min. Final pH, 7.0. Use this formula for both sample dilution and blank for V. parahaemolyticus.

R34. **0.2 M Sodium Chloride Solution**

NaCl	11.7 g
Distilled water to make	1 litre

Dispense in suitable containers. Autoclave 15 min at 121°C.

R35. **Sodium Citrate, 3.8%**

Sodium citrate	38 g
Distilled water	1 litre

Dissolve 38 g sodium citrate in water. Autoclave 15 min at 121°C. Cool to room temperature and store at 4°C.

R36. **Sodium Hippurate Reagent**

Sodium hippurate	0.1 g
Distilled water	10 ml

Dissolve in distilled water and filter sterilize through 0.45 μm membrane. Store refrigerated or in 0.4 ml aliquots at -20°C.

R37. **1 N Sodium Hydroxide Solution**

NAOH	40 g
Distilled water to make	1 litre

Use for adjusting pH of culture media.

R38. **Tergitol Anionic 7**

This reagent is a sodium sulfate derivative of 3,9-diethyl tridecanol-6. Its recommended use is for wetting and emulsifying when the electrolyte is below 1% in textiles, emulsion polymers, rubber latices, leather, and pharmaceuticals. Tergitol-7™ is an anionic wetting agent manufactured by Union Carbide Corp., Chemicals and Plastics, 270 Park Avenue, New York, NY 10017, USA.

R39. **Triton X-100™**

This reagent is the registered trademark for octylphenoxy polyethoxy ethanol. Its recommended uses include wetting agent, detergent, dispersant, emulsifier in household and industrial cleaners, textile processing, wool scouring, emulsifying agent for insecticides and herbicides, etc. Triton X-100 is a nonionic preparation manufactured by Rohm and Haas Company, Independence Mall West, Philadelphia, PA 19105, USA.

R40. **Voges-Proskauer (VP) Test Reagents**

Solution 1

a-Naphtol	5 g
Alcohol (absolute)	100 ml

Solution 2

Potassium hydroxide	40 g
Distilled water to make	100 ml

Voges-Proskauer (VP) test. At room temperature, transfer 1 ml of 48 hr culture to test tube and add 0.6 ml solution 1 and 0.2 ml solution 2. Shake after adding each solution. To intensify and speed reaction, add a few creatine crystals to mixture. Read results 4 hr after adding reagents. Development of eosin pink color is a positive test.

STAINS AND DYES

S1. Basic Fuchsin Staining Solution

Dissolve 0.5 g basic fuschsin dye in 20 ml 95% ethanol. Dilute to 100 ml with distilled water. Filter, if necessary, with Whatman No. 31 filter paper to remove any undissolved dye. (TB Carbolfuchsin ZN staining solution, available from Difco Laboratories, is satisfactory.)

S2. Brilliant Green Dye Solution, 1%

Brilliant green dye	1 g
Distilled water (sterile)	100 ml

Dissolve 1 g dye in sterile water. Dilute to 100 ml. Before use, test all batches of dye for toxicity with known positive and negative test microorganisms.

S3. Bromocresol Purple Dye Solution, 0.2%

Bromcresol purple dye	0.2 g
Distilled water (sterile)	100 ml

Dissolve 0.2 g dye in sterile water and dilute to 100 ml.

S4. Carbol fuchsin, 0.5%

Carbol fuchsin	0.5 g
Distilled water	1 litre

Dissolve 0.5 g dye in sterile water. Dilute to 100 ml.

S5. Crystal Violet Stain (for Bacteria)

1. Crystal violet in dilute alcohol

Crystal violet (90% dye content)	2 g
Ethanol (95%)	20 ml
Distilled water	80 ml

2. Ammonium oxalate crystal violet (Hucker's) (See S6)

Either solution is generally considered suitable as a simple stain to observe morphology.

S6. **Gram Stain**

(Commercial staining solutions are satisfactory)

1. Hucker's crystal violet

Solution A

Crystal violet (90% dye content)	2 g
Ethanol, 95%	20 ml

Solution B

Ammonium oxalate	0.8 g
Distilled water	80 ml

Mix solutions A and B. Store 24 hr and filter through coarse filter paper.

Gram's iodine

Iodine	1 g
Potassium iodide (KI)	2 g
Distilled water	300 ml

Place KI in mortar, add iodine, and grind with pestle for 5-10 s. Add 1 ml water and grind; then add 5 ml of water and grind, then 10 ml and grind. Pour this solution into reagent bottle. Rinse mortar and pestle with amount of water needed to bring total volume to 300 ml.

2. Hucker's counterstain (stock solution)

Safranin O (certified)	2.5 g
Ethanol, 95%	100 ml

Add 10 ml stock solution to 90 ml distilled water.

3. Staining procedure (Gram stain)

Fix air-dried films of food sample in moderate heat. Stain films 1 min with crystal violet-ammonium oxalate solution. Wash briefly in tap water and drain. Apply Gram's iodine for 1 min. Wash in tap water and drain. Decolorize with 95% ethanol until washings are no longer blue (about 30 s). Alternatively, flood slides with ethanol, pour off immediately, and reflood with ethanol for 10 s. Wash briefly with water, drain, and apply Hucker's counterstain (safranin solution) for 10-30 s. Wash briefly with water, drain, blot or air-dry, and examine. Gram-positive bacteria stain blue; Gram-negative bacteria stain red. Some Gram-negative bacteria do not destain readily after staining with Hucker's crystal violet. To avoid this difficulty, dilute crystal violet solution up to 10-fold before mixing with equal parts of ammonium oxalate solution. Stain reference Gram-positive and Gram-negative bacteria to ensure valid staining results.

S7. **Methylene Blue Stain (Loeffler's)**

Solution A

Methylene blue (90% dye content)	0.3 g
Ethanol (95%)	30 ml

Solution B

Diluted potassium hydroxide (0.01%)	100 ml

Mix solutions A and B.

S8. **Thiazine Red R Stain**
(for S. aureus enterotoxin gel diffusion technique)

Dissolve Thiazine Red R (Colour Index No. 14780), 0.1%, in 1.0% acetic acid.

CHAPTER 21

MOST PROBABLE NUMBER DETERMINATION

The most probable number (MPN) technique is an estimate of the density of viable organisms in a sample. To obtain this estimate, the sample must be diluted in such a manner that a more dilute sample will result in fewer positive tubes, which are indicated by the presence of gas or microbial growth. The number of dilutions to be prepared should be based on the expected population of the sample. The most reliable results are obtained when all the tubes at the lowest dilution are positive (i.e., microbial growth present) and all tubes at the highest dilution are negative (i.e., microbial growth absent). If a high microbial count is expected, the sample must be diluted to the extent at which the MPN can be obtained. Serial 10-fold dilutions of the sample homogenate are used in either a 3- or 5- tube MPN series. As the number of tubes inoculated for each dilution increases, the confidence limits for the MPN are narrowed.

Table 28 demonstrates how a 3-tube MPN is obtained, assuming that each of the 5 sets of 3 tubes has been properly incubated and inoculated with 1.0, 0.1, 0.01, 0.001, and 0.0001 g of the sample and that the numbers of positive and negative tubes (showing growth or no growth) have been determined. Tables 29-32, for the 3- and 5- tube MPNs, are supplied here for the analyst's use.

Table 28

Example of selecting and recording MPN data

Example	Dilution (g)					Combination of positive tubes	Table used	MPN/g	95% Confidence limits	
	1.0	0.1	0.01	0.001	0.0001				Lower	Upper
a	3/3*	3/3	1/3			3-3-1	2	46	7.1	240
b	3/3	3/3	2/3	1/3	0/3	3-2-1	3	150	30	440
c	3/3	2/3	0/3	0/3	0/3	3-2-0	2	9.3	1.5	38
d	2/3	2/3	0/3			2-2-0	2	2.1	0.4	4.7
e	3/3	3/3	2/3	1/3	1/3	3-2-2	3	210	35	470
f	3/3	2/3	0/3	1/3	0/3	3-2-1	2	15	3	44

*Numerator = number of positive tubes; denominator = number of tubes inoculated.

Select highest dilution in which all tubes are positive (3, 5, or 10) and next 2 highest dilutions (examples a, b, c, and e). If only 3 dilutions are made (examples a and d), use those 3 dilutions to obtain an MPN. Where there are skips, as in examples e and f, add highest dilution to next lower dilution and derive "Index Number" from 3 dilutions as before. Since Table 1 is devised for portions of 10, 1 and 0.1 g, Tables 2-5 are provided for MPNs determined by higher dilutions of the sample.

Table 29

MPN index and 95% confidence limits for various combinations of positive results when various numbers of tubes are used (Inocula of 10, 1.0, and 0.1 g)

	3 Tubes per dilution			5 Tubes per dilution		
		95% Confidence limits			95% Confidence limits	
Combination of positives	MPN index per g	Lower	Upper	MPN index per g	Lower	Upper
0-0-0	<0.03	<0.005	<0.09	<0.02	<0.005	<0.07
0-0-1	0.03	<0.005	0.09	0.02	<0.005	0.0
0-1-0	0.03	<0.005	0.13	0.02	<0.005	0.0
0-2-0	--	--	--	0.04	<0.005	0.1
1-0-0	0.04	<0.005	0.20	0.02	<0.005	0.0
1-0-1	0.07	0.01	0.21	0.04	<0.005	0.1
1-1-0	0.07	0.01	0.23	0.04	<0.005	0.1
1-1-1	0.11	0.03	0.36	0.06	<0.005	0.1
1-2-0	0.11	0.03	0.36	0.06	<0.005	0.1
2-0-0	0.09	0.01	0.36	0.05	<0.005	0.1
2-0-1	0.14	0.03	0.37	0.07	0.01	0.1
2-1-0	0.15	0.03	0.44	0.07	0.01	0.1
2-1-1	0.2	0.07	0.89	0.09	0.02	0.2
2-2-0	0.21	0.04	0.47	0.09	0.02	0.2
2-2-1	0.28	0.10	1.50	--	--	--
2-3-0	--	--	--	0.12	0.03	0.2
3-0-0	0.23	0.04	1.20	0.08	0.01	0.1
3-0-1	0.39	0.07	1.30	0.11	0.02	0.2
3-0-2	0.64	0.15	3.80	--	--	--
3-1-0	0.43	0.07	2.10	0.11	0.02	0.2
3-1-1	0.75	0.14	2.30	0.14	0.04	0.3
3-1-2	1.20	0.30	3.80	--	--	--
3-2-0	0.93	0.15	3.80	0.14	0.04	0.3
3-2-1	1.5	0.30	4.40	0.17	0.05	0.4
3-2-2	2.1	0.35	4.70	--	--	--
3-3-0	2.4	0.36	13.0	--	--	--
3-3-1	4.6	0.71	24.0	--	--	--
3-3-2	11.0	1.50	48.0	--	--	--
3-3-3	>11.0	>1.50	>48.0	--	--	--

Table 29 (continued)

Combination of positives	5 Tubes per dilution		
	MPN index per g	95% Confidence limits	
		Lower	Upper
4-0-0	0.13	0.03	0.31
4-0-1	0.17	0.05	0.46
4-1-0	0.17	0.05	0.46
4-1-1	0.21	0.07	0.63
4-1-2	0.26	0.09	0.78
4-2-0	0.22	0.07	0.67
4-2-1	0.26	0.09	0.78
4-3-0	0.27	0.09	0.80
4-3-1	0.33	0.11	0.93
4-4-0	0.34	0.12	0.93
5-0-0	0.23	0.07	0.70
5-0-1	0.31	0.11	0.89
5-0-2	0.43	0.15	1.14
5-1-0	0.33	0.11	0.93
5-1-1	0.46	0.16	1.2
5-1-2	0.63	0.21	1.5
5-2-0	0.49	0.17	1.3
5-2-1	0.70	0.23	1.7
5-2-2	0.94	0.28	2.2
5-3-0	0.79	0.25	1.9
5-3-1	1.10	0.31	2.5
5-3-2	1.40	0.37	3.4
5-3-3	1.80	0.44	5.0
5-4-0	1.30	0.35	3.0
5-4-1	1.70	0.43	4.9
5-4-2	2.20	0.57	7.0
5-4-3	2.80	0.90	8.5
5-4-4	3.50	1.20	10.0
5-5-0	2.40	0.68	7.5
5-5-1	3.50	1.20	10.0
5-5-2	5.40	1.80	14.0
5-5-3	9.20	3.00	32.0
5-5-4	16.09	6.40	58.0
5-5-5	--	--	--

Table 30

MPN index and 95% confidence limits for various combinations of positive results when various numbers of tubes are used. (Inocula of 1, 0.1, and 0.01 g)

	3 Tubes per dilution			5 Tubes per dilution		
		95% Confidence limits			95% Confidenc limits	
Combination of positives	MPN index per g	Lower	Upper	MPN index per g	Lower	Upp
0-0-0	<0.3	<0.05	<0.9	<0.2	<0.05	<0.
0-0-1	0.3	<0.05	0.9	0.2	<0.05	0.7
0-1-0	0.3	<0.05	1.3	0.2	<0.05	0.7
0-2-0	--[b]	--	--	0.4	<0.05	1.1
1-0-0	0.4	<0.05	2	0.2	<0.05	0.7
1-0-1	0.7	0.1	2.0	0.4	<0.05	1.1
1-1-0	0.7	0.1	2.3	0.4	<0.05	1.1
1-1-1	1.1	0.3	3.6	0.6	<0.05	1.5
1-2-0	1.1	0.3	3.6	0.6	<0.05	1.5
2-0-0	0.9	0.1	3.6	0.5	<0.05	1.3
2-0-1	1.4	0.3	3.7	0.7	0.1	1.7
2-1-0	1.5	0.3	4.4	0.7	0.1	1.7
2-1-1	2.0	0.7	8.9	0.9	0.2	2.1
2-2-0	2.1	0.4	4.7	0.9	0.2	2.1
2-2-1	2.8	1.0	15	--	--	--
2-3-0	--	--	--	1.2	0.3	2.8
3-0-0	2.3	0.4	12	0.8	0.1	1.9
3-0-1	3.9	0.7	13	1.1	0.2	2.5
3-0-2	6.4	1.5	38	--	--	--
3-1-0	4.3	0.7	21	1.1	0.2	2.5
3-1-1	7.5	1.4	23	1.4	0.4	3.4
3-1-2	12	3	38	--	--	--
3-2-0	9.3	1.5	38	1.4	0.4	3.4
3-2-1	15	3	44	1.7	0.5	4.6
3-2-2	21	3.5	47	--	--	--
3-3-0	24	3.6	130	--	--	--
3-3-1	46	7.1	240	--	--	--
3-3-2	110	15	480	--	--	--
3-3-3	>110	>15	>480	--	--	--

Table 30 (continued)

Combination of positives	5 Tubes per dilution		
	MPN index per g	95% Confidence limits	
		Lower	Upper
4-0-0	1.3	0.3	3.1
4-0-1	1.7	0.5	4.6
4-1-0	1.7	0.5	4.6
4-1-1	2.1	0.7	6.3
4-1-2	2.6	0.9	7.8
4-2-0	2.2	0.7	6.7
4-2-1	2.6	0.9	7.8
4-3-0	2.7	0.9	8.0
4-3-1	3.3	1.1	9.3
4-4-0	3.4	1.2	9.3
5-0-0	2.3	0.7	7.0
5-0-1	3.1	1.1	8.9
5-0-2	4.3	1.5	11.4
5-1-0	3.3	1.1	9.3
5-1-1	4.6	1.6	12
5-1-2	6.3	2.1	15
5-2-0	4.9	1.7	13
5-2-1	7.0	2.3	17
5-2-2	9.4	2.8	22
5-3-0	7.9	2.5	19
5-3-1	11	3.1	25
5-3-2	14	3.7	34
5-3-3	18	4.4	50
5-4-0	13	3.5	30
5-4-1	17	4.3	49
5-4-2	22	5.7	70
5-4-3	28	9.0	85
5-4-4	35	12	100
5-5-0	24	6.8	75
5-5-1	35	12	100
5-5-2	54	18	140
5-5-3	92	30	320
5-5-4	161	64	580
5-5-5	>161	>64	>580

Table 31

MPN index and 95% confidence limits for various combinations of positive results when various numbers of tubes are used. (Inocula of 0.1, 0.01, and 0.001 g)

	3 Tubes per dilution			5 Tubes per dilution		
		95% Confidence limits			95% Confidenc limits	
Combination of positives	MPN index per g	Lower	Upper	MPN index per g	Lower	Upp
0-0-0	<3	<0.5	<9	<2	<0.5	<7
0-0-1	3	<0.5	9	2	<0.5	7
0-1-0	3	<0.5	13	2	<0.5	7
0-2-0	--	--	--	4	<0.5	11
1-0-0	4	<0.5	20	2	<0.5	7
1-0-1	7	1	21	4	<0.5	11
1-1-0	7	1	23	4	<0.5	11
1-1-1	11	3	36	6	<0.5	15
1-2-0	11	3	36	6	<0.5	15
2-0-0	9	1	36	5	<0.5	13
2-0-1	14	3	37	7	1	17
2-1-0	15	3	44	7	1	17
2-1-1	20	7	89	9	2	21
2-2-0	21	4	47	9	2	21
2-2-1	28	10	150	--	--	--
2-3-0	--	--	--	12	3	28
3-0-0	23	4	120	8	1	19
3-0-1	39	7	130	11	2	25
3-0-2	64	15	380	--	--	--
3-1-0	43	7	210	11	2	25
3-1-1	75	14	230	14	4	34
3-1-2	120	30	380	--	--	--
3-2-0	93	15	380	14	4	34
3-2-1	150	30	440	17	5	46
3-2-2	210	35	470	--	--	--
3-3-0	240	36	1,300	--	--	--
3-3-1	460	71	2,400	--	--	--
3-3-2	1,100	150	4,800	--	--	--
3-3-3	>1,100	>150	>4,800	--	--	--

Table 31 (continued)

Combination of positives	MPN index per g	5 Tubes per dilution: 95% Confidence limits Lower	Upper
4-0-0	13	3	31
4-0-1	17	5	46
4-1-0	17	5	46
4-1-1	21	7	63
4-1-2	26	9	78
4-2-0	22	7	67
4-2-1	26	9	78
4-3-0	27	9	80
4-3-1	33	11	93
4-4-0	34	12	93
5-0-0	23	7	70
5-0-1	31	11	89
5-0-2	43	15	114
5-1-0	33	11	93
5-1-1	46	16	120
5-1-2	63	21	150
5-2-0	49	17	130
5-2-1	70	23	170
5-2-2	94	28	220
5-3-0	79	25	190
5-3-1	110	31	250
5-3-2	140	37	340
5-3-3	180	44	500
5-4-0	130	35	300
5-4-1	170	43	490
5-4-2	220	57	700
5-4-3	280	90	850
5-4-4	350	120	1,000
5-5-0	240	68	750
5-5-1	350	120	1,000
5-5-2	540	180	1,400
5-5-3	920	300	3,200
5-5-4	1,600	640	5,800
5-5-5	>1,600	>640	>5,800

Table 32

MPN index and 95% confidence limits for various combinations of positive results when various numbers of tubes are used. (Inocula of (0.01, 0.001, and 0.0001 g)

Combination of positives	3 Tubes per dilution			5 Tubes per dilution		
	MPN index per g	95% Confidence limits		MPN index per g	95% Confidenc limits	
		Lower	Upper		Lower	Upp
0-0-0	<30	<5	<90	<20	<5	<70
0-0-1	30	<5	90	20	<5	70
0-1-0	30	<5	130	20	<5	70
0-2-0	--	--	--	40	<5	110
1-0-0	40	<5	200	20	<5	70
1-0-1	70	10	210	40	<5	110
1-1-0	70	10	230	40	<5	110
1-1-1	110	30	360	60	<5	150
1-2-0	110	30	360	60	<5	150
2-0-0	90	10	360	50	<5	130
2-0-1	140	30	370	70	10	170
2-1-0	150	30	440	70	10	170
2-1-1	200	70	890	90	20	210
2-2-0	210	40	470	90	20	210
2-2-1	280	100	1,500	--	--	--
2-3-0	--	--	--	120	30	280
3-0-0	230	40	1,200	80	10	190
3-0-1	390	70	1,300	110	20	250
3-0-2	640	150	3,800	--	--	--
3-1-0	430	70	2,100	110	20	250
3-1-1	750	140	2,300	140	40	340
3-1-2	1,200	300	3,800	--	--	--
3-2-0	930	150	3,800	140	40	340
3-2-1	1,500	300	4,400	170	50	460
3-2-2	2,100	350	4,700	--	--	--
3-3-0	2,400	360	13,000	--	--	--
3-3-1	4,600	710	24,000	--	--	--
3-3-2	11,000	1,500	48,000	--	--	--
3-3-3	>11,000	>1,500	>48,000	--	--	--

Table 32 (continued)

Combination of positives	5 Tubes per dilution		
	MPN index per g	95% Confidence limits	
		Lower	Upper
4-0-0	130	30	310
4-0-1	170	50	460
4-1-0	170	50	460
4-1-1	210	70	630
4-1-2	260	90	780
4-2-0	220	70	670
4-2-1	260	90	780
4-3-0	270	90	800
4-3-1	330	110	930
4-4-0	340	120	930
5-0-0	230	70	700
5-0-1	310	110	890
5-0-2	430	150	1,140
5-1-0	330	110	930
5-1-1	460	160	1,200
5-1-2	630	210	1,500
5-2-0	490	170	1,300
5-2-1	700	230	1,700
5-2-2	940	280	2,200
5-3-0	790	250	1,900
5-3-1	1,100	310	2,500
5-3-2	1,400	370	3,400
5-3-3	1,800	440	5,000
5-4-0	1,300	350	3,000
5-4-1	1,700	430	4,900
5-4-2	2,200	570	7,000
5-4-3	2,800	900	8,500
5-4-4	3,500	1,200	10,000
5-5-0	2,400	680	7,500
5-5-1	3,500	1,200	10,000
5-5-2	5,400	1,800	14,000
5-5-3	9,200	3,000	32,000
5-5-4	16,000	6,400	58,000
5-5-5	>16,000	>6,400	>58,000

CHAPTER 22

INTERPRETATION OF DATA

In determining the acceptability of a particular food, sampling is necessary, because it is usually not feasible to subject an entire lot to microbiological examination. A lot is a quantity of food produced and handled under uniform conditions. In practice, this usually means food produced within a limited period of time. Examination of every subunit in the lot would be destructive and too expensive. Results from the portion of the lot represented by the sample are used to draw conclusions about the whole lot.

A. FRESH, FROZEN AND DRIED FOODS

1. Two-class attribute plan

A simple way to make a decision to accept or reject a food lot is the following. The decision will be based on some microbiological test performed on several (n) sample units. This might be a test for the presence or absence of an organism (positive or negative); or it could be a plate count, to see whether or not the count is above some critical number, m. The decision-making process is defined by 2 numbers. The first of these is the number of sample units represented by the letter n. If n = 5, one tests 5 sample units of material. If n = 15, one tests 15 sample units. The second number is the maximum allowable number of sample units yielding unsatisfactory test results, e.g., the presence of the organism, or a count above m. This acceptable number is given the letter c. Thus, in a presence/absence type of decision on a food lot, the sample plan n = 10, c = 2 means the following: take a sample of 10 sample units and make a test on each: then, if 2 or fewer show the presence of the organism, accept the lot; but if 3 or more of the 10 show the presence of the organism, reject the lot.

2. Three-class attribute plan

This is a type of plan devised for a situation where the quality of the product, in terms of microbial criteria, can be divided into 3 classes, where any single sample unit may be: wholly acceptable, marginally acceptable, or defective. There are 2 levels of sample counts, m and M, that one uses to classify a sample unit into one of the 3 categories above. A count above M for any sample unit is unacceptable. If there are any such counts among the n sample units from a lot, then this lot is withheld pending further investigation. Counts between m and M are undesirable, but some such counts can be accepted if there are not too many of them.

3. Recommended microbiological limits

The International Commission on Microbiological Specifications for Foods (1) has published recommendations for microbiological limits of selected fresh, frozen, and dried foods (Table 33). Thus, these recommendations may be useful in determining acceptability of a particular lot of food.

B. CANNED FOODS

Determining the acceptability of a lot of canned foods is based on 3 factors: culture analysis, organoleptic examination, and microleak and seam analysis. The final determination regarding the acceptability of a particular lot of canned food must be made based on a combination of results from all 3 analyses. Tables 34-37 provide a key to the probable cause of spoilage in foods based on culture analysis and organoleptic examination.

Wherever possible, an attempt should be made to confirm the diagnosis in Tables 2-5 with an analysis for seam defects. Seam defects are much like can dents in that, although the defect may be substantial, the product in the can may not be contaminated. A can with a seam defect will have a better chance of leaking, but if the cooling water is chlorinated properly and the cans are handled properly after cooling, leakage and contamination still may not occur. Conversely, a can with apparently adequate seams, if handled improperly, may leak and, regardless of the cooling water quality, may become contaminated.

Because seam defects do not always lead to product contamination, a lot of canned food cannot be rejected based on seam defects only. It must be demonstrated that the product is adulterated (swollen, leaking, contains viable microorganisms or is decomposed) in addition to having seam defects. The can seam examination and microleak test serve only to document an apparent reason for the adulteration.

REFERENCE

1. International Commission on Microbiological Specifications for Foods of the International Union of Microbiological Societies. 1986. Microorganisms in Foods. 2. Sampling for Microbiological Analysis: Principles and Specific Applications, 2nd ed. University of Toronto Press, Toronto, Canada.

Table 33. Recommended microbiological limits for various foods

Product	Test	n	c	Limit per g	
				m	M
Raw meats					
1. Carcass meat, chilled	APC[a]	5	3	10^6	10^7
2. Edible offal, chilled	APC	5	3	10^6	10^7
3. Carcass meat, frozen	APC	5	3	5×10^5	10^7
4. Boneless meat, frozen (beef, pork, veal, mutton)	APC	5	3	5×10^5	10^7
5. Comminuted meat, frozen	APC	5	3	10^6	10^7
6. Edible offal, frozen	APC	5	3	5×10^5	10^7
Processed meats					
1. Roast beef	Salmonella	20	0	0	-
2. Pate	Salmonella	20	0	0	-

Table 33. Recommended microbiological limits for various foods (continued)

Product	Test	n	c	Limit per g	
				m	M
Poultry and poultry products					
1. Cooked poultry meat, frozen; to be reheated before eating (e.g., prepared frozen meals)	Staphylococcus aureus	5	1	10^3	10^4
	Salmonella	5	0	0	-
2. Cooked poultry meat, frozen, ready-to-eat (e.g., turkey rolls)	Staphylococcus aureus	5	1	10^3	10^4
	Salmonella	10	0	0	-
3. Cured and/or smoked poultry meat	Staphylococcus aureus	10	1	10^3	10^4
	Salmonella	10	0	0	-
4. Dehydrated poultry products	Salmonella	10	0	0	-
5. Raw chicken (fresh or frozen), during processing	APC	5	3	5×10^5	10^7

Table 33. Recommended microbiological limits for various foods (continued)

Product	Test	n	c	Limit per g m	Limit per g M
Milk and milk products					
1. Dried milk	APC	5	2	3×10^4	3×10^5
	Total coliforms	5	1	10	10^2
	Salmonella, normal routine	30	0	0	-
	Salmonella for high-risk population	60	0	0	-
2. Cheese, hard and semi-soft	Staphylococcus aureus	5	0	10^4	-
Egg and egg products					
Pasteurized liquid, frozen, and dried egg products	APC	5	2	5×10^4	10^6
	Total coliforms	5	2	10	10^3
	Salmonella, routine	30	0	0	-
	Salmonella, for high-risk population	60	0	0	-

Table 33. Recommended microbiological limits for various foods (continued)

Product	Test	n	c	Limit per g	
				m	M
Fish and shellfish					
1. Fresh and frozen fish and cold-smoked	APC	5	3	5×10^5	10^7
	Escherichia coli	5	3	11	500
2. Precooked breaded fish	APC	5	2	5×10^5	10^7
	Escherichia coli	5	2	11	500
3. Frozen raw crustaceans	APC	5	3	10^6	10^7
	Escherichia coli	5	3	11	500
4. Frozen cooked crustaceans	APC	5	2	5×10^5	10^7
	Escherichia coli	5	2	11	500
	Staphylococcus aureus	5	0	10^3	-

Table 33. Recommended microbiological limits for various foods (continued)

Product	Test	n	c	Limit per g	
				m	M
5. Cooked, chilled, and frozen crabmeat	APC	5	2	10^5	10^6
	Escherichia coli	5	1	11	500
	Staphylococcus aureus	5	0	10^3	-
6. Fresh and frozen bivalve molluscs	APC	5	0	5×10^5	-
	Esherichia coli	5	0	16	-
7. Fresh and frozen fish and cold-smoked	Salmonella	5	0	0	-
	Vibrio parahaemolyticus	5	2	10^2	10^3
	Staphylococcus aureus	5	2	10^3	10^4
8. Precooked breaded fish	Staphylococcus aureus	5	1	10^3	10^4

Table 33. Recommended microbiological limits for various foods (continued)

Product	Test	n	c	Limit per g	
				m	M
9. Frozen raw crustaceans	Salmonella	5	0	0	-
	Vibrio parahaemolyticus	5	1	10^2	10^3
	Staphylococcus aureus	5	2	10^3	10^4
10. Frozen cooked crustaceans	Salmonella	10	0	0	-
	Vibrio parahaemolyticus	5	1	10^2	10^3
11. Cooked, chilled, and frozen crabmeat	Vibrio parahaemolyticus	10	1	10^2	10^3
12. Fresh and frozen bivalve molluscs	Salmonella	20	0	0	-
	Vibrio parahaemolyticus	10	1	10^2	10^3

Table 33. Recommended microbiological limits for various foods (continued)

Product	Test	n	c	Limit per g	
				m	M
Vegetables, fruits, nuts, and yeast					
1. Frozen vegetables and fruits (pH >4.5)	*Escherichia coli*	5	2	10^2	10^3
2. Dried vegetables	*Escherichia coli*	5	2	10^2	10^3
3. Coconut	*Salmonella*	20	0	0	-
4. Yeast	*Salmonella*	20	0	0	-
Cereals and cereal products					
1. Cereals	Moulds	5	2	10^2-10^4	10^5
2. Soya flours, concentrates, and isolates	Moulds	5	2	10^2-10^4	10^5
	Salmonella	5	0	0	-
3. Frozen bakery products (ready to eat) with low acid or high water activity fillings or toppings	*Staphylococcus aureus*	5	1	10^2	10^4
	Salmonella	20	0	0	-

Table 33. Recommended microbiological limits for various foods (continued)

Product	Test	n	c	Limit per g	
				m	M
4. Frozen bakery products (to be cooked) with low acid or high water activity fillings or toppings (e.g., meat pies, pizzas)	Staphylococcus aureus	5	1	10^2	10^4
	Salmonella	5	0	0	-
5. Frozen entrees containing rice or corn flour as a main ingredient	Bacillus cereus	5	1	10^3	10^4
6. Frozen and dried products	Staphylococcus aureus	5	1	10^2	10^4
	Salmonella	5	0	0	-
Cocoa, chocolate and confectionery					
1. Cocoa	Salmonella	10	0	0	-
2. Chocolate and other confectionery	Salmonella	10	0	0	-

Table 33. Recommended microbiological limits for various foods (continued)

Product	Test	n	c	Limit per g	
				m	M
Formulated foods					
1. Coated or filled, dried shelf-stable biscuits	Total coliforms	5	2	10	10^2
	Salmonella	30	0	0	-
2. Dried and instant products requiring reconstitution	APC	5	1	10^4	10^5
	Total coliforms	5	1	10	10^2
	Salmonella	60	0	0	-
3. Dried products requiring heating to boiling before consumption	APC	5	3	10^5	10^6
	Total coliforms	5	3	10	10^2
	Salmonella	15	0	0	-
Bottled water					
Non-carbonated natural mineral waters and bottled non-carbonated waters, not classified as mineral waters	Total coliforms	5	0	0[b]	-[b]

[a] Aerobic plate count

[b] Limit per 250 ml.

Table 34. Key to probable cause of spoilage in canned foods: low acid foods (pH 5.0 - 8.0)

Condition	Odor	Appearance	Gas	pH	Smear	Cultures	Diagnosis
Swells	Normal to "metallic"	Normal to frothy; (cans usually etched or corroded)	More than 20% H_2	Normal	Negative to occasional organisms	Negative	Hydrogen swells
	Sour	Frothy; possibly ropy brine	Mostly CO_2	Below normal	Pure or mixed cultures of rods, coccoids, cocci yeasts or mold	Growth aerobically and/or anaerobically at 30°C and possibly at 50°C	Leakage
	Sour	Frothy; possibly ropy brine; food particles firm with uncooked appearance	Mostly CO_2	Below normal	Pure or mixed cultures of rods, coccoids, cocci, and yeasts	Growth aerobically and/or anaerobically at 30°C and possibly at 50°C (if product received high exhaust, only spore formers may be recovered)	No process given
	Normal to sour	Frothy	H_2 and CO_2	Slightly to definitely below normal	Rods, med.short to med. long, usually granular; spores seldom seen	Gas, anaerobically at 50°C and possibly slowly at 30°C	Underprocessing-thermophilic anaerobes
	Cheesy to putrid	Usually frothy with disintegration of solid particles	Mostly CO_2; possibly some H_2	Slightly to definitely below normal	Rods; usually spores present	Gas, anaerobically at 30°C	Underprocessing-mesophilic anaerobes (possibility of <u>C</u>. <u>botulinum</u>)
	Slightly "off"; possibly ammonical	Normal to frothy		Slightly to definitely below normal	Rods; spores occasionally seen	Growth aerobically and/or anaerobically with gas at at 30°C and possibly at 50°C; pellicle in aerobic broth tubes; spores formed on agar and in pellicle	Underprocessing - <u>B</u>. <u>subtilis</u> type
No vacuum and/or cans buckled	Normal	Normal	No H_2	Normal to slightly below normal	Negative to moderate number or organisms	Negative	Insufficient vacuum, caused by: 1. Incipient spoilage 2. Insufficient exhaust 3. Insufficent blanch 4. Improper retort cooling procedures 5. Over fill
Flat cans (0 to normal vacuum)	Normal to sour	Normal to cloudy brine		Slightly to definitely below normal	Rods, generally granular in appearance; spores seldom seen	Growth without gas at 50°C; spore formation on nutrient agar	Underprocessing - thermophilic flat sours
	Normal to sour	Normal to cloudy brine; possibly moldy		Slightly to defintely below normal	Pure or mixed cultures of rods, coccoids, cocci	Growth aerobically and/or anaerobically at 30°C and possibly at 50°C	Leakage

Table 35. Key to probable cause of spoilage in canned foods: semi-acid foods (pH 4.6-5.0)

Condition	Odor	Appearance	Gas	pH	Smear	Cultures	Diagnosis
Swells	Normal to "metallic"	Normal to frothy (cans usually etched or corroded)	More than 20% H_2	Normal	Negative to occasional organisms	Negative	Hydrogen swells
	Sour	Frothy; possibly ropy brine	Mostly CO_2	Below normal	Pure or mixed cultures of rods coccoids, cocci, yeasts and mold	Growth aerobically and/or anaerobically at 30°C and possibly at 50°C	Leakage
	Sour	Frothy, possibly ropy brine; food particles firm with uncooked appear-ance	Mostly CO_2	Below normal	Pure or mixed cultures of rods, coccoids, cocci and yeasts	Growth aerobically and/or anaerobically at 30°C and possibly at 50°C (if product received high exhaust, only spore formers may be recovered)	No process given
	Normal to sour cheesy	Frothy	H_2 and CO_2	Slightly to definitely below normal	Rods, med. short to med. long, usually granular; spores seldom seen	Gas, anaerobically at 50°C and possibly slowly at 30°C	Underprocessing-thermophilic anaerobes
(Note: Cans are some-times flat)	Normal to cheesy to putrid	Normal to frothy disintegration of solid particles	Mostly CO_2; possibly some H_2	Normal to slightly below normal	Rods; possibly spores present	Gas, anaerobically at 30°C; putrid odor	Underprocessing-mesophilic anaerobes (possibility of C. botulinum)
	Slightly "off"; possibly ammonical	Normal to frothy		Slightly to definitely below normal	Rods; spores occasionally seen	Growth aerobically and/or anaerobically with gas at 30°C and possibly at 50°C; pellicle in aerobic broth tubes; spores formed on agar and in pellicle	Underprocessing - B. subtilis type
	Butyric acid	Frothy, large volume gas	H_2 and CO_2	Definitely below normal	Rods; bipolar staining possibly spores	Gas anaerobically at 30°C; butyric acid color	Underprocessing-butyric acid anaerobe
No vacuum and/or cans buckled	Normal	Normal	No H_2	Normal to slightly below normal	Negative to moderate number of organisms	Negative	1. Incipient spoilage 2. Insufficient exhaust 3. Insufficient blanch 4. Improper retort cooling procedures 5. Over fill
Flat cans (0 to normal vacuum)	Sour to "medicinal"	Normal to cloudy brine		Slightly to definitely below normal	Rods, possibly granular in appearance	Growth without gas as 50°C; growth on thermoacidurans agar	Underprocessing-B coagulans; flat sours
	Normal to sour	Normal to cloudy brine; possibly moldy		Slightly to definitely below normal	Pure or mixed cultures of rods, coccoids, cocci or mold	Growth aerobically and/or anaerobically at 30°C and possibly at 50°C	Leakage

Table 36. Key to probable cause of spoilage in canned foods: acid foods (pH 4.0-4.6)

Condition	Odor	Appearance	Gas	pH	Smear	Cultures	Diagnosis
Swells	Normal to "metallic"	Normal to frothy (cans usually etched or corroded)	More than 20% H_2	Normal	Negative to occasional organisms	Negative	Hydrogen swells
	Sour	Frothy; possibly ropy brine	Mostly CO_2	Below normal	Pure or mixed cultures of rods, coccoids, coci yeasts and mold	Growth aerobically and/or anaerobically at 30°C and possibly at 50°C	Leakage or gross processing
	Sour	Frothy; possibly ropy brine; food particles firm	Mostly CO_2	Below normal	Pure or mixed cultures of rods, coccoids, cocci and yeasts	Growth aerobically and/or anaerobically at 30°C and possibly at 50°C (if product received high exhaust, only spore formers may be recovered)	No process given
	Normal to sour/cheesy	Frothy	H_2 and CO_2	Normal to slightly below normal	Rods, med. short to med. long, usually granular; spores seldom seen	Gas, anaerobically at 50°C and possibly slowly at 30°C	Underprocessing-thermophilic anaerobes
	Butyric acid	Frothy; large volume gas; solid particles	H_2 and CO_2; some H_2	Below normal	Rods-bipolar staining; possibly spores	Gas, anaerobically at 30°C; butyric acid odor	Underprocessing-butyric acid anaerobes
	Sour	Frothy	Mostly CO_2	Below normal	Short to long rods	Growth aerobically; acid and broth gas aerobically in broth tubes are 30°C; possible growth at 50°C	Gross underprocessing-lactobacilli
No vacuum and/or cans buckled	Normal	Normal	No H_2	Normal to slightly below normal	Negative to moderate number or organisms	Negative	Insufficient vacuum caused by: 1. Incipient spoilage 2. Insufficient exhaust 3. Insufficient blanch 4. Improper retort cooling procedures 5. Over fill
Flat cans (0 to normal vacuum	Sour to "medicinal"	Normal		Slighhtly to definitely below normal	Rods, possibly granular in appearance	Growthh without gas at 50°C and possibly at 30°C; growth on thermoacidurans agar	Underprocessing <u>B. coagulans</u> (Spoilage of this type usually limited to tomato juice)
	Normal to sour	Normal to cloudy brine; possibly moldy		Slightly to definitely below normal	Pure or mixed cultures of rods, coccoids, cocci or mold	Growth aerobically and/or anaerobically at 30°C and possibly at 50°C	Leakage or no process given

Table 37. Key to probable cause of spoilage in canned foods: high acid foods (pH below 4.0)

Condition	Odor	Appearance	Gas	pH	Smear	Cultures	Diagnosis
Swells	Normal to "metallic"	Normal to frothy (cans usually etched or corroded)	More than 20% H_2	Normal	Negative to occasional organisms	Negative	Hydrogen swells
	Normal	Normal to frothy	All CO_2	Below normal	Negative to occasional organisms	Negative	"Frothy fermentation" (This spoilage is limited to concentrated products)
	Normal to sour to cheesy	Normal to frothy; possibly surface growth	Mostly CO_2	Normal to below normal	(1) Short to long rods (2) Pure or mixed cultures of short to long rods, cocci, coccoids or yeasts	(1) and (2) Growth aerobically and/or anaerobically with gas production at 30°C and possibly at 50°C; spore formers may be recovered.	(1) Underprocessing or leakage (2) Gross underprocessing or leakage
	Sour	Frothy, possibly ropy brine; food particles firm with uncooked appearance	Mostly CO_2	Below normal	Pure or mixed cultures of rods, coccoids, cocci or yeasts	Gas, aerobically and/or anaerobically at 30°C and possibly at 50°C	No process given
No vacuum and/or cans buckled	Normal	Normal	No H_2	Normal to slightly below normal	Negative to moderate number of organisms	Negative	Insufficient vacuum caused by: (1) Incipient spoilage (2) Insufficient exhaust (3) Insufficient blanch (4) Improper retort cooling procedure (5) Over fill
Flat cans (0 to normal vacuum)	Normal to sour	Normal to cloudy brine; possibly moldy		Normal to definitely below normal	Pure or mixed cultures of rods, coccoids, cocci or mold	Growth aerobically and/or anaerobically at 30°C and possibly at 50°C	Leakage or underprocessing (If mold present, leakage)

Annex 1
Aerobic Plate Count Record

Sample number ______________ Food type ______________

Date received ______________ Date analysis initiated ______________

Aerobic plate count/g ______________ Date reported ______________

Dilution	Aerobic plate count	
	Plate 1	Plate 2
10^{-1}		
10^{-2}		
10^{-3}		
10^{-4}		
10^{-5}		
10^{-6}		
10^{-7}		

Annex 2
Coliform Record — Conventional Method

Sample number ____________________ Food type ____________________

Date received ____________________ Date analysis initiated ____________________

Total coliform MPN/g ____________________ Fecal coliform MPN/g ____________________

Escherichia coli MPN/g ____________________ Date reported ____________________

Amount (g)	LT		BGB		EC		L-EMB	IMViC				LT	Gram stain	E. coli
	24 hr	48 hr	24 hr	48 hr	24 hr	48 hr		Indole	Methyl red	Voges-Proskauer	Citrate			
10^{-1}														
10^{-2}														
10^{-3}														
10^{-4}														
10^{-5}														
10^{-6}														

Annex 3
Escherichia coli Record – MUG Method

Sample number ______________________ Food type ______________________

Date received ______________________ Date analysis initiated ______________________

Escherichia coli MPN/g ______________________ Date reported ______________________

Amount (g)	LT-MUG			L-EMB	IMViC				LT	Gram stain	
	Growth	Gas	Fluorescence		Indole	Methyl red	Voges-Proskauer	Citrate			E. coli
10^{-1}											
10^{-2}											
10^{-3}											
10^{-4}											
10^{-5}											
10^{-6}											

Annex 4
Salmonella Record

Sample number ______________________ Food type ______________________ Date received ______________________

Date analysis initiated ______________________ Results ______________________

Date reported ______________________

				TSI				LI				Biochemical screening																Serological tests			
Analytical unit	Preenrichment	Selective enrichment	Selective agar	Slant	Butt	H_2S	Gas	Slant	Butt	H_2S	Gas	Glucose	Lysine broth	Urease	Dulcitol	KCN	Malonate	Indole	Methyl red	Voges-Proskauer	Citrate	Lactose	Sucrose					Polyvalent flagellar	Polyvalent somatic	Somatic group	Salmonella
		Tetrathionate	BS																												
			BS																												
			HE																												
			HE																												
			XLD																												
			XLD																												
		Selenite cystine	BS																												
			BS																												
			HE																												
			HE																												
			XLD																												
			XLD																												

Annex 5
Shigella Record

Sample number ______________________ Food type ______________________ Date received ______________________

Date analysis initiated ______________________ Results ______________________

Date reported ______________________

			TSI				Biochemical screening																		Serological somatic tests								
Analytical unit	Selective enrichment	Selective agar	Slant	Butt	H_2S	Gas	Glucose	Lactose	Sucrose	Urease	Lysine	Ornithine	Motility	Indole	Methyl red	Voges-Proskauer	Citrate	KCN	Malonate	Adonitol	Salicin	Mannitol	Inositol		A	A_1	B	C	C_1	C_2	D	A-D	Shigella
	Shigella broth (0.5 μg novobiocin)	MacConkey																															
	Shigella broth (3 μg novobiocin)	MacConkey																															

Annex 6
Vibrio cholerae Record

Sample number ______________________ Food type ______________________ Date received ______________________

Date analysis initiated ______________________ Results ______________________ Date reported ______________________

				KI				TSI				LI				1% Tryptone		Hugh-Leifson					Dihydrolase/ Decarboxylases			Fermentation reactions										Somatic O serological tests					Other tests		
Analytical unit	Enrichment	Selective agar	T_1N_1 agar	Slant	Butt	H_2S	Gas	Slant	Butt	H_2S	Gas	Slant	Butt	H_2S	Gas	With 3% NaCl	Without 3% NaCl	Oxid	Ferm	TSA	Oxidase	Urease	Lysine	Arginine	Ornithine	Glucose	Lactose	Sucrose	Mannose	Mannitol	Salicin	Arabinose	Esculin	Inositol	Citrate	Polyvalent	Inaba	Ogawa	Hikojima	Gram stain			V. cholerae
	Alkaline peptone water	TCBS																																									
		TCBS																																									
		TCBS																																									
		TCBS																																									
		Gel																																									
		Gel																																									
		Gel																																									
		Gel																																									
		GPS																																									
		GPS																																									
		GPS																																									
		GPS																																									
	GPS medium	TCBS																																									
		TCBS																																									
		TCBS																																									
		TCBS																																									
		Gel																																									
		Gel																																									
		Gel																																									
		Gel																																									
		GPS																																									
		GPS																																									
		GPS																																									
		GPS																																									

Annex 7
Vibrio parahaemolyticus Record

Sample number ______________________ Food type ________________ Date received ______________________

Date analysis initiated __________________ *Vibrio parahaemolyticus* MPN/g ______________________________________

Date reported ______________________

Amount (g)		GSTB	TCBS	TSI				Motility	Hugh Leifson		Cytochrome oxidase	Arginine	Lysine	Ornithine	Growth in NaCl (%)				Growth at 42°C	Voges-Proskauer	Sucrose	Lactose	Mannitol	Arabinose	Kanagawa	Serology		Gram stain	V. para-haemolyticus
				Slant	Butt	H_2S	Gas		Overlay	Open					0	6	8	10								0 group	K type		
1	A																												
	B																												
	C																												
– 1	A																												
	B																												
	C																												
– 2	A																												
	B																												
	C																												
– 3	A																												
	B																												
	C																												
– 4	A																												
	B																												
	C																												

Annex 8
Motile Aeromonas Record

Sample number ______________ Food type ________________ Date received ______________

Date analysis initiated __________ Results __________________ Date reported ______________

Analytical unit	Enrichment	Selective agar	AHM					Maintenance medium	Growth in 1% tryptone wo NaCl	Growth in 0/129 (μg/ml)			Gram stain	Indole	Cytochrome oxidase	TSI				Arginine dihydrolase					Voges-Proskauer	Methyl red	Salicin	Arabinose	Glucose	KCN	Esculin hydrolysis	Beta-hemolysin	Aeromonas hydrophila
			Top	Butt	H_2S (top)	H_2S (butt)	Motility			10	150	Control				Slant	Butt	H_2S	Gas	1 day	2 days	3 days	4 days	Control									
	Tryptic soy ampicillin broth	MAC																															
		MAC																															
		MAC																															
		PBG																															
		PBG																															
		PBG																															
		YSA																															
		YSA																															
		YSA																															

Annex 9
Campylobacter Record

Sample number ______________ Food type ______________ Date received ______________

Date analysis initiated ______________ Results ______________ Date reported ______________

Analytical unit	Enrichment	Direct plating	Dilution of sample/pellet before incubation	Treatment of incubated enrichment		Selective agar		Microscopic examination of wet mount	Catalase	Colony pick	Antibiotic sensitivity		Gram stain	Hippurate hydrolysis	TSI			O-F		Oxidase	Growth (°C)			MacConkey	1% Glycine	3.5% NaCl	H_2S from cysteine	Nitrate reduction	Campylobacter
				Antibiotic supplement	Filtered (F)/unfiltered (U)	Colony pick	Typical morphology				Cephalothin	Nalidixic acid			Slant	Butt	H_2S	Glucose	Control		25	35-37	42						
	Campylobacter enrichment broth		Undiluted	1	F					1																			
										2																			
					U					1																			
										2																			
				2	F					1																			
										2																			
					U					1																			
										2																			
				3	F					1																			
										2																			
					U					1																			
										2																			
			1:10	1	F					1																			
										2																			
					U					1																			
										2																			
				2	F					1																			
										2																			
					U					1																			
										2																			
				3	F					1																			
										2																			
					U					1																			
										2																			
			1:100	1	F					1																			
										2																			
					U					1																			
										2																			
				2	F					1																			
										2																			
					U					1																			
										2																			
				3	F					1																			
										2																			
					U					1																			
										2																			
	None	Yes	Undiluted							1																			
										2																			
			1:10							1																			
										2																			

Annex 10
Yersinia enterocolitica Record

Sample number ____________ Food type ____________ Date received ____________

Date analysis initiated ____________ Results ____________ Date reported ____________

Analytical unit	Enrichment	Alkali treatment	Selective agar	LAIA				Urea	Bile esculin	Anaerobic egg yolk agar	Lipase	Oxidase	Lysine decarboxylase	Arginine dihydrolase	Ornithine decarboxylase	Phenylalanine deaminase	Motility, 22-26 C	Motility, 35-37 C	Indole	Autoagglutination	Voges-Proskauer	Mannitol	Sorbitol	Cellobiose	Adonitol	Inositol	Sucrose	Rhamnose	Raffinose	Melibiose	Salicin	Trehalose	Xylose	Alpha methyl-D glucoside	Pyrazinamidase	Citrate	Gram stain	Pathogenicity testing		Y. enterocolitica
				Slant	Butt	H_2S	Gas																															Calcium dependency	Crystal violet binding	
	Peptone sorbitol bile broth	No Alkali	MAC																																					
			MAC																																					
			MAC																																					
			MAC																																					
			CIN																																					
			CIN																																					
			CIN																																					
			CIN																																					
		Alkali	MAC																																					
			MAC																																					
			MAC																																					
			MAC																																					
			CIN																																					
			CIN																																					
			CIN																																					
			CIN																																					

Annex 11
Listeria monocytogenes Record

Sample number ______________ Food type ______________ Date received ______________

Date analysis initiated ______________ Results ______________ Date reported ______________

Analytical unit	Enrichment	Selective agar	Henry illumination	TSI				Micro-scopic		Biochemical screening												Sheep blood stab	Serological tests					Mouse pathogenicity			CAMP Test	Other tests				L. monocytogenes
				Slant	Butt	H_2S	Gas	Wet mount	Gram stain	Catalase	Urease	Nitrate reductase	SIM	Methyl red	Voges-Proskauer	Glucose	Esculin	Maltose	Rhamnose	Mannitol	Xylose		Polyvalent somatic	Group 1	Group 2	Group 3	Group 4	Test culture	Positive Control	Negative Control						
	Listeria enrichment broth	LPM																																		
		LPM																																		
		LPM																																		
		LPM																																		
		LPM																																		
		OX																																		
		OX																																		
		OX																																		
		OX																																		
		OX																																		

Annex 12

Staphylococcus aureus Record- Direct Plate Method

Sample number ____________________ Food type ____________________

Date received____________________ Date analysis initiated ____________________

No. *Staphylococcus aureus* cells/g ________ Date reported ____________________

Analytical unit	Dilution	Baird-Parker agar plate number	Colony type on Baird-Parker agar	Colony pick	Coagulase	Catalase	Glucose	Mannitol	Lysostaphin sensitivtity	Nuclease production	Gram stain	S. aureus	Number S. aureus colonies picked
		I (0.4 ml inoculum)	A	1									
				2									
				3									
			B	1									
				2									
				3									
			C	1									
				2									
				3									
		II (0.3 ml inoculum)	A	1									
				2									
				3									
			B	1									
				2									
				3									
			C	1									
				2									
				3									
		III (0.3 ml inoculum)	A	1									
				2									
				3									
			B	1									
				2									
				3									
			C	1									
				2									
				3									

Annex 13
Staphylococcus aureus Record-
Most Probable Number Method

Sample number ______________ Food type ______________

Date received______________ Date analysis intitiated ______________

Staphylococcus aureus MPN/g ______________ Date reported ______________

Amount (g)		TSB/ 10% NaCl/ 1% pyruvate	Baird-Parker agar	Coagulase	Catalase	Glucose	Mannitol	Lysostaphin sensitivity	Nuclease production	Gram stain	S. aureus
−1	A										
	B										
	C										
−2	A										
	B										
	C										
−3	A										
	B										
	C										
−4	A										
	B										
	C										
−5	A										
	B										
	C										
−6	A										
	B										
	C										

Annex 14
Staphylococcal Enterotoxin Record

Sample number ____________ Food type ______________ Date received______________

Date analysis initiated __________ Enterotoxigenic staphylococci/g _____ Date reported______________

Dilution	Staphylococcus 110 agar plate number	Morphological type											
		1			2			3			4		
		Count	Pick	Toxi-genic	Count	Pick	Toxi-genic	Count	Pick	Toxi-genic	Count	Pick	Toxi-genic
	1		1			1			1			1	
			2			2			2			2	
	2		1			1			1			1	
			2			2			2			2	
	1		1			1			1			1	
			2			2			2			2	
	2		1			1			1			1	
			2			2			2			2	
	1		1			1			1			1	
			2			2			2			2	
	2		1			1			1			1	
			2			2			2			2	
	1		1			1			1			1	
			2			2			2			2	
	2		1			1			1			1	
			2			2			2			2	

Reference enterotoxins tested (circle as many as appropriate): A B C D E

Enterotoxins detected (circle as many as appropriate): A B C D E

Annex 15
Low Acid Canned Food Microbiological Analysis Record

Sample number ______________ Can number ________________ Food type ________________

Date received ______________________________ Date analysis initiated ______________________

Results ________________________________ Date reported __________________________

Incubation temperature(C)	Enrichment	Tube number	Growth	LVA, NA			Agar slant (growth)	Gram stain	CMM			LVA, NA		Other tests
				Atmosphere	Growth	Pick			Atmosphere	Growth	Gram stain	Atmosphere	Growth	
35	CMM	1		A		1			A			AN		
						2								
				AN		1			AN			A		
						2								
		2		A		1			A			AN		
						2								
				AN		1			AN			A		
						2								
	BCP	1		A		1			A			AN		
						2								
				AN		1			AN			A		
						2								
		2		A		1			A			AN		
						2								
				AN		1			AN			A		
						2								
55	CMM	1		A		1			A			AN		
						2								
				AN		1			AN			A		
						2								
		2		A		1			A			AN		
						2								
				AN		1			AN			A		
						2								
	BCP	1		A		1			A			AN		
						2								
				AN		1			AN			A		
						2								
		2		A		1			A			AN		
						2								
				AN		1			AN			A		
						2								

LVA, liver veal agar; NA, nutrient agar; CMM, cooked meat medium; BCP, bromcresol purple broth; A, aerobic; AN, anaerobic.

Annex 16
Acid Canned Food Microbiological Analysis Record

Sample number ______________ Can number ________________ Food type ________________

Date received ______________________________ Date analysis initiated ________________

Results ______________________________ Date reported ______________________

Incubation temperature(C)	Enrichment	Tube number	Growth	SDA, NA			Agar slant (growth)	Gram stain	AB, MEB			SDA, NA		Other tests
				Atmosphere	Growth	Pick			Atmosphere	Growth	Gram stain	Atmosphere	Growth	
30	AB	1		A		1			A			AN		
						2								
				AN		1			AN			A		
						2								
		2		A		1			A			AN		
						2								
				AN		1			AN			A		
						2								
	MEB	1		A		1			A			AN		
						2								
				AN		1			AN			A		
						2								
		2		A		1			A			AN		
						2								
				AN		1			AN			A		
						2								
55	AB	1		A		1			A			AN		
						2								
				AN		1			AN			A		
						2								
		2		A		1			A			AN		
						2								
				AN		1			AN			A		
						2								

SDA, Sabouraud's dextrose agar; NA, nutrient agar; AB, acid broth; MEB, malt extract broth; A, aerobic; AN, anaerobic.

Annex 17
Double Seam Measurements Record

Sample number ____________ Food type ____________ Date received ____________

Date analysis initiated ________ Results ____________ Date reported ____________

Sub no.	Code	Visual	Width (w)		Thickness (Th)			Countersink (CS)		Cover hook (CH)		Body hook (BH)		Overlap (OL)				Tightness rating % (Ti)	Junction rating %	Pressure ridge	Remarks
														In		%					
			Min	Max	Min	Max	X-O	Min	Max	1	2	1	2	1	2	1	2				

Measurement system	Required	Optional
Micrometer	CH, BH, W, Th, Ti	OL, CS
Projector	BH, OL, Th, Ti	W, CH, CS

Reference specification (for size ×):

W____________ CH____________

Th____________ BH____________

CS____________ OL____________

Ti____________

Container vacuum, in Hg:

Minimum____________

Maximum____________

Other____________

Annex 18
Container Integrity Examination Record

Sample number ______________ Food type ______________ Manufacturer ______________

Container code ______________ Container type and size ______________

Date received ______________ Number of subs received ______________

Date examined ______________ Number of subs examined ______________

Results ______________ Date reported ______________

I. HISTORICAL DATA

1. Extent of spoilage
2. Number of cans involved
3. Canning procedure used
4. Scheduled process
5. Method of cooling
6. Deviations from normal processing and handling
7. Other pertinent information

II. PRELIMINARY OBSERVATIONS

1. Flat____ Flipper____ Springer____
   Soft swell____ Hard swell____
2. Other conditions (e.g., evidence of damage, leakage, etc.)

III. VISUAL DEFECT EXAMINATION

IV. VACUUM

V. MICROLEAK EXAMINATIONS

1. Air pressure test
2. Fluorescein dye test
3. Vacuum leak test
4. Mead jar test

VI. DOUBLE SEAM MEASUREMENTS

VII. SIDE SEAM EXAMINATION

VIII. OTHER TESTS

Annex 19
Bacillus cereus Record

Sample number ______________ Food type ______________ Date received ______________

Date analysis initiated ______________ Results ______________ Date reported ______________

Amount (g)		TSP	MYP	Nutrient agar pick	Gram stain			Confirmation							Differentiation				Other tests			Bacillus cereus
					Positive or negative	Bacilli or cocci	Spores	Phenol red glucose	Nitrate reduction	Voges-Proskauer	Tyrosine	Lysozyme	MYP	Motility	Rhizoid growth	Hemolysis	Protein toxin crystals				
	A			1																		
				2																		
				3																		
				4																		
				5																		
	B			1																		
				2																		
				3																		
				4																		
				5																		
	C			1																		
				2																		
				3																		
				4																		
				5																		

Annex 20
Clostridium perfringens Record

Sample number ____________________ Food type ____________________

Date received ____________________ Date analysis initiated ____________________

Results ____________________ Date reported ____________________

Analytical unit	Dilution	Selective agar	Colony pick number	Thioglycollate	Iron milk	Motility	Nitrate reduction	Lactose	Gelatin	Sporulation broth	Thioglycollate	Spray's medium			Gram stain			C. perfringens
												Salicin	Raffinose	None	Positive or negative	Bacilli or cocci	Spores	
			1															
			2															
			3															
			4															
			5															
		TSC	6															
			7															
			8															
			9															
			10															
			1															
			2															
			3															
			4															
			5															
		TSC	6															
			7															
			8															
			9															
			10															

Annex 21
Clostridium botulinum Record

Sample number ______________________ Food type ______________________

Date received ______________________ Date analysis initiated ______________________

Results ______________________ Date reported ______________________

Dilution	Cooked meat medium[a]				TPGY	
	Trypsin treated		Untreated		Untreated	
	Tube 1	Tube 2	Tube 1	Tube 2	Tube 1	Tube 2
Undiluted						
1:2						
1:10						
1:100						

[a]Or food supernatant fluid or liquid food.

Heated, untreated controls:

Tube 1 ______________

Tube 2 ______________

Annex 22
Yeast and Mold Count Record

Sample number ____________________ Food type ____________________

Date received ____________________ Date analysis initiated ____________________

Total yeast and mold count/g ____________________ Date reported ____________________

Dilution	Yeast and mold count		
	Plate 1	Plate 2	Plate 3
10^{-1}			
10^{-2}			
10^{-3}			
10^{-4}			
10^{-5}			
10^{-6}			

FIAT PANIS